Advances on Fractional
Dynamic Inequalities
on Time Scales

Advances on Fractional Dynamic Inequalities on Time Scales

Svetlin G Georgiev

Sofia University, Bulgaria

Khaled Zennir

Qassim University, Saudi Arabia

World Scientific

NEW JERSEY · LONDON · SINGAPORE · BEIJING · SHANGHAI · HONG KONG · TAIPEI · CHENNAI · TOKYO

Published by

World Scientific Publishing Co. Pte. Ltd.

5 Toh Tuck Link, Singapore 596224

USA office: 27 Warren Street, Suite 401-402, Hackensack, NJ 07601

UK office: 57 Shelton Street, Covent Garden, London WC2H 9HE

Library of Congress Cataloging-in-Publication Data
Names: Georgiev, Svetlin, author. | Zennir, Khaled, author.
Title: Advances on fractional dynamic inequalities on time scales : fractional dynamic
    inequalities on time scales / Svetlin G Georgiev, Sofia University, Bulgaria,
    Khaled Zennir, Qassim University, Saudi Arabia.
Description: New Jersey : World Scientific, [2024] | Includes bibliographical references and index.
Identifiers: LCCN 2023016967 | ISBN 9789811275463 (hardcover) |
    ISBN 9789811275470 (ebook) | ISBN 9789811275487 (ebook other)
Subjects: LCSH: Integral inequalities. | Fractional calculus. | Calculus, Integral. |
    Fractional integrals. | Differentiable dynamical systems.
Classification: LCC QA295 .G465 2024 | DDC 515/.46--dc23/eng20230722
LC record available at https://lccn.loc.gov/2023016967

British Library Cataloguing-in-Publication Data
A catalogue record for this book is available from the British Library.

For any available supplementary material, please visit
https://www.worldscientific.com/worldscibooks/10.1142/13386#t=suppl

Desk Editors: Nimal Koliyat/Ana Ovey

Typeset by Stallion Press
Email: enquiries@stallionpress.com

Preface

This book is devoted to the recent developments in linear and nonlinear fractional Riemann–Liouville and Caputo integral inequalities on time scales. It is intended for use in the fields of fractional dynamic calculus on time scales and fractional dynamic equations on time scales. Suitable for graduate courses in the above fields, this book is designed for those who have a mathematical background in dynamic calculus and fractional dynamic equations on time scales.

This book contains 10 chapters. In Chapter 1, we introduce some basic notations, definitions and facts on the time scale calculus. Chapter 2 provides the basic definitions and facts on the Riemann–Liouville fractional integral and derivative and the Caputo fractional derivative on time scales. Chapter 3 is devoted to the Riemann–Liouville fractional delta integral inequalities. In this chapter, the fractional admissible triple is defined, and using this, some Gronwall-type Riemann–Liouville fractional inequalities are deducted. Some classes of Volterra-type fractional inequalities as well as simultaneous Riemann–Liouville fractional inequalities are investigated, and some analogues of the Pachpatte inequalities are deducted. As applications, we investigate the Cauchy problem for Riemann–Liouville fractional dynamic equations on arbitrary time scales and the dependency of its solution on the initial data. The main aim of Chapter 4 is to investigate the fractional analogues of some Hölder and Young inequalities and their reverse inequalities. In Chapter 5, we introduce the notion of exponentially convex functions on time scales and then establish Hermite–Hadamard-type inequalities for this class of functions. As a special case, we derive this double inequality in the

v

context of the classical notion of exponentially convex functions and convex functions. Moreover, we prove some new integral inequalities for n-times continuously differentiable functions with exponentially convex first-order Δ-derivative. Some Ostrowski-type inequalities on arbitrary time scales are deducted. In this chapter, strongly r-convex functions are defined and some upper bounds for the delta-Riemann–Liouville fractional integral are deducted. Next, we attempt to prove some upper bounds for the delta-Riemann–Liouville fractional integral of functions which are n-times rd-continuously Δ-differentiable with exponentially s-convexity property in the second sense on an interval in some time scales. In Chapter 6, some opial-type inequalities on arbitrary time scales are introduced. Using the classical Young and Hölder inequalities and their refinements, we deduct opial-type inequalities for Riemann–Liouville fractional derivatives, Poincaré inequalities and Ostrowski inequalities. In Chapter 7, we deduct some Chebyshev-type inequalities for two or more functions. Chapter 8 is devoted to some Hardy-type inequalities on time scales. Some fractional analogues of the Copson, Leindler and Bennett inequalities are deducted. In Chapter 9, we investigate some reverse Copson-type, Leindler-type and Bennett-type fractional inequalities on arbitrary time scales. Finally, in Chapter 10, we study some inequalities for generalized Riemann–Liouville fractional integral operators. In this chapter, some fractional analogues of Minkowski and Grüss-type inequalities are deducted.

The aim of this book is to present a clear and well-organized treatment of the concept behind the development of mathematics as well as solution techniques. The contents of the book are hence presented in an easy-to-read and mathematically solid format.

About the Authors

 **Svetlin G. Georgiev** is a mathematician who has worked in various areas of mathematics. His research currently focuses on harmonic analysis, functional analysis, partial differential equations, ordinary differential equations, Clifford and quaternion analysis, integral equations, and dynamic calculus on time scales. He is the author of the book series, *Foundations of Iso-Differential Calculus* (Vols. I–VI), and the author of several books, including *Real Quaternion Calculus, Theory of Distributions, Fractional Dynamic Calculus and Fractional Dynamic Equations on Time Scales, Fuzzy Dynamic Equations, Dynamic Inclusions and Optimal Control Problems on Time Scales, Functional Dynamic Equations on Time Scales, Multiple Fixed-Point Theorems and Applications in the Theory of ODEs, FDEs and PDEs,* and *Boundary Value Problems on Time Scales* (Vols. I and II). He also co-authored *Conformable Dynamic Equations on Time Scales.*

 **Khaled Zennir** was born in Skikda, Algeria, in 1982. He received his PhD in Mathematics in 2013 from Sidi Bel Abbès University, Algeria. He became Associate Professor at Constantine University, Algeria, in May 2015. His research interests include nonlinear hyperbolic partial differential equations: global existence, blow-up, and long-time behaviour. Khaled Zennir currently works at Qassim University (KSA).

Contents

Chapter 1

Elements of Time Scale Calculus

This chapter provides a brief introduction to some basic notions and concepts of time scales. For detailed information on time scale calculus, we refer the reader to Refs. [4, 5, 8, 16].

1.1 Introduction to Time Scales

Definition 1.1 ([4, 5]). A time scale is an arbitrary non-empty closed subset of the real numbers.

We will denote a time scale by the symbol $\mathbb{T}$.

Example 1.1. The set $\mathbb{T} = [0, 1]$, where $[0, 1]$ is the real-valued interval, is a time scale.

Example 1.2. The set $\mathbb{T} = \left\{ -\frac{1}{3}, -\frac{1}{8}, 0, \frac{1}{9}, \frac{1}{2}, 1, 3, 4 \right\}$ is a time scale.

Definition 1.2 ([4, 5]).

1. The forward jump operator $\sigma : \mathbb{T} \to \mathbb{T}$ is defined by

$$\sigma(t) = \inf\{s \in \mathbb{T} : s > t\} \quad \text{for } t \in \mathbb{T}.$$

2. The backward jump operator $\rho : \mathbb{T} \to \mathbb{T}$ is defined by

$$\rho(t) = \sup\{s \in \mathbb{T} : s < t\} \quad \text{for } t \in \mathbb{T}.$$

3. The forward graininess function $\mu : \mathbb{T} \to [0, \infty)$ is defined by

$$\mu(t) = \sigma(t) - t \quad \text{for } t \in \mathbb{T}.$$

4. The backward graininess function $\nu : \mathbb{T} \to \mathbb{R}$ is defined as follows:

$$\nu(t) = t - \rho(t), \quad t \in \mathbb{T}.$$

We set

$$\inf \emptyset = \sup \mathbb{T}, \quad \sup \emptyset = \inf \mathbb{T}.$$

We note that $\sigma(t) \geq t$ for any $t \in \mathbb{T}$ and $\rho(t) \leq t$ for any $t \in \mathbb{T}$.

Example 1.3. Let $\mathbb{T} = \left\{0, 1, \frac{3}{2}, \frac{7}{4}, 2, 4\right\}$. Then

$$\sigma(0) = 1,$$

$$\sigma(1) = \frac{3}{2},$$

$$\sigma\left(\frac{3}{2}\right) = \frac{7}{4},$$

$$\sigma(2) = 4,$$

$$\sigma(4) = 4,$$

$$\rho(0) = 0,$$

$$\rho(1) = 0,$$

$$\rho\left(\frac{3}{2}\right) = 1,$$

$$\rho\left(\frac{7}{4}\right) = \frac{3}{2},$$

$$\rho(2) = \frac{7}{4},$$

$$\rho(4) = 2.$$

Example 1.4. Let $\mathbb{T} = \left\{\sqrt[3]{2n + 7} : n \in \mathbb{N}_0\right\}$. For $n \in \mathbb{N}$, set $t = \sqrt[3]{2n + 7}$. Then $t^3 - 7 = 2n$ or $n = \frac{1}{2}(t^3 - 7)$. Hence,

$$\sigma(n) = \inf\left\{l \in \mathbb{N} : \sqrt[3]{2l + 7} > \sqrt[3]{2n + 7}\right\}$$

$$= \sqrt[3]{2n + 9}$$

$$= \sqrt[3]{2\left(\frac{1}{2}(t^3 - 7)\right) + 9}$$

$$= \sqrt[3]{t^3 - 7 + 9}$$

$$= \sqrt[3]{t^3 + 2}.$$

For $n = 0$, we have $\rho\left(\sqrt[3]{7}\right) = \sqrt[3]{7}$. Let $n \neq 0$. Then

$$\rho(n) = \sup\left\{l \in \mathbb{N} : \sqrt[3]{2l + 7} < \sqrt[3]{2n + 7}\right\}$$

$$= \sqrt[3]{2n + 5}$$

$$= \sqrt[3]{2\left(\frac{1}{2}(t^3 - 7)\right) + 5}$$

$$= \sqrt[3]{t^3 - 7 + 5}$$

$$= \sqrt[3]{t^3 - 2}.$$

Exercise 1.1. Let $\mathbb{T} = \left\{-1, 0, 1, \frac{5}{2}, 3, 8, 15, 31\right\}$. Find $\sigma(t)$, $\rho(t)$, $t \in \mathbb{T}$.

Definition 1.3 ([4,5]). For $a, b \in \mathbb{T}$, $a < b$, we define the interval $[a, b]$ in $\mathbb{T}$ as follows:

$$[a, b] = \{t \in \mathbb{T} : a \leq t \leq b\}.$$

Open intervals and half-open intervals are defined accordingly.

Let $\mathbb{T}$ be a time scale with forward jump operator and backward jump operator σ and ρ, respectively.

Definition 1.4 ([4,5]). We define the set

$$\mathbb{T}^\kappa = \begin{cases} \mathbb{T} \backslash (\rho(\sup \mathbb{T}), \sup \mathbb{T}] & \text{if} \quad \sup \mathbb{T} < \infty, \\ \mathbb{T} & \text{otherwise.} \end{cases}$$

Example 1.5. Let $\mathbb{T} = 2^{\mathbb{N}_0}$. Then $\mathbb{T}^\kappa = 2^{\mathbb{N}_0}$.

Example 1.6. Let $\mathbb{T} = \{1, 3, 5, 7, 8\}$. Then $\mathbb{T}^\kappa = \{1, 3, 5, 7\}$.

Definition 1.5 ([4, 5]). We define

$$\mathbb{T}_\kappa = \begin{cases} \mathbb{T}\backslash[\inf\mathbb{T}, \sigma(\inf\mathbb{T})) & \text{if} \quad \inf\mathbb{T} > -\infty, \\ \mathbb{T} & \text{otherwise.} \end{cases}$$

Example 1.7. Let $\mathbb{T} = 2^{\mathbb{N}_0}$. Then

$$[2, 32]_\kappa = [4, 32].$$

Definition 1.6. For $t \in \mathbb{T}$, we have the following cases:

1. If $\sigma(t) > t$, then we say that t is right-scattered.
2. If $t < \sup\mathbb{T}$ and $\sigma(t) = t$, then we say that t is right-dense.
3. If $\rho(t) < t$, then we say that t is left-scattered.
4. If $t > \inf\mathbb{T}$ and $\rho(t) = t$, then we say that t is left-dense.
5. If t is left-scattered and right-scattered at the same time, then we say that t is isolated.
6. If t is left-dense and right-dense at the same time, then we say that t is dense.

Example 1.8. Let $\mathbb{T} = \{-1, 0, 2, 7, 15\}$. Then $\inf\mathbb{T} = -1$, $\sup\mathbb{T} = 15$ and

$$\sigma(-1) = 0,$$
$$\sigma(0) = 2,$$
$$\sigma(2) = 7,$$
$$\sigma(7) = 15,$$
$$\sigma(15) = 15,$$
$$\rho(-1) = -1,$$
$$\rho(0) = -1,$$
$$\rho(2) = 0,$$
$$\rho(7) = 2,$$
$$\rho(15) = 7.$$

Therefore, the points $-1, 0, 2, 7$ are right-scattered, the points $0, 2, 7, 15$ are left-scattered and the points $0, 2, 7$ are isolated.

Example 1.9. Let $\mathbb{T} = \{0\} \cup \left\{\frac{1}{n}\right\}_{n \in \mathbb{N}} \cup \left\{\frac{2n}{n+1}\right\}_{n \in \mathbb{N}_0}$. Here, $\inf \mathbb{T} = 0$, $\sup \mathbb{T} = 2$ and

$$\sigma(0) = 0,$$

$$\sigma\left(\frac{1}{n}\right) = \frac{1}{n-1}, \quad n \in \mathbb{N}\setminus\{1\},$$

$$\sigma(1) = \frac{4}{3},$$

$$\sigma\left(\frac{2n}{n+1}\right) = \frac{2n+2}{n}, \quad n \in \mathbb{N},$$

$$\sigma(2) = 2,$$

$$\rho(0) = 0,$$

$$\rho(2) = 2,$$

$$\rho\left(\frac{1}{n}\right) = \frac{1}{n+1}, \quad n \in \mathbb{N},$$

$$\rho(1) = \frac{1}{2},$$

$$\rho\left(\frac{2n}{n+1}\right) = \frac{2n-2}{n+2}, \quad n \in \mathbb{N}.$$

Therefore, the points $\frac{1}{n}$, $n \in \mathbb{N}\setminus\{1\}$, $\frac{2n}{n+1}$, $n \in \mathbb{N}$, are right-scattered, the points $\frac{1}{n}$, $n \in \mathbb{N}$, $\frac{2n}{n+1}$, $n \in \mathbb{N}_0$, are left-scattered, and the points $\frac{1}{n}$, $n \in \mathbb{N}\setminus\{1\}$, $\frac{2n}{n+1}$, $n \in \mathbb{N}$, are isolated.

Exercise 1.2. Classify the points of $\mathbb{T} = \left\{\frac{1}{2n+1}\right\}_{n \in \mathbb{N}_0} \cup \{0\} \cup \left\{\frac{3}{n}\right\}_{n \in \mathbb{N}}$.

Definition 1.7 ([4, 5]). For a function $f : \mathbb{T} \to \mathbb{R}$, define

$$f^{\sigma}(t) = f(\sigma(t)), \quad f^{\rho}(t) = f(\rho(t)), \quad t \in \mathbb{T}.$$

Example 1.10. Let $\mathbb{T} = 2\mathbb{Z}$ and $f : \mathbb{T} \to \mathbb{R}$ be defined by

$$f(t) = \frac{1+t}{1+t^2}, \quad t \in \mathbb{T}.$$

Then $\sigma(t) = t + 2$, $\rho(t) = t - 2$, $t \in \mathbb{T}$, and

$$f^\sigma(t) = f(\sigma(t))$$
$$= \frac{1 + \sigma(t)}{1 + (\sigma(t))^2}$$
$$= \frac{1 + t + 2}{1 + (t + 2)^2}$$
$$= \frac{t + 3}{t^2 + 4t + 5},$$
$$f^\rho(t) = f(\rho(t))$$
$$= \frac{1 + \rho(t)}{1 + (\rho(t))^2}$$
$$= \frac{1 + t - 2}{1 + (t - 2)^2}$$
$$= \frac{t - 1}{t^2 - 4t + 5}, \quad t \in \mathbb{T}.$$

Exercise 1.3. Let $\mathbb{T} = 2^{\mathbb{N}_0}$ and $f, g : \mathbb{T} \to \mathbb{R}$ be defined by

$$f(t) = 1 + t - t^2, \quad g(t) = \frac{1 + t}{2 + t}, \quad t \in \mathbb{T}.$$

Find $f^\sigma(t)$, $f^\rho(t)$, $g^\sigma(t)$, $g^\rho(t)$, $f^\sigma(t) - g^\rho(t)$, $t \in \mathbb{T}$.

1.2   Differentiation on Time Scales

Definition 1.8 ([4, 5]). Assume that $f : \mathbb{T} \to \mathbb{R}$ is a function and let $t \in \mathbb{T}^\kappa$. We define $f^\Delta(t)$ to be the number, provided that it exists, such that for any $\varepsilon > 0$, there is a neighbourhood U of t, $U = (t - \delta, t + \delta) \cap \mathbb{T}$ for some $\delta > 0$, for which we have

$$|f(\sigma(t)) - f(s) - f^\Delta(t)(\sigma(t) - s)| \leq \varepsilon |\sigma(t) - s| \quad \text{for all } s \in U.$$

We call $f^\Delta(t)$ the delta or Hilger derivative of f at t. We say that f is delta or Hilger differentiable, shortly differentiable, in $\mathbb{T}^\kappa$ if $f^\Delta(t)$ exists for all $t \in \mathbb{T}^\kappa$. The function $f^\Delta : \mathbb{T} \to \mathbb{R}$ is said to be the delta derivative or Hilger derivative, shortly derivative, of f in $\mathbb{T}^\kappa$.

Remark 1.1. If $\mathbb{T} = \mathbb{R}$, then the delta derivative coincides with the classical derivative.

Note that the delta derivative is well defined.

Example 1.11. Let $f(t) = \alpha \in \mathbb{R}$. We will prove that $f^{\Delta}(t) = 0$ for any $t \in \mathbb{T}^{\kappa}$. Indeed, for $t \in \mathbb{T}^{\kappa}$ and for any $\varepsilon > 0$, there exists $\delta > 0$ such that $s \in (t - \delta, t + \delta) \cap \mathbb{T}$ implies

$$|f(\sigma(t)) - f(s) - 0(\sigma(t) - s)| = |\alpha - \alpha|$$
$$= 0$$
$$\leq \varepsilon |\sigma(t) - s|.$$

Theorem 1.1 ([4,5]). *Assume that $f : \mathbb{T} \to \mathbb{R}$ is a function and let $t \in \mathbb{T}^{\kappa}$. Then we have the following:*

1. *If f is differentiable at t, then f is continuous at t.*
2. *If f is continuous at t and t is right-scattered, then f is differentiable at t with*

$$f^{\Delta}(t) = \frac{f(\sigma(t)) - f(t)}{\mu(t)}.$$

3. *If t is right-dense, then f is differentiable at t if and only if the limit*

$$\lim_{s \to t} \frac{f(t) - f(s)}{t - s}$$

exists as a finite number. In this case,

$$f^{\Delta}(t) = \lim_{s \to t} \frac{f(t) - f(s)}{t - s}.$$

4. *If f is differentiable at t, then*

$$f(\sigma(t)) = f(t) + \mu(t) f^{\Delta}(t).$$

Example 1.12. Let $f(t) = t^2$. By Theorem 1.1, if $t \in \mathbb{T}$ is right-scattered, we have

$$f^\Delta(t) = \frac{(\sigma(t))^2 - t^2}{\sigma(t) - t}$$

$$= \frac{(\sigma(t) - t)(\sigma(t) + t)}{\sigma(t) - t}$$

$$= \sigma(t) + t.$$

If $t \in \mathbb{T}$ is right-dense, then $\sigma(t) = t$ and

$$f^\Delta(t) = \lim_{s \to t} \frac{s^2 - t^2}{s - t}$$

$$= \lim_{s \to t}(s + t)$$

$$= 2t$$

$$= \sigma(t) + t.$$

Therefore, $f^\Delta(t) = \sigma(t) + t$, $t \in \mathbb{T}^\kappa$.

Theorem 1.2 ([4, 5]). *Let $f, g : \mathbb{T} \to \mathbb{R}$ be delta differentiable at $t \in \mathbb{T}^\kappa$. Then*

1. *$f + g$ is delta differentiable at t and*
$$(f + g)^\Delta(t) = f^\Delta(t) + g^\Delta(t),$$

2. *for any $\alpha \in \mathbb{R}$, the function αf is delta differentiable at t and*
$$(\alpha f)^\Delta(t) = \alpha f^\Delta(t),$$

3. *the product fg is delta differentiable at t and*
$$(fg)^\Delta(t) = f^\Delta(t)g(t) + f(\sigma(t))g^\Delta(t)$$
$$= f^\Delta(t)g(\sigma(t)) + f(t)g^\Delta(t),$$

4. *if $g(t) \neq 0$, then $\frac{f}{g}$ is delta differentiable at t and*
$$\left(\frac{f}{g}\right)^\Delta(t) = \frac{f^\Delta(t)g(t) - f(t)g^\Delta(t)}{g(t)g(\sigma(t))}.$$

Definition 1.9 ([4, 5]). Assume that $f : \mathbb{T} \to \mathbb{R}$ is a function and let $t \in \mathbb{T}_\kappa$. We define $f^\nabla(t)$ to be the number, provided that it exists, such that for any $\varepsilon > 0$, there is a neighbourhood U of t, $U = (t - \delta, t + \delta) \cap \mathbb{T}$ for some $\delta > 0$, for which we have

$$|f(\rho(t)) - f(s) - f^\nabla(t)(\rho(t) - s)| \leq \varepsilon |\rho(t) - s| \quad \text{for all} \quad s \in U.$$

We call $f^\nabla(t)$ the nabla derivative of f at t. We say that f is nabla differentiable in $\mathbb{T}_\kappa$ if $f^\nabla(t)$ exists for all $t \in \mathbb{T}_\kappa$. The function $f^\nabla : \mathbb{T} \to \mathbb{R}$ is said to be the nabla derivative of f in $\mathbb{T}_\kappa$.

Remark 1.2. If $\mathbb{T} = \mathbb{R}$, then the nabla derivative coincides with the classical derivative.

Note that the nabla derivative is well defined.

Theorem 1.3 ([4,5]). *Assume that $f : \mathbb{T} \to \mathbb{R}$ is a function and let $t \in \mathbb{T}_\kappa$. Then we have the following:*

1. *If f is nabla differentiable at t, then f is continuous at t.*
2. *If f is continuous at t and t is left-scattered, then f is nabla differentiable at t with*

$$f^\nabla(t) = \frac{f(t) - f(\rho(t))}{\nu(t)}.$$

3. *If t is left-dense, then f is nabla differentiable at t if and only if the limit*

$$\lim_{s \to t} \frac{f(t) - f(s)}{t - s}$$

exists as a finite number. In this case,

$$f^\nabla(t) = \lim_{s \to t} \frac{f(t) - f(s)}{t - s}.$$

4. *If f is nabla differentiable at t, then*

$$f(\rho(t)) = f(t) - \nu(t) f^\nabla(t).$$

Example 1.13. Let $\mathbb{T} = 2\mathbb{Z}$ and $f(t) = t^2 + 2t + 1$, $t \in \mathbb{T}$. Then, any point of $\mathbb{T}$ is left-scattered and

$$\rho(t) = t - 2,$$
$$f^\nabla(t) = \rho(t) + t + 2$$
$$= t - 2 + t + 2$$
$$= 2t, \quad t \in \mathbb{T}.$$

Theorem 1.4 ([4, 5]). *Let $f, g : \mathbb{T} \to \mathbb{R}$ be nabla differentiable at $t \in \mathbb{T}_\kappa$. Then*

1. *$f + g$ is nabla differentiable at t and*

$$(f + g)^\nabla(t) = f^\nabla(t) + g^\nabla(t),$$

2. *for any $\alpha \in \mathbb{R}$, the function αf is nabla differentiable at t and*

$$(\alpha f)^\nabla(t) = \alpha f^\nabla(t),$$

3. *the product fg is nabla differentiable at t and*

$$(fg)^\nabla(t) = f^\nabla(t)g(t) + f(\rho(t))g^\nabla(t)$$
$$= f^\nabla(t)g(\rho(t)) + f(t)g^\nabla(t),$$

4. *if $g(t) \neq 0$, then $\frac{f}{g}$ is nabla differentiable at t and*

$$\left(\frac{f}{g}\right)^\nabla(t) = \frac{f^\nabla(t)g(t) - f(t)g^\nabla(t)}{g(t)g(\rho(t))}.$$

Definition 1.10. Suppose that $\alpha \in [0, 1]$ and f is delta and nabla differentiable at $t \in \mathbb{T}$. The diamond-α derivative of f at t is defined as follows:

$$f^{\diamond\alpha}(t) = \alpha f^\Delta(t) + (1 - \alpha)f^\nabla(t).$$

Example 1.14. Let $\mathbb{T} = 2\mathbb{N}$ and

$$f(t) = t^2 - t, \quad t \in \mathbb{T}.$$

We have

$$\sigma(t) = t + 2,$$
$$\rho(t) = t - 2, \quad t \in \mathbb{N},$$

and

$$f^\Delta(t) = \sigma(t) + t - 1$$
$$= t + 2 + t - 1$$
$$= 2t - 1,$$

$$f^{\nabla}(t) = \rho(t) + t - 1$$
$$= t - 2 + t - 1$$
$$= 2t - 3, \quad t \in \mathbb{T}.$$

Hence,

$$f^{\diamond\frac{1}{2}}(t) = \frac{1}{2}f^{\Delta}(t) + \frac{1}{2}f^{\nabla}(t)$$
$$= \frac{1}{2}(2t + 1) + \frac{1}{2}(2t - 3)$$
$$= \frac{1}{2}(4t - 2)$$
$$= 2t - 1, \quad t \in \mathbb{T}.$$

Exercise 1.4. Let $\mathbb{T} = 2^{\mathbb{N}_0}$ and

$$f(t) = t^2 - 3t + 1, \quad t \in \mathbb{T}.$$

Find

1. $f^{\diamond\frac{1}{3}}(t), \quad t \in \mathbb{T},$
2. $f^{\diamond\frac{1}{4}}(t), \quad t \in \mathbb{T}.$

Answer:

1. $2t - 3, \quad t \in \mathbb{T},$
2. $\frac{15}{8}t - 3, \quad t \in \mathbb{T}.$ □

Definition 1.11 ([16]). Let $\alpha \in (0,1]$ and $t \in \mathbb{T}$. Define conformable α-fractional derivative of f at t, denoted by $D_{\alpha}^{\Delta}f(t)$ or $f^{\Delta\alpha}(t)$, to be the number with the property that for any $\epsilon > 0$ there is a neighbourhood U of t such that

$$\left| (f^{\sigma}(t) - f(s))\sigma^{1-\alpha}(s) - f^{\Delta\alpha}(t)(\sigma(t) - s) \right| \leq \epsilon |\sigma(t) - s|$$

for any $s \in U$.

Theorem 1.5 ([16]). *Let $\alpha \in (0,1]$ and $f, g : \mathbb{T} \to \mathbb{R}$ be conformable α-fractional differentiable at $t \in \mathbb{T}$. Then*

1. *$f + g$ is conformable α-fractional differentiable at t and*

$$(f + g)^{\Delta\alpha}(t) = f^{\Delta\alpha}(t) + g^{\Delta\alpha}(t),$$

2. *for any $a \in \mathbb{R}$, the function af is conformable α-fractional differentiable at t and*

$$(af)^{\Delta_\alpha}(t) = af^{\Delta_\alpha}(t),$$

3. *the product fg is conformable α-fractional differentiable at t and*

$$(fg)^{\Delta_\alpha}(t) = f^{\Delta_\alpha}(t)g(t) + f(\sigma(t))g^{\Delta_\alpha}(t)$$
$$= f^{\Delta_\alpha}(t)g(\sigma(t)) + f(t)g^{\Delta_\alpha}(t),$$

4. *if $g(t) \neq 0$, then $\frac{f}{g}$ is conformable α-fractional differentiable at t and*

$$\left(\frac{f}{g}\right)^{\Delta_\alpha}(t) = \frac{f^{\Delta_\alpha}(t)g(t) - f(t)g^{\Delta_\alpha}(t)}{g(t)g(\sigma(t))}.$$

1.3   Integration on Time Scales

Definition 1.12 ([4, 5]).

1. A function $f : \mathbb{T} \to \mathbb{R}$ is called regulated, provided that its right-sided limits exist (finite) at all right-dense points in $\mathbb{T}$ and its left-sided limits exist (finite) at all left-dense points in $\mathbb{T}$.
2. A continuous function $f : \mathbb{T} \to \mathbb{R}$ is called pre-differentiable with region of differentiation D, provided that

   a. $D \subset \mathbb{T}^\kappa$,
   b. $\mathbb{T}^\kappa \backslash D$ is countable and contains no right-scattered elements of $\mathbb{T}$,
   c. f is differentiable at each $t \in D$.

Theorem 1.6 ([4, 5]). *Let $t_0 \in \mathbb{T}$, $x_0 \in \mathbb{R}$, $f : \mathbb{T}^\kappa \to \mathbb{R}$ be a given regulated map. Then there exists exactly one pre-differentiable function F, satisfying*

$$F^\Delta(t) = f(t) \quad \text{for all } t \in D, \quad F(t_0) = x_0.$$

Definition 1.13 ([4, 5]).

1. Assume $f : \mathbb{T} \to \mathbb{R}$ is a regulated function. Any function F in Theorem 1.6 is called a pre-antiderivative of f. We define the

indefinite integral of a regulated function f by

$$\int f(t)\Delta t = F(t) + c,$$

where c is an arbitrary constant and F is a pre-antiderivative of f. We define the Cauchy integral by

$$\int_\tau^s f(t)\Delta t = F(s) - F(\tau) \quad \text{for all } \tau, \quad s \in \mathbb{T}.$$

2. A function $F : \mathbb{T} \to \mathbb{R}$ is called an antiderivative of $f : \mathbb{T} \to \mathbb{R}$, provided that

$$F^\Delta(t) = f(t) \quad \text{holds for all } t \in \mathbb{T}^\kappa.$$

Example 1.15. Let $\mathbb{T} = \mathbb{Z}$. Then $\sigma(t) = t + 1$, $t \in \mathbb{T}$. Define $f(t) = 3t^2 + 5t + 2$, $g(t) = t^3 + t^2$, $t \in \mathbb{T}$. Since

$$\begin{aligned}
g^\Delta(t) &= (\sigma(t))^2 + t\sigma(t) + t^2 + \sigma(t) + t \\
&= (t+1)^2 + t(t+1) + t^2 + t + 1 + t \\
&= t^2 + 2t + 1 + t^2 + t + t^2 + 2t + 1 \\
&= 3t^2 + 5t + 2, \quad t \in \mathbb{T},
\end{aligned}$$

we conclude that

$$\begin{aligned}
\int f(t)\Delta t &= \int (3t^2 + 5t + 2)\Delta t \\
&= g(t) + c \\
&= t^3 + t^2 + c, \quad t \in \mathbb{T}.
\end{aligned}$$

Here, c is a constant.

Example 1.16. Let $\mathbb{T} = 3^{\mathbb{N}_0}$. In this case, we have $\sigma(t) = 3t$. Let $f(t) = 4t + 2$ and $g(t) = t^2 + 2t$, for $t \in \mathbb{T}$. Since

$$g^\Delta(t) = \sigma(t) + t + 2$$

$$= 3t + t + 2$$
$$= 4t + 2, \quad t \in \mathbb{T},$$

then we compute

$$\int f(t)\Delta t = \int (4t + 2)\Delta t$$
$$= g(t) + c$$
$$= t^2 + 2t + c, \quad t \in \mathbb{T}.$$

Here, c is a constant.

Definition 1.14 ([4, 5]). A function $f : \mathbb{T} \to \mathbb{R}$ is called rd-continuous (ld-continuous), provided that it is continuous at right-dense (left-dense) points in $\mathbb{T}$ and its left-sided (right-sided) limits exist (finite) at left-dense (right-dense) points in $\mathbb{T}$. The set of rd-continuous (ld-continuous) functions $f : \mathbb{T} \to \mathbb{R}$ is denoted by $C_{rd}(\mathbb{T})$ $(C_{ld}(\mathbb{T}))$. The set of functions $f : \mathbb{T} \to \mathbb{R}$ that are delta (nabla) differentiable and whose delta (nabla) derivative is rd-continuous (ld-continuous) is denoted by $C_{rd}^1(\mathbb{T})(C_{ld}^1(\mathbb{T}))$.

Note that if f is rd-continuous, then f is regulated.

Theorem 1.7 ([4,5]). *If* $a, b, c \in \mathbb{T}$, $\alpha \in \mathbb{R}$ *and* $f, g \in C_{rd}(\mathbb{T})$, *then*

(i) $\displaystyle\int_a^b (f(t) + g(t))\Delta t = \int_a^b f(t)\Delta t + \int_a^b g(t)\Delta t,$

(ii) $\displaystyle\int_a^b (\alpha f)(t)\Delta t = \alpha \int_a^b f(t)\Delta t,$

(iii) $\displaystyle\int_a^b f(t)\Delta t = -\int_b^a f(t)\Delta t,$

(iv) $\displaystyle\int_a^b f(t)\Delta t = \int_a^c f(t)\Delta t + \int_c^b f(t)\Delta t,$

(v) $\displaystyle\int_a^b f(\sigma(t))g^\Delta(t)\Delta t = (fg)(b) - (fg)(a) - \int_a^b f^\Delta(t)g(t)\Delta t,$

(vi) $\displaystyle\int_a^b f(t)g^\Delta(t)\Delta t = (fg)(b) - (fg)(a) - \int_a^b f^\Delta(t)g(\sigma(t))\Delta t,$

(vii) $\displaystyle\int_a^a f(t)\Delta t = 0,$

(viii) *if* $|f(t)| \le g(t)$ *on* $[a,b)$, *then*

$$\left| \int_a^b f(t)\Delta t \right| \le \int_a^b g(t)\Delta t,$$

(ix) *if* $f(t) \ge 0$ *for all* $a \le t < b$, *then* $\displaystyle\int_a^b f(t)\Delta t \ge 0.$

Definition 1.15. A function $F : \mathbb{T} \to \mathbb{R}$ is called nabla antiderivative of the function $f : \mathbb{T} \to \mathbb{R}$ if $F^\nabla(t) = f(t)$ for any $t \in \mathbb{T}_\kappa$. Then
we define the nabla integral of f as follows:

$$\int_a^t f(s)\nabla s = F(t) - F(a), \quad t \in \mathbb{T}.$$

Note that any *ld*-continuous function has a nabla antiderivative.

Theorem 1.8 ([4,5]). *If* $a, b, c \in \mathbb{T}$, $\alpha \in \mathbb{R}$ *and* $f, g \in C_{ld}(\mathbb{T})$, *then*

(i) $\displaystyle\int_a^b (f(t) + g(t))\nabla t = \int_a^b f(t)\nabla t + \int_a^b g(t)\nabla t,$

(ii) $\displaystyle\int_a^b (\alpha f)(t)\nabla t = \alpha \int_a^b f(t)\nabla t,$

(iii) $\displaystyle\int_a^b f(t)\nabla t = - \int_b^a f(t)\nabla t,$

(iv) $\displaystyle\int_a^b f(t)\nabla t = \int_a^c f(t)\nabla t + \int_c^b f(t)\nabla t,$

(v) $\displaystyle\int_a^b f(\rho(t))g^\nabla(t)\nabla t = (fg)(b) - (fg)(a) - \int_a^b f^\nabla(t)g(t)\Delta t,$

(vi) $\displaystyle\int_a^b f(t)g^\nabla(t)\nabla t = (fg)(b) - (fg)(a) - \int_a^b f^\nabla(t)g(\rho(t))\nabla t,$

(vii) $\int_a^a f(t)\nabla t = 0,$

(viii) *if* $|f(t)| \le g(t)$ *on* $[a, b)$, *then*

$$\left| \int_a^b f(t)\nabla t \right| \le \int_a^b g(t)\nabla t,$$

(ix) *if* $f(t) \ge 0$ *for all* $a \le t < b$, *then* $\int_a^b f(t)\nabla t \ge 0.$

Definition 1.16. Let f be δ- and ∇-integrable on $[a, b] \subseteq \mathbb{T}$ and $\alpha \in [0, 1]$. The diamond-α integral of f on $[a, b]$ is defined as follows:

$$\int_a^b f(t) \diamond_\alpha t = \alpha \int_a^b f(t)\Delta t + (1 - \alpha) \int_a^b f(t)\nabla t.$$

Exercise 1.5.

(1) Find $\sigma(t)$, $\rho(t)$ and $\mu(t)$ for the following time scales:

    a. $\mathbb{T} = \mathbb{N}^2 = \{1, 4, 9, \ldots\}$.
    b. $\mathbb{T} = a\mathbb{Z}$, where a is a positive real number.
    c. $\mathbb{T} = q^{\mathbb{N}_0} \cup \{0\} = \{0, 1, q, q^2, q^3, \ldots\}$, where $q > 1$.

(2) Find the delta and nabla derivatives of the following functions:

    a. $f(t) = \dfrac{1}{t^2}$, where $\mathbb{T} = a\mathbb{Z}$.

    b. $f(t) = t^3 - 2t^2 + t$, where $\mathbb{T} = q^{\mathbb{N}_0} = \{1, q, q^2, q^3, \ldots\}$ and $q > 1$.

(3) Compute the following integrals:

    a. $\int (t^2 - 3t + 1)\Delta t$, where $\mathbb{T} = \mathbb{Z}$.
    b. $\int (t^2 - 3t + 1)\nabla t$, where $\mathbb{T} = \mathbb{Z}$.
    c. $\int (t^2 - 3t + 1) \diamond_{\frac{1}{2}} t$, where $\mathbb{T} = \mathbb{Z}$.
    d. $\int (t^2 - 3t + 1) \diamond_{\frac{1}{4}} t$, where $\mathbb{T} = \mathbb{Z}$.
    e. $\int (t + 2\sqrt{t} + 1)\Delta t$, where $\mathbb{T} = \mathbb{N}^2$.

f. $\int (t + 2\sqrt{t} + 1)\nabla t$, where $\mathbb{T} = \mathbb{N}^2$.

g. $\int (t + 2\sqrt{t} + 1) \diamond_{\frac{1}{6}} t$, where $\mathbb{T} = \mathbb{N}^2$.

h. $\int (t + 2\sqrt{t} + 1) \diamond_{\frac{1}{8}} t$, where $\mathbb{T} = \mathbb{N}^2$.

Definition 1.17 ([16]). Let $\alpha \in (0, 1]$ and f be δ-integrable on $[a, b]$. Define conformable α-fractional integral of f on $[a, b]$ as follows:

$$\int_a^b f(s)\Delta_\alpha s = \int_a^b f(s)s^{1-\alpha}\Delta s.$$

Theorem 1.9 ([16]). *If* $a, b, c \in \mathbb{T}$, $\alpha \in (0, 1]$, $\beta \in \mathbb{R}$ *and* $f, g \in C_{\mathrm{rd}}(\mathbb{T})$, *then*

(i) $\displaystyle\int_a^b (f(t) + g(t))\Delta_\alpha t = \int_a^b f(t)\Delta_\alpha t + \int_a^b g(t)\Delta_\alpha t,$

(ii) $\displaystyle\int_a^b (\beta f)(t)\Delta_\alpha t = \beta \int_a^b f(t)\Delta_\alpha t,$

(iii) $\displaystyle\int_a^b f(t)\Delta_\alpha t = -\int_b^a f(t)\Delta_\alpha t,$

(iv) $\displaystyle\int_a^b f(t)\Delta_\alpha t = \int_a^c f(t)\Delta_\alpha t + \int_c^b f(t)\Delta_\alpha t,$

(v) $\displaystyle\int_a^b f(\sigma(t))g^{\Delta_\alpha}(t)\Delta_\alpha t = (fg)(b) - (fg)(a) - \int_a^b f^{\Delta_\alpha}(t)g(t)\Delta_\alpha t,$

(vi) $\displaystyle\int_a^b f(t)g^{\Delta_\alpha}(t)\Delta_\alpha t = (fg)(b) - (fg)(a) - \int_a^b f^{\Delta_\alpha}(t)g(\sigma(t))\Delta_\alpha t,$

(vii) $\displaystyle\int_a^a f(t)\Delta_\alpha t = 0,$

(viii) *if* $|f(t)| \le g(t)$ *on* $[a, b)$, *then*

$$\left| \int_a^b f(t)\Delta_\alpha t \right| \le \int_a^b g(t)\Delta_\alpha t,$$

(ix) *if* $f(t) \ge 0$ *for all* $a \le t < b$, *then* $\displaystyle\int_a^b f(t)\Delta_\alpha t \ge 0.$

Definition 1.18. Let $\delta > 0$, f be δ-integrable on $[a,b] \subseteq \mathbb{T}$, $\phi \in \mathcal{C}^1([a,b])$, $\phi^\Delta \neq 0$ on $[a,b]$. Define generalized fractional integral of order δ of f as follows:

$$_\phi I^\delta_{\Delta,a}(f)(b) = \frac{1}{\Gamma(\delta)} \int_a^b (\phi(b) - \phi(s))^{\delta-1} \phi^\Delta(s) f(s) \delta s.$$

Here, Γ is the classical gamma function.

1.4   The Taylor Formula

First, we will define the monomials on general time scales.

Definition 1.19. [4,5]. Let $s,t \in \mathbb{T}$. The time scale monomials are defined as

$$g_0(t,s) = h_0(t,s) = 1,$$

$$g_{k+1}(t,s) = \int_s^t g_k(\sigma(\tau),s)\Delta\tau,$$

$$h_{k+1}(t,s) = \int_s^t h_k(\tau,s)\Delta\tau, \quad k \in \mathbb{N}_0.$$

It is clear that

$$g_1(t,s) = \int_s^t g_0(\sigma(\tau),s)\Delta\tau$$

$$= \int_s^t \Delta\tau$$

$$= t - s,$$

$$g_2(t,s) = \int_s^t g_1(\sigma(\tau),s)\Delta\tau$$

$$= \int_s^t (\sigma(\tau) - s)\Delta\tau,$$

$$h_1(t, s) = \int_s^t h_0(\tau, s)\Delta\tau$$

$$= \int_s^t \Delta\tau$$

$$= t - s,$$

$$h_2(t, s) = \int_s^t h_1(\tau, s)\Delta\tau$$

$$= \int_s^t (\tau - s)\Delta\tau,$$

and so on.

Example 1.17. Let $\mathbb{T} = 2^{\mathbb{N}_0}$. Then $\sigma(t) = 2t$, $\mu(t) = t$, $t \in \mathbb{T}$. Hence,

$$g_1(t, s) = h_1(t, s)$$

$$= t - s,$$

$$g_2(t, s) = \int_s^t (2\tau - s)\Delta\tau$$

$$= \frac{(t - s)(2t - s)}{3},$$

$$h_2(t, s) = \int_s^t (\tau - s)\Delta\tau$$

$$= \frac{(t - s)(t - 2s)}{3}, \quad s, t \in \mathbb{T}.$$

Example 1.18. Let $\mathbb{T} = \mathbb{Z}$. Then $\sigma(t) = t + 1$, $\mu(t) = 1$, $t \in \mathbb{T}$. Hence,

$$g_1(t, s) = h_1(t, s)$$

$$= t - s,$$

$$g_2(t, s) = \int_s^t (\tau + 1 - s)\Delta\tau$$

$$= \frac{(t - s)(t - s + 1)}{2},$$

$$h_2(t,s) = \int_s^t (\tau - s)\Delta\tau$$

$$= \frac{(t-s)(t-s-1)}{2}.$$

Note that

$$g_k^\Delta(t,s) = g_{k-1}(\sigma(t),s),$$
$$h_k^\Delta(t,s) = h_{k-1}(t,s), \quad k \in \mathbb{N}.$$

Now, we give some intermediate results related to the properties of the monomials h_k and g_k.

Lemma 1.1 ([4, 5]). *Let $n \in \mathbb{N}$. If f is n times differentiable and p_k, $0 \le k \le n-1$, are differentiable at some $t \in \mathbb{T}$ with*

$$p_{k+1}^\Delta(t) = p_k^\sigma(t) \quad for \ all \ 0 \le k \le n-2, \quad n \ge 2,$$

then

$$\left(\sum_{k=0}^{n-1} (-1)^k f^{\Delta^k} p_k\right)^\Delta (t) = (-1)^{n-1} f^{\Delta^n}(t)p_{n-1}^\sigma(t) + f(t)p_0^\Delta(t).$$

We also have the following result.

Lemma 1.2 ([4, 5]). *The relationship*

$$g_n(\rho^k(t),t) = 0 \quad for \ all \ n \in \mathbb{N} \ and \ all \ 0 \le k \le n-1$$

holds for all $t \in \mathbb{T}$.

Lemma 1.3 ([4,5]). *Let $n \in \mathbb{N}$, and suppose that f is $(n-1)$ times differentiable at $\rho^{n-1}(t)$. Then*

$$f(t) = \sum_{k=0}^{n-1} (-1)^k f^{\Delta^k}(\rho^{n-1}(t))g_k(\rho^{n-1}(t),t).$$

Now, employing the above results, we will formulate the Taylor formula on time scales with the monomials g_k.

Theorem 1.10 ([4, 5] (The Taylor Formula)). *Let $n \in \mathbb{N}$ and let f be n times differentiable on $\mathbb{T}^{\kappa^n}$. Let $\alpha \in \mathbb{T}^{\kappa^{n-1}}$, $t \in \mathbb{T}$. Then*

$$f(t) = \sum_{k=0}^{n-1} (-1)^k g_k(\alpha, t) f^{\Delta^k}(\alpha)$$

$$+ \int_{\alpha}^{\rho^{n-1}(t)} (-1)^{n-1} g_{n-1}(\sigma(\tau), t) f^{\Delta^n}(\tau) \Delta \tau.$$

Theorem 1.11 ([4, 5]). *The functions g_n and h_n satisfy the relationship*

$$h_n(t, s) = (-1)^n g_n(s, t)$$

for all $t \in \mathbb{T}$ and all $s \in \mathbb{T}^{\kappa^n}$.

Another version of the Taylor formula with the monomials h_k follows directly from Theorems 1.10 and 1.11.

Theorem 1.12 ([4, 5] (The Taylor Formula)). *Let $n \in \mathbb{N}$. Suppose f is n times differentiable on $\mathbb{T}^{\kappa^n}$. Let $\alpha \in \mathbb{T}^{\kappa^{n-1}}$ and $t \in \mathbb{T}$. Then*

$$f(t) = \sum_{k=0}^{n-1} (-1)^k h_k(t, \alpha) f^{\Delta^k}(\alpha)$$

$$+ \int_{\alpha}^{\rho^{n-1}(t)} (-1)^{n-1} h_{n-1}(t, \sigma(\tau)) f^{\Delta^n}(\tau) \Delta \tau.$$

Next, we give the Leibnitz formula on arbitrary time scales.

Theorem 1.13 ([4, 5] (The Leibnitz Formula)). *Let $S_k^{(n)}$ be the set consisting of all possible strings of length n, containing exactly k times σ and $n - k$ times Δ. If*

$$f^{\Lambda} \text{ exists for all } \Lambda \in S_k^{(n)},$$

then

$$(fg)^{\Delta^n} = \sum_{k=0}^{n} \left(\sum_{\Lambda \in S_k^{(n)}} f^{\Lambda} \right) g^{\Delta^k}.$$

1.5   Definition of the Laplace Transform: Properties

Let $\mathbb{T}$ be a time scale with forward jump operator, backward jump operator, delta differentiation operator and nabla differentiation operator σ, ρ, Δ and ∇, respectively. Let also $\sup \mathbb{T} = \infty$. Take $s \in \mathbb{T}$. If $z \in \mathbb{C}$ and $1 + \mu(t)z \neq 0$ for all $t \in \mathbb{T}^{\kappa}$, then

$$\ominus z = -\frac{z}{1 + \mu(t)z}$$

and

$$1 + \mu(t)(\ominus z) = 1 - \frac{\mu(t)z}{1 + \mu(t)z}$$

$$= \frac{1}{1 + \mu(t)z}$$

for any $t \in \mathbb{T}^{\kappa}$, and hence, $e_{\ominus z}(\cdot, 0)$ is well defined on $\mathbb{T}^{\kappa}$.

Definition 1.20 ([4,5]). Suppose that $s \in \mathbb{T}$, $f : \mathbb{T} \to \mathbb{R}$ is regulated. Then the Laplace transform of f is defined by

$$\mathcal{L}(f)(z, s) = \int_{s}^{\infty} e_{\ominus z}^{\sigma}(t, s) f(t) \Delta t \tag{1.1}$$

for $z \in \mathbb{C}$ for which $1 + \mu(t)z \neq 0$ for any $t \in \mathbb{T}^{\kappa}$ and the improper integral (1.1) exists. When $s = 0$, we will write $\mathcal{L}(f)(z)$.

Remark 1.3. Let $f : \mathbb{T} \to \mathbb{R}$ be regulated. With $\mathcal{D}\{f\}$, we will denote the set of all complex numbers z for which $1 + \mu(t)z \neq 0$ for any $t \in \mathbb{T}^{\kappa}$ and the improper integral (1.1) exists.

Theorem 1.14 ([4,5] (Linearity)). *Assume that $f, g : \mathbb{T} \to \mathbb{R}$ are regulated. Then for any constants α and β, we have*

$$\mathcal{L}(\alpha f + \beta g)(z, s) = \alpha \mathcal{L}(f)(z, s) + \beta \mathcal{L}(g)(z, s)$$

for $z \in \mathcal{D}\{f\} \bigcap \mathcal{D}\{g\}$, $t \in \mathbb{T}$.

Lemma 1.4 ([4,5]). *Let $z \in \mathbb{C}$. Then*

$$e_{\ominus z}^{\sigma}(t, s) = \frac{e_{\ominus z}(t, s)}{1 + \mu(t)z}$$

$$= -\frac{(\ominus z)(t)}{z} e_{\ominus z}(t, s), \quad t \in \mathbb{T}.$$

Theorem 1.15 ([4, 5]). *Assume* $f : \mathbb{T} \to \mathbb{C}$ *such that* f^{Δ^l}, $l \in \{0, \dots, k\}$, *is regulated. Then*

$$\mathcal{L}\left(f^{\Delta^k}\right)(z, s) = z^k \mathcal{L}(f)(z, s) - \sum_{l=0}^{k-1} z^l f^{\Delta^{k-1-l}}(s) \tag{1.2}$$

for those $z \in \mathcal{D}\{f\} \cap \mathcal{D}\{f^\Delta\} \cap \dots \cap \mathcal{D}\{f^{\Delta^k}\}$, *satisfying*

$$\lim_{t \to \infty} \left(f^{\Delta^l}(t) e_{\ominus z}(t, s)\right) = 0, \quad l = 0, \dots, k-1, \tag{1.3}$$

for any $s \in \mathbb{T}$.

Example 1.19. Let $z \in \mathbb{C}$ be such that

$$1 + \mu(t)z \neq 0 \quad \text{and} \quad \lim_{t \to \infty} e_{\ominus z}(t, s) = 0, \quad s \in \mathbb{T}.$$

Then

$$
\begin{aligned}
\mathcal{L}(1)(z) &= \int_s^\infty e_{\ominus z}^\sigma(t, s) \Delta t \\
&= \int_s^\infty \left(1 + \mu(t)\left(\ominus z\right)(t)\right) e_{\ominus z}(t, s) \Delta t \\
&= \int_s^\infty \left(1 - \frac{z\mu(t)}{1 + z\mu(t)}\right) e_{\ominus z}(t, s) \Delta t \\
&= -\frac{1}{z} \int_s^\infty \left(-\frac{z}{1 + z\mu(t)}\right) e_{\ominus z}(t, s) \Delta t \\
&= -\frac{1}{z} \int_s^\infty \left(\ominus z\right)(t) e_{\ominus z}(t, s) \Delta t \\
&= -\frac{1}{z} \int_s^\infty e_{\ominus z}^\Delta(t, s) \Delta t \\
&= -\frac{1}{z} e_{\ominus z}(t, s) \Big|_{t=s}^{t \to \infty} \\
&= \frac{1}{z}.
\end{aligned}
$$

Example 1.20. Let $\mathbb{T} = \mathbb{Z}$, $s = 0$ and

$$f(t) = t^2 + 1, \quad t \in \mathbb{T}.$$

We will find $\mathcal{L}(f)(z)$ for $z \in \mathcal{D}\{f\}$. We have

$$\sigma(t) = t + 1, \quad \mu(t) = 1, \quad t \in \mathbb{T}.$$

We set

$$f_1(t) = t, \quad t \in \mathbb{T}.$$

By (1.2), we have

$$\mathcal{L}\left(f_1^\Delta\right)(z) = z\mathcal{L}(f_1)(z) - f_1(0)$$

or

$$\mathcal{L}(1)(z) = z\mathcal{L}(f_1)(z)$$

or

$$\frac{1}{z} = z\mathcal{L}(f_1)(z)$$

or

$$\mathcal{L}(f_1)(z) = \frac{1}{z^2}.$$

Note that

$$\begin{aligned}
f^\Delta(t) &= \sigma(t) + t \\
&= t + 1 + t \\
&= 2t + 1, \quad t \in \mathbb{T}.
\end{aligned}$$

Therefore,

$$\begin{aligned}
\mathcal{L}\left(f^\Delta\right)(z) &= 2\mathcal{L}(f_1)(z) + \mathcal{L}(1)(z) \\
&= 2\frac{1}{z^2} + \frac{1}{z} \\
&= \frac{z + 2}{z^2}.
\end{aligned}$$

Now, we use (1.2) to get

$$\mathcal{L}\left(f^{\Delta}\right)(z) = z\mathcal{L}(f)(z) - f(0)$$

or

$$\frac{z+2}{z^2} = z\mathcal{L}(f)(z) - 1$$

or

$$\frac{z+2}{z^2} + 1 = z\mathcal{L}(f)(z)$$

or

$$\frac{z^2 + z + 2}{z^2} = z\mathcal{L}(f)(z)$$

or

$$\mathcal{L}(f)(z) = \frac{z^2 + z + 2}{z^3}.$$

Example 1.21. Let $\mathbb{T} = \mathbb{N}_0^2$, $s = 0$ and

$$f(t) = \frac{2t + 2\sqrt{t} + 3}{(t+1)^2 \left(t + 2\sqrt{t} + 2\right)^2}, \quad t \in \mathbb{T}.$$

We will find $\mathcal{L}(f)(z)$ for $z \in \mathcal{D}\{f\}$. Let

$$h(t) = \frac{1}{(t+1)^2}, \quad t \in \mathbb{T}.$$

We have

$$\sigma(t) = t + 2\sqrt{t} + 1, \quad \mu(t) = 2\sqrt{t} + 1, \quad t \in \mathbb{T}.$$

Note that

$$e^{\sigma}_{\ominus z}(t,0) = e_{-\frac{z}{1+z\mu(t)}}(\sigma(t),0)$$

$$= e^{-\int_0^{\sigma(t)} \log(1+z\mu(\tau))\Delta\tau}$$

$$= e^{-\sum_{\tau\in[0,t]} \mu(\tau)\log(1+z\mu(\tau))}$$

$$= \prod_{\tau\in[0,t]} \frac{1}{(1+z\mu(\tau))^{\mu(\tau)}}$$

$$= \prod_{\tau\in[0,t]} \frac{1}{(1+z(1+2\sqrt{\tau}))^{1+2\sqrt{\tau}}}, \quad t\in\mathbb{T}.$$

Then

$$\mathcal{L}(h)(z) = \int_0^{\infty} \frac{1}{(1+t)^2} e^{\sigma}_{\ominus z}(t,0)\,\Delta t$$

$$= \sum_{t\in[0,\infty)} \frac{2\sqrt{t}+1}{(t+1)^2} \prod_{\tau\in[0,t]} \frac{1}{(1+z(1+2\sqrt{\tau}))^{1+2\sqrt{\tau}}}, \quad t\in\mathbb{T}.$$

Next,

$$h^{\Delta}(t) = -\frac{\sigma(t)+t+2}{(t+1)^2(\sigma(t)+1)^2}$$

$$= -\frac{2t+2\sqrt{t}+3}{(t+1)^2(t+2\sqrt{t}+2)^2}$$

$$= -f(t), \quad t\in\mathbb{T}.$$

Hence, from (1.2), we obtain

$$\mathcal{L}(h^{\Delta})(z) = z\mathcal{L}(h)(z) - h^{\Delta}(0)$$

or

$$-\mathcal{L}(f)(z) = z\mathcal{L}(h)(z) - \frac{3}{4}$$

or

$$\mathcal{L}(f)(z) = -z \sum_{t\in[0,\infty)} \frac{2\sqrt{t}+1}{(t+1)^2} \prod_{\tau\in[0,t]} \frac{1}{(1+z(1+2\sqrt{\tau}))^{1+2\sqrt{\tau}}} + \frac{3}{4}.$$

Example 1.22. Let $\mathbb{T} = 3^{\mathbb{N}_0} \bigcup \{0\}$, $s = 0$ and

$$f(t) = \begin{cases} \dfrac{4}{(t+1)(3t+1)(9t+1)}, & t \neq 0, \quad t \in \mathbb{T}, \\ \dfrac{3}{8}, & t = 0. \end{cases}$$

We will find $\mathcal{L}(f)(z)$ for $z \in \mathcal{D}\{f\}$. Let

$$h(t) = \frac{1}{1+t}, \quad t \in \mathbb{T}.$$

We have

$$\sigma(t) = 3t, \quad \mu(t) = 2t, \quad \sigma(0) = 1, \quad \mu(0) = 1, \quad t \in 3^{\mathbb{N}_0}.$$

Next,

$$\begin{aligned}
e^{\sigma}_{\ominus z}(t,0) &= e^{\int_0^{\sigma(t)} \log(1+(\ominus z)(t)\mu(\tau))\Delta\tau} \\
&= e^{\int_0^{\sigma(t)} \log\left(1-\frac{z\mu(\tau)}{1+z\mu(\tau)}\right)\Delta\tau} \\
&= e^{\int_0^{\sigma(t)} \log\frac{1}{1+z\mu(\tau)}\Delta\tau} \\
&= e^{\sum_{\tau\in[0,t]} \mu(\tau)\log\frac{1}{1+z\mu(\tau)}} \\
&= \prod_{\tau\in[0,t]} \frac{1}{(1+z\mu(\tau))^{\mu(\tau)}} \\
&= \begin{cases} \dfrac{1}{1+z} \prod_{\tau\in[1,t]} \dfrac{1}{(1+2z\tau)^{2\tau}}, & t \in 3^{\mathbb{N}_0}, \\ \dfrac{1}{z+1}, & t = 0. \end{cases}
\end{aligned}$$

Hence,

$$\mathcal{L}(h)(z) = \int_0^\infty \frac{1}{t+1} e_{\ominus z}^\sigma(t,0)\,\Delta t$$

$$= \sum_{t\in[0,\infty)} \frac{\mu(t)}{t+1} e_{\ominus z}^\sigma(t,0)$$

$$= \frac{1}{1+z}\left(1 + \sum_{t\in[1,\infty)} \frac{2t}{t+1} \prod_{\tau\in[1,t]} \frac{1}{(1+2z\tau)^{2\tau}}\right).$$

Note that

$$h^\Delta(t) = -\frac{1}{(1+t)(1+\sigma(t))}$$

$$= \begin{cases} -\dfrac{1}{2}, & t = 0, \\[2mm] -\dfrac{1}{(1+t)(1+3t)}, & t \neq 0, \end{cases}$$

for $t = 0$, we get

$$h^{\Delta^2}(0) = \frac{h^\Delta(\sigma(0)) - h^\Delta(0)}{\sigma(0) - 0}$$

$$= h^\Delta(1) + \frac{1}{2}$$

$$= -\frac{1}{8} + \frac{1}{2}$$

$$= \frac{3}{8}$$

$$= f(0),$$

for $t \neq 0$, we get

$$h^{\Delta^2}(t) = \frac{h^\Delta(\sigma(t)) - h^\Delta(t)}{\sigma(t) - t}$$

$$= \frac{-\dfrac{1}{(3t+1)(9t+1)} + \dfrac{1}{(t+1)(3t+1)}}{2t}$$

$$= \frac{4}{(t+1)(3t+1)(9t+1)}$$

$$= f(t), \quad t \in \mathbb{T}.$$

Hence, from (1.2), we get

$$\mathcal{L}\left(h^{\Delta^2}\right)(z) = \mathcal{L}(f)(z)$$

$$= z^2 \mathcal{L}(h)(z) - h^{\Delta}(0) - zh(0)$$

$$= \frac{z^2}{1+z}\left(1 + \sum_{t\in[1,\infty)} \frac{2t}{1+t} \prod_{\tau\in[1,t]} \frac{1}{(1+2z\tau)^{2\tau}}\right)$$

$$+ \frac{1}{2} - z.$$

Exercise 1.6. Let $\mathbb{T} = 2^{\mathbb{N}_0} \bigcup \{0\}$, $s = 0$ and

$$f(t) = \frac{1}{2t^3 + 1}, \quad t \in \mathbb{T}.$$

Find $\mathcal{L}(f)(z)$ for $z \in \mathcal{D}\{f\}$.

Answer:

$$\mathcal{L}(f)(z) = \frac{1}{1+z}\left(1 + \sum_{t\in[1,\infty)} \frac{t}{2t^3+1} \prod_{\tau\in[1,t]} \frac{1}{(1+\tau z)^{\tau}}\right).$$

Theorem 1.16 ([4, 5]). *Assume $f : \mathbb{T} \to \mathbb{C}$ is regulated. If*

$$F(x) = \int_s^x f(y)\Delta y$$

for $x \in \mathbb{T}$, then

$$\mathcal{L}(F)(z,s) = \frac{1}{z}\mathcal{L}(f)(z,s)$$

for those $z \in \mathcal{D}\{f\}\backslash\{s\}$, satisfying

$$\lim_{x\to\infty}\left(e_{\ominus z}(x,s)\int_s^x f(y)\Delta y\right) = 0.$$

Theorem 1.17 ([4, 5]). *For all $n \in \mathbb{N}_0$, we have*

$$\mathcal{L}(h_n(x, s))(z) = \frac{1}{z^{n+1}}, \quad x \in \mathbb{T}, \tag{1.4}$$

for $z \in \mathbb{C}\backslash\{s\}$ such that $1 + z\mu(x) \neq 0$, $x \in \mathbb{T}$, and

$$\lim_{x \to \infty} (h_n(x, s)e_{\ominus z}(x, s)) = 0. \tag{1.5}$$

Theorem 1.18 ([4, 5]). *Let $\alpha \in \mathbb{C}$ and $1 + \alpha\mu(x) \neq 0$ for $x \in \mathbb{T}$. Then*

$$\mathcal{L}(e_\alpha(x, s))(z, s) = \frac{1}{z - \alpha}, \quad x \in \mathbb{T},$$

provided that

$$\lim_{x \to \infty} e_{\alpha \ominus z}(x, s) = 0.$$

Corollary 1.1 ([4, 5]). *We have*

1. $\mathcal{L}(\cos_\alpha(x, s))(z, s) = \dfrac{z}{z^2 + \alpha^2}$,
2. $\mathcal{L}(\sin_\alpha(x, s))(z, s) = \dfrac{\alpha}{z^2 + \alpha^2}$,

provided that

$$\lim_{x \to \infty} e_{i\alpha \ominus z}(x, s) = \lim_{x \to \infty} e_{-i\alpha \ominus z}(x, s) = 0.$$

Definition 1.21 ([4, 5]). *Let $f : \mathbb{N}_0 \to \mathbb{R}$ and let $z \in \mathbb{R}$. Then the $\mathcal{Z}$-transform is defined by*

$$\mathcal{Z}(f)(z) = \sum_{t=0}^{\infty} \frac{f(t)}{z^t},$$

provided that the series converges.

Theorem 1.19. *Let $\mathbb{T} = \mathbb{N}_0$. Then*

$$(z + 1)\mathcal{L}(f)(z) = \mathcal{Z}(f)(z + 1)$$

for every $f : \mathbb{T} \to \mathbb{R}$ and every $z \in \mathcal{D}\{f\}$.

Exercise 1.7. Let $\mathbb{T} = \mathbb{N}_0$ and $\alpha > 0$. Prove that

$$\mathcal{Z}\left(\alpha^t\right)(z) = \frac{z}{z - \alpha}$$

for every $z \in \mathcal{D}\{\alpha^t\}$ such that $|z| > \alpha$.

Exercise 1.8. Let $f : \mathbb{N}_0 \to \mathbb{R}$. Prove that

1. $\mathcal{Z}(f^\sigma)(z) = z\left[\mathcal{Z}(f)(z) - f(0)\right],$
2. $\mathcal{Z}(f^{\sigma\sigma})(z) = z^2\mathcal{Z}(f)(z) - z^2 f(0) - z f(1),$
3. $\mathcal{Z}\left(f^{\sigma^l}\right)(z) = z\left[z^{l-1}\mathcal{Z}(f)(z) - \sum_{k=0}^{l-1} z^{l-1-k} f(k)\right], \quad l \in \mathbb{N},$

for every $z \in \mathcal{D}\{f\}$.

1.6 Shifts and Convolutions

Let $\mathbb{T}$ be a time scale with forward jump operator and delta differentiation operator σ and Δ, respectively, such that $\sup \mathbb{T} = \infty$. Let also $t_0 \in \mathbb{T}$.

Definition 1.22 ([8] (Shift (Delay) of a Function)). For a given function $f : [t_0, \infty) \to \mathbb{C}$, the solution of the shifting problem

$$u^{\Delta t}(t, \sigma(s)) = -u^{\Delta s}(t, s), \quad t, s \in \mathbb{T}, \quad t \geq s \geq t_0, \tag{1.6}$$
$$u(t, t_0) = f(t), \quad t \in \mathbb{T}, \quad t \geq t_0, \tag{1.7}$$

is denoted by $\hat{f}$ and it is called the shift or delay of f.

Example 1.23. Let $\mathbb{T} = \mathbb{R}$. Then the problem (1.6), (1.7) takes the following form:

$$\frac{\partial u}{\partial t}(t, s) = -\frac{\partial u}{\partial s}(t, s), \quad t, s \in \mathbb{T}, \quad t \geq s \geq t_0,$$
$$u(t, t_0) = f(t), \quad t \in \mathbb{T}, \quad t \geq t_0,$$

where $f \in \mathcal{C}^1([t_0, \infty))$. Its unique solution is

$$u(t, s) = f(t - s + t_0), \quad t, s \in \mathbb{T}, \quad t \geq s \geq t_0.$$

Really, we have

$$u(t, t_0) = f(t - t_0 + t_0)$$
$$= f(t),$$

$$\frac{\partial u}{\partial t}(t, s) = f'(t - s + t_0),$$

$$\frac{\partial u}{\partial s}(t, s) = -f'(t - s + t_0), \quad t, s \in \mathbb{T}, \quad t \geq s \geq t_0.$$

Therefore,

$$\frac{\partial u}{\partial t}(t, s) = -\frac{\partial u}{\partial s}(t, s), \quad t, s \in \mathbb{T}, \quad t \geq s \geq t_0.$$

Example 1.24. Let $\mathbb{T} = \mathbb{Z}$. Then the problem (1.6), (1.7) takes the form

$$u(t + 1, s + 1) - u(t, s + 1) = -u(t, s + 1) + u(t, s), \quad t, s \in \mathbb{T},$$
$$t \geq s \geq t_0,$$
$$u(t, t_0) = f(t), \quad t \in \mathbb{T}, \quad t \geq t_0.$$

Its unique solution is

$$u(t, s) = f(t - s + t_0), \quad t, s \in \mathbb{T}, \quad t \geq s \geq t_0.$$

Really, we have

$$u(t + 1, s + 1) = f(t + 1 - s - 1 + t_0)$$
$$= f(t - s + t_0),$$
$$u(t, s + 1) = f(t - s - 1 + t_0),$$
$$u(t + 1, s + 1) - u(t, s + 1) = f(t + s + t_0) - f(t - s - 1 + t_0)$$
$$= -u(t, s + 1) + u(t, s), \quad t, s \in \mathbb{T},$$
$$t \geq s \geq t_0,$$
$$u(t, t_0) = f(t - t_0 + t_0)$$
$$= f(t), \quad t \in \mathbb{T}, \quad t \geq t_0.$$

Example 1.25. Consider the problem

$$u^{\Delta_t}(t, \sigma(s)) = -u^{\Delta_s}(t, s), \quad \forall t, s \in \mathbb{T}, \quad \text{independent of } t_0,$$

$$u(t, t_0) = e_\lambda(t, t_0), \quad \forall t \in \mathbb{T},$$

where $\lambda \in \mathbb{C}$ is a constant such that $1 + \lambda\mu(t) \neq 0$, $t \in \mathbb{T}$. We will prove that its unique solution is

$$u(t, s) = e_\lambda(t, s), \quad t, s \in \mathbb{T}, \quad \text{independent of } t_0.$$

Really, we have

$$u^{\Delta_t}(t, \sigma(s)) = \lambda e_\lambda(t, \sigma(s)),$$

$$u^{\Delta_s}(t, s) = e_\lambda^{\Delta_s}(t, s)$$

$$= \left(\frac{1}{e_{\lambda(s,t)}}\right)^{\Delta_s}$$

$$= -\frac{e_\lambda^{\Delta_s}(s, t)}{e_\lambda(s, t)e_\lambda(\sigma(s), t)}$$

$$= -\frac{\lambda e_\lambda(s, t)}{e_\lambda(s, t)e_\lambda(\sigma(s), t)}$$

$$= -\frac{\lambda}{e_\lambda(\sigma(s), t)}$$

$$= -\lambda e_\lambda(t, \sigma(s))$$

$$= -u^{\Delta_t}(t, \sigma(s)), \quad \forall t, s \in \mathbb{T}, \quad \text{independent of } t_0,$$

$$u(t, t_0) = e_\lambda(t, t_0), \quad t \in \mathbb{T}.$$

Consequently,

$$\widehat{e_\lambda(\cdot, t_0)}(t, s) = e_\lambda(t, s), \quad \forall t, s \in \mathbb{T}, \quad \text{independent of } t_0.$$

1.7 Investigation of the Shifting Problem

Suppose that $\mathbb{T}$ is an arbitrary time scale with forward jump operator and delta differentiation operator σ and Δ, respectively, such that $\sup \mathbb{T} = \infty$. Let also $t_0 \in \mathbb{T}$ be fixed.

Theorem 1.20 ([8]). *Let $f \in \mathcal{P}$ be so that*

$$f(t) = \sum_{k=0}^{\infty} a_k h_k(t, t_0), \quad t \in [t_0, \infty), \quad (1.8)$$

where the coefficients a_k, $k \in \mathbb{N}_0$, satisfy

$$|a_k| \leq M R^k, \quad k \in \mathbb{N}_0, \quad (1.9)$$

for some constants $M > 0$ and $R > 0$. Then the problem (1.6), (1.7) has a solution u of the form

$$u(t, s) = \sum_{k=0}^{\infty} a_k h_k(t, s), \quad t, s \in \mathbb{T}, \quad t \geq s \geq t_0. \quad (1.10)$$

This solution is unique in the class of functions u for which

$$A_k(s) = u^{\Delta_t^k}(t, s)\Big|_{t=s}, \quad k \in \mathbb{N}_0, \quad (1.11)$$

are delta differentiable of $s \in \mathbb{T}$ and

$$|A_k(s)| \leq A|s|^k, \quad \left|A_k^{\Delta}(s)\right| \leq B|s|^k \quad (1.12)$$

for some constants $A > 0$ and $B > 0$ for all $t, s \in \mathbb{T}$, $t \geq s \geq t_0$.

Theorem 1.21 ([8]). *Suppose that the function $f : [t_0, \infty) \to \mathbb{C}$ has the form (1.8) and (1.9) holds. Then the function*

$$u(t, s) = \int_{\Omega} \phi(\lambda) e_\lambda(t, s) dw(\lambda), \quad t, s \in \mathbb{T}, \quad t \geq s \geq t_0, \quad (1.13)$$

has first-order partial delta derivatives with respect to t and s for $t \geq s \geq t_0$ and satisfies (1.6), (1.7).

By Theorem 1.21, it follows that the solution of the problem (1.6), (1.7) has the form

$$u(t, s) = \sum_{k=1}^{n} c_k e_{\lambda_k}(t, s), \quad t, s \in \mathbb{T}, \quad t \geq s \geq t_0.$$

Example 1.26. Let $\mathbb{T} = 2^{\mathbb{N}_0}$, $t_0 = 1$, $f(t) = t^2 + 2$, $t \in \mathbb{T}$. We will find $\hat{f}(t, s)$, $t, s \in \mathbb{T}$, $t \geq s \geq 1$. Consider the problem

$$u^{\Delta_t}(t, \sigma(s)) = -u^{\Delta_s}(t, s), \quad t, s \in \mathbb{T}, \quad t \geq s \geq 1,$$
$$u(t, 1) = t^2 + 2, \quad t \in \mathbb{T}, \quad t \geq 1.$$

Here,

$$\sigma(t) = 2t, \quad t \in \mathbb{T}.$$

We have

$$f(1) = 3,$$
$$f^{\Delta}(t) = \sigma(t) + t$$
$$= 2t + t$$
$$= 3t,$$
$$f^{\Delta}(1) = 3,$$
$$f^{\Delta^2}(t) = 3,$$
$$f^{\Delta^2}(1) = 3, \quad t \in \mathbb{T}, \quad t \geq 1.$$

Hence,

$$f(t) = 3h_0(t, 1) + 3h_1(t, 1) + 3h_2(t, 1), \quad t \in \mathbb{T}, \quad t \geq 1.$$

Let

$$g(t) = \frac{t^2}{3} - st, \quad t, s \in \mathbb{T}, \quad t \geq s \geq t_0.$$

Then

$$g^{\Delta}(t) = \frac{1}{3}(\sigma(t) + t) - s$$
$$= \frac{1}{3}(2t + t) - s$$
$$= t - s.$$

Hence, from Theorem 1.20, we get

$$u(t, s) = 3h_0(t, s) + 3h_1(t, s) + 3h_2(t, s)$$

$$= 3 + 3(t - s) + 3 \int_s^t h_1(\tau, s) \Delta \tau$$

$$= 3 + 3(t - s) + 3 \int_s^t g^{\Delta}(\tau) \Delta \tau$$

$$= 3 + 3(t - s) + 3g(\tau) \Big|_{\tau=s}^{\tau=t}$$

$$= 3 + 3t - 3s + 3 \left(\frac{\tau^2}{3} - s\tau \right) \Big|_{\tau=s}^{\tau=t}$$

$$= 3 + 3t - 3s + 3 \left(\frac{t^2}{3} - st - \frac{s^2}{3} + s^2 \right)$$

$$= 3 + 3t - 3s + t^2 - 3st + 2s^2, \quad t, s \in \mathbb{T}, \quad t \geq s \geq 1.$$

Consequently,

$$u(t, s) = t^2 + 2s^2 - 3st + 3t - 3s + 3, \quad t, s \in \mathbb{T}, \quad t \geq s \geq 1.$$

Example 1.27. Let $\mathbb{T} = 3^{\mathbb{N}_0}$, $t_0 = 1$, $f(t) = \frac{1}{t}$, $t \in \mathbb{T}$. We will find the shift of f. Here,

$$\sigma(t) = 3t, \quad t \in \mathbb{T}.$$

We have

$$f(1) = 1,$$

$$f^{\Delta}(t) = -\frac{1}{t\sigma(t)}$$

$$= -\frac{1}{3t^2},$$

$$f^{\Delta}(1) = -\frac{1}{3},$$

$$f^{\Delta^2}(t) = \frac{\sigma(t) + t}{3t^2(\sigma(t))^2}$$

$$= \frac{3t + t}{3t^2(3t)^2}$$

$$= \frac{4t}{27t^4}$$

$$= \frac{4}{27t^3},$$

$$f^{\Delta^2}(1) = \frac{4}{27},$$

$$f^{\Delta^3}(t) = -\frac{4}{27}\frac{(\sigma(t))^2 + t\sigma(t) + t^2}{t^3(\sigma(t))^3}$$

$$= -\frac{4}{27}\frac{(3t)^2 + 3t^2 + t^2}{t^3(3t)^3}$$

$$= -\frac{4}{27}\frac{9t^2 + 3t^2 + t^2}{27t^6}$$

$$= -\frac{52}{729t^4},$$

$$f^{\Delta^3}(1) = -\frac{52}{729}, \quad t \in \mathbb{T},$$

and so on. Therefore,

$$f(t) = h_0(t, 1) - \frac{1}{3}h_1(t, 1) + \frac{4}{27}h_2(t, 1)$$

$$- \frac{52}{729}h_3(t, 1) + \cdots, \quad t \in \mathbb{T}, \quad t \geq 1.$$

Hence, from Theorem 1.20, we obtain

$$u(t, s) = h_0(t, s) - \frac{1}{3}h_1(t, s) + \frac{4}{27}h_2(t, s)$$

$$- \frac{52}{729}h_3(t, s) + \cdots, \quad t, s \in \mathbb{T}, \quad t \geq s \geq 1.$$

Example 1.28. Let $\mathbb{T} = \mathbb{N}_0^2$, $t_0 = 0$, $f(t) = \frac{1}{t+1}$, $t \in \mathbb{T}$. We will find the shift of f. Here,

$$\sigma(t) = \left(\sqrt{t} + 1\right)^2, \quad t \in \mathbb{T}.$$

We have

$$f(0) = 1,$$

$$f^{\Delta}(t) = -\frac{1}{(t+1)(\sigma(t)+1)}$$

$$= -\frac{1}{(t+1)\left(\left(\sqrt{t}+1\right)^2+1\right)}$$

$$= -\frac{1}{(t+1)\left(t+2\sqrt{t}+2\right)},$$

$$f^{\Delta}(0) = -\frac{1}{2},$$

$$f^{\Delta^2}(t) = -\frac{1}{\sigma(t)-t}\left(\frac{1}{(\sigma(t)+1)\left(\sigma(t)+2\sqrt{\sigma(t)}+2\right)}\right.$$

$$\left. -\frac{1}{(t+1)\left(t+2\sqrt{t}+2\right)}\right)$$

$$= -\frac{1}{\left(\sqrt{t}+1\right)^2-t}\left(\frac{1}{\left(\left(\sqrt{t}+1\right)^2+1\right)\left(\left(\sqrt{t}+1\right)^2+2(\sqrt{t}+1)+2\right)}\right.$$

$$\left. -\frac{1}{(t+1)\left(t+2\sqrt{t}+2\right)}\right)$$

$$= -\frac{1}{1+2\sqrt{t}}\left(\frac{1}{\left(t+2\sqrt{t}+2\right)\left(t+2\sqrt{t}+1+2\sqrt{t}+2+2\right)}\right.$$

$$\left. -\frac{1}{(t+1)\left(t+2\sqrt{t}+2\right)}\right)$$

$$= -\frac{1}{1+2\sqrt{t}}\left(\frac{1}{\left(t+2\sqrt{t}+2\right)\left(t+4\sqrt{t}+5\right)}\right.$$

$$\left. -\frac{1}{(t+1)\left(t+2\sqrt{t}+2\right)}\right),$$

$$f^{\Delta^2}(0) = -\left(\frac{1}{10}-\frac{1}{2}\right)$$

$$= \frac{2}{5}, \quad t \in \mathbb{T},$$

and so on. Then

$$f(t) = h_0(t,0) - \frac{1}{2}h_1(t,0) + \frac{2}{5}h_2(t,0) + \cdots, \quad t \in \mathbb{T}.$$

Therefore, using Theorem 1.20, we get

$$\hat{f}(t,s) = h_0(t,s) - \frac{1}{2}h_1(t,s) + \frac{2}{5}h_2(t,s) + \cdots, \quad t,s \in \mathbb{T}, \quad t \geq s \geq 0.$$

Exercise 1.9. Let $\mathbb{T} = 3^{\mathbb{N}_0}$, $t_0 = 1$,

$$f(t) = t^3 - 7t^2 + 4t + 5, \quad t \in \mathbb{T}.$$

Find the shift of f.

Answer:

$$\hat{f}(t,s) = 3h_0(t,s) - 11h_1(t,s) + 24h_2(t,s)$$
$$+ 52h_3(t,s), \quad t,s \in \mathbb{T}, \quad t \geq s \geq 1.$$

1.8   Convolutions

Suppose that $\mathbb{T}$ is an arbitrary time scale with forward jump operator and delta differentiation operator σ and Δ, respectively, and $t_0 \in \mathbb{T}$. We start with the following useful lemma.

Lemma 1.5 ([8]). *For a given $f : \mathbb{T} \to \mathbb{C}$, we have*

$$\hat{f}(t,t) = f(t_0)$$

for any $t \in \mathbb{T}$.

Definition 1.23 ([8]). For given functions $f, g : \mathbb{T} \to \mathbb{R}$, their convolution $f \star g$ is defined by

$$(f \star g)(t) = \int_{t_0}^{t} \hat{f}(t, \sigma(s))g(s)\Delta s, \quad t \in \mathbb{T}, \quad t \geq t_0.$$

Example 1.29. Let $\mathbb{T} = \mathbb{Z}$, $t_0 = 0$, $f(t) = t^2$, $g(t) = t$, $t \in \mathbb{T}$. We will find $(f \star g)(t)$, $t \in \mathbb{T}$, $t \geq t_0$. We have

$$\sigma(t) = t + 1,$$
$$f(0) = 0,$$
$$f^{\Delta}(t) = \sigma(t) + t$$
$$= t + 1 + t$$
$$= 2t + 1,$$
$$f^{\Delta}(0) = 1,$$
$$f^{\Delta^2}(t) = 2,$$
$$f^{\Delta^2}(0) = 2, \quad t \in \mathbb{T}.$$

Therefore,

$$f(t) = h_1(t, 0) + 2h_2(t, 0), \quad t \in \mathbb{T}.$$

Hence, from Theorem 1.20, we get

$$\hat{f}(t, s) = h_1(t, s) + 2h_2(t, s), \quad t, s \in \mathbb{T}, \quad t \geq s \geq 0.$$

We set

$$g(t) = \frac{1}{2}t^2 - \frac{1}{2}t - st, \quad t \in \mathbb{T},$$

for some $s \in \mathbb{T}$. Then

$$g^{\Delta}(t) = \frac{1}{2}(\sigma(t) + t) - \frac{1}{2} - s$$
$$= \frac{1}{2}(t + 1 + t) - \frac{1}{2} - s$$
$$= \frac{1}{2}(2t + 1) - \frac{1}{2} - s$$
$$= t - s, \quad t \in \mathbb{T}.$$

Therefore,

$$h_2(t,s) = \int_s^t h_1(\tau,s)\Delta\tau$$

$$= \int_s^t (\tau - s)\Delta\tau$$

$$= \int_s^t g^\Delta(\tau)\Delta\tau$$

$$= g(\tau)\Big|_{\tau=-s}^{\tau=t}$$

$$= \left(\frac{1}{2}\tau^2 - \frac{1}{2}\tau - s\tau\right)\Big|_{\tau=s}^{\tau=t}$$

$$= \frac{1}{2}t^2 - \frac{1}{2}t - st - \frac{1}{2}s^2 + \frac{1}{2}s + s^2$$

$$= \frac{1}{2}t^2 - \frac{1}{2}t - st + \frac{1}{2}s + \frac{1}{2}s^2$$

and

$$\hat{f}(t,s) = h_1(t,s) + 2h_2(t,s)$$

$$= t - s + 2\left(\frac{1}{2}t^2 - \frac{1}{2}t - st + \frac{1}{2}s + \frac{1}{2}s^2\right)$$

$$= t - s + t^2 - t - 2st + s + s^2$$

$$= t^2 - 2st + s^2,$$

$$\hat{f}(t,\sigma(s)) = t^2 - 2\sigma(s)t + (\sigma(s))^2$$

$$= t^2 - 2(s+1)t + (s+1)^2$$

$$= t^2 - 2st - 2t + s^2 + 2s + 1$$

$$= (t-1)^2 - 2s(t-1) + s^2, \quad t,s \in \mathbb{T}, \quad t \geq s \geq 0.$$

Let

$$l(s) = (t-1)^2\left(\frac{1}{2}s^2 - \frac{1}{2}s\right) - 2(t-1)\left(\frac{1}{3}s^3 - \frac{1}{2}s^2 + \frac{1}{6}s\right)$$

$$+ \frac{1}{4}s^4 - \frac{1}{2}s^3 + \frac{1}{4}s^2, \quad s \in \mathbb{T},$$

for some $t \in \mathbb{T}$. Then

$$l^{\Delta}(s) = (t-1)^2 \left(\frac{1}{2}(\sigma(s) + s) - \frac{1}{2} \right)$$

$$- 2(t-1) \left(\frac{1}{3} \left((\sigma(s))^2 + s\sigma(s) + s^2 \right) - \frac{1}{2}(\sigma(s) + s) + \frac{1}{6} \right)$$

$$+ \frac{1}{4} \left((\sigma(s))^3 + s(\sigma(s))^2 + s^2\sigma(s) + s^3 \right)$$

$$- \frac{1}{2} \left((\sigma(s))^2 + s\sigma(s) + s^2 \right) + \frac{1}{4}(\sigma(s) + s)$$

$$= (t-1)^2 \left(\frac{1}{2}(s + 1 + s) - \frac{1}{2} \right)$$

$$- 2(t-1) \left(\frac{1}{3} \left((s+1)^2 + s(s+1) + s^2 \right) - \frac{1}{2}(s+1+s) + \frac{1}{6} \right)$$

$$+ \frac{1}{4} \left((s+1)^3 + s(s+1)^2 + s^2(s+1) + s^3 \right)$$

$$- \frac{1}{2} \left((s+1)^2 + s(s+1) + s^2 \right) + \frac{1}{4}(s+1+s)$$

$$= (t-1)^2 \left(s + \frac{1}{2} - \frac{1}{2} \right)$$

$$- 2(t-1) \left(\frac{1}{3} \left(s^2 + 2s + 1 + s^2 + s + s^2 \right) - s - \frac{1}{2} + \frac{1}{6} \right)$$

$$+ \frac{1}{4} \left(s^3 + 3s^2 + 3s + 1 + s^3 + 2s^2 + s + s^3 + s^2 + s^3 \right)$$

$$- \frac{1}{2} \left(s^2 + 2s + 1 + s^2 + s + s^2 \right) + \frac{1}{4}(2s + 1)$$

$$= (t-1)^2 s - 2(t-1) \left(\frac{1}{3} \left(3s^2 + 3s + 1 \right) - s - \frac{1}{3} \right)$$

$$+ \frac{1}{4} \left(4s^3 + 6s^2 + 4s + 1 \right) - \frac{1}{2} \left(3s^2 + 3s + 1 \right) + \frac{1}{2}s + \frac{1}{4}$$

$$= (t-1)^2 s - 2(t-1)\left(s^2 + s + \frac{1}{3} - s - \frac{1}{3}\right)$$

$$+ s^3 + \frac{3}{2}s^2 + s + \frac{1}{4} - \frac{3}{2}s^2 - \frac{3}{2}s - \frac{1}{2} + \frac{1}{2}s + \frac{1}{4}$$

$$= (t-1)^2 s - 2(t-1)s^2 + s^3, \quad s \in \mathbb{T},$$

for some $t \in \mathbb{T}$. Consequently,

$$(f \star g)(t) = \int_0^t \hat{f}(t, \sigma(s))g(s)\Delta s$$

$$= \int_0^t \left((t-1)^2 - 2s(t-1) + s^2\right)s\Delta s$$

$$= \int_0^t \left(s(t-1)^2 - 2s^2(t-1) + s^3\right)\Delta s$$

$$= \int_0^t l^\Delta(s)\Delta s$$

$$= l(s)\Big|_{s=0}^{s=t}$$

$$= \left((t-1)^2\left(\frac{1}{2}s^2 - \frac{1}{2}s\right) - 2(t-1)\left(\frac{1}{3}s^3 - \frac{1}{2}s^2 + \frac{1}{6}s\right)\right.$$

$$\left. + \frac{1}{4}s^4 - \frac{1}{2}s^3 + \frac{1}{4}s^2\right)\Bigg|_{s=0}^{s=t}$$

$$= (t-1)^2\left(\frac{1}{2}t^2 - \frac{1}{2}t\right) - 2(t-1)\left(\frac{1}{3}t^3 - \frac{1}{2}t^2 + \frac{1}{6}t\right)$$

$$+ \frac{1}{4}t^4 - \frac{1}{2}t^3 + \frac{1}{4}t^2$$

$$= \frac{1}{2}t(t-1)^3 - \frac{1}{3}(t-1)\left(2t^3 - 3t^2 + t\right)$$

$$+ \frac{1}{4}t^2\left(t^2 - 2t + 1\right)$$

$$= \frac{1}{2}t(t-1)^3 - \frac{1}{3}(t-1)\left(2t^2(t-1) - t(t-1)\right)$$

$$+ \frac{1}{4}t^2(t-1)^2$$

$$= \frac{1}{2}t(t-1)^3 - \frac{1}{3}t(t-1)^2(2t-1) + \frac{1}{4}t^2(t-1)^2$$

$$= \frac{1}{12}t(t-1)^2\left(6(t-1) - 4(2t-1) + 3t\right)$$

$$= \frac{1}{12}t(t-1)^2(6t - 6 - 8t + 4 + 3t)$$

$$= \frac{1}{12}t(t-1)^2(t-2), \quad t \in \mathbb{T}, \quad t \geq 0.$$

Example 1.30. Let $\mathbb{T} = 2^{\mathbb{N}_0}$, $t_0 = 1$, $f(t) = g(t) = t$, $t \in \mathbb{T}$. We will find $(f \star g)(t)$, $t \in \mathbb{T}$, $t \geq 1$. Here,

$$\sigma(t) = 2t, \quad t \in \mathbb{T},$$

and

$$f(1) = 1,$$
$$f^{\Delta}(t) = 1,$$
$$f^{\Delta}(1) = 1,$$
$$f^{\Delta^k}(t) = 0, \quad k \in \mathbb{N}, \quad k \geq 2.$$

Then

$$f(t) = h_0(t,1) + h_1(t,1), \quad t \in \mathbb{T}, \quad t \geq 1.$$

Hence, from Theorem 1.20, we get

$$\hat{f}(t,s) = h_0(t,s) + h_1(t,s)$$
$$= 1 + t - s,$$
$$\hat{f}(t,\sigma(s)) = 1 + t - \sigma(s)$$
$$= 1 + t - 2s, \quad t, s \in \mathbb{T}, \quad t \geq s \geq 1.$$

Let

$$g(s) = \frac{1}{3}(1+t)s^2 - \frac{2}{7}s^3, \quad s \in \mathbb{T},$$

for some $t \in \mathbb{T}$. We have

$$g^\Delta(s) = \frac{1}{3}(1+t)\left(\sigma(s) + s\right) - \frac{2}{7}\left((\sigma(s))^2 + s\sigma(s) + s^2\right)$$

$$= \frac{1}{3}(1+t)(2s+s) - \frac{2}{7}\left((2s)^2 + s(2s) + s^2\right)$$

$$= \frac{1}{3}(1+t)(3s) - \frac{2}{7}\left(4s^2 + 2s^2 + s^2\right)$$

$$= (1+t)s - 2s^2, \quad s \in \mathbb{T},$$

for some $t \in \mathbb{T}$. Therefore,

$$(f \star g)(t) = \int_1^t \hat{f}(t, \sigma(s))g(s)\Delta s$$

$$= \int_1^t (1 + t - 2s)s\Delta s$$

$$= \int_1^t g^\Delta(s)\Delta s$$

$$= g(s)\Big|_{s=1}^{s=t}$$

$$= \left(\frac{1}{3}(1+t)s^2 - \frac{2}{7}s^3\right)\Big|_{s=1}^{s=t}$$

$$= \frac{1}{3}(1+t)t^2 - \frac{2}{7}t^3 - \frac{1}{3}(1+t) + \frac{2}{7}$$

$$= \frac{1}{3}(1+t)\left(t^2 - 1\right) - \frac{2}{7}\left(t^3 - 1\right)$$

$$= \frac{1}{3}(1+t)(t-1)(t+1) - \frac{2}{7}(t-1)\left(t^2 + t + 1\right)$$

$$= \frac{1}{21}(t-1)\left(7(t+1)^2 - 6\left(t^2 + t + 1\right)\right)$$

$$= \frac{1}{21}(t-1)\left(7t^2 + 14t + 7 - 6t^2 - 6t - 6\right)$$

$$= \frac{1}{21}(t-1)\left(t^2 + 8t + 1\right), \quad t \in \mathbb{T}, \quad t \geq 1.$$

Example 1.31. Let $\mathbb{T} = 3^{\mathbb{N}_0}$, $t_0 = 1$,

$$f(t) = \sum_{k=0}^{\infty} \frac{1}{k^2+1} h_k(t,1), \quad g(t) = \sum_{k=0}^{\infty} \frac{1}{k^2+k+1} h_k(t,1), \quad t \in \mathbb{T}.$$

We will find $(f \star g)(t)$, $t \in \mathbb{T}$, $t \geq 1$. Here,

$$\sigma(t) = 3t, \quad t \in \mathbb{T}.$$

By Theorem 1.20, we get

$$\hat{f}(t,s) = \sum_{k=0}^{\infty} \frac{1}{k^2+1} h_k(t,s),$$

$$\hat{f}(t,\sigma(s)) = \sum_{k=0}^{\infty} \frac{1}{k^2+1} h_k(t,\sigma(s))$$

$$= \sum_{k=0}^{\infty} \frac{1}{k^2+1} h_k(t,3s), \quad t,s \in \mathbb{T}, \quad t \geq s \geq 1.$$

Then

$$(f \star g)(t) = \int_1^t \hat{f}(t,\sigma(s)) g(s) \Delta s$$

$$= \int_1^t \left(\sum_{k=0}^{\infty} \frac{1}{k^2+1} h_k(t,3s) \right) \left(\sum_{k=0}^{\infty} \frac{k}{k^2+k+1} h_k(s,1) \right) \Delta s$$

$$= \int_1^t \sum_{k=0}^{\infty} \sum_{l=0}^{k} \frac{1}{(k-l)^2+1} h_{k-l}(t,3s) \frac{l}{l^2+l+1} h_l(s,1) \Delta s$$

$$= \sum_{k=0}^{\infty} \sum_{l=0}^{k} \left(\frac{1}{(k-l)^2+1} \right) \left(\frac{l}{l^2+l+1} \right) \int_1^t h_{k-l}(t,3s) h_l(s,1) \Delta s$$

$$= \sum_{k=0}^{\infty} \sum_{l=0}^{k} \left(\frac{1}{(k-l)^2+1} \right) \left(\frac{l}{l^2+l+1} \right) h_{k+1}(t,1), \quad t \in \mathbb{T}, \quad t \geq 1.$$

Exercise 1.10. Let $\mathbb{T} = 4^{\mathbb{N}_0}$, $t_0 = 4$,

$$f(t) = \sum_{k=0}^{\infty} e^{-k^2} h_k(t,4), \quad g(t) = \sum_{k=0}^{\infty} e^{-k^3} h_k(t,4), \quad t \in \mathbb{T}, \quad t \geq 4.$$

Find $(f \star g)(t)$, $t \in \mathbb{T}$, $t \geq 4$.

Answer:

$$(f \star g)(t) = \sum_{k=0}^{\infty} \sum_{l=0}^{k} e^{-(k-l)^2 - l^3} h_{k+1}(t, 4), \quad t \in \mathbb{T}, \quad t \geq 4.$$

Theorem 1.22 ([8]). *Let $f, g : \mathbb{T} \to \mathbb{C}$ be given functions. The shift of the convolution $f \star g$ is given by the formula*

$$\left(\widehat{f \star g}\right)(t, s) = \int_{s}^{t} \hat{f}(t, \sigma(u)) \hat{g}(u, s) \Delta u, \quad t, s \in \mathbb{T}, \quad t \geq s \geq t_0.$$

Theorem 1.23 ([8] (Associativity of the Convolution)). *The convolution is associative, i.e.,*

$$(f \star g) \star h = f \star (g \star h).$$

Theorem 1.24 ([8]). *Let f be delta differentiable, then*

$$(f \star g)^{\Delta} = f^{\Delta} \star g + f(t_0)g. \tag{1.14}$$

In addition, if g is delta differentiable, then

$$(f \star g)^{\Delta} = f \star g^{\Delta} + fg(t_0). \tag{1.15}$$

Corollary 1.2 ([8]). *Let f be delta differentiable. Then*

$$\int_{t_0}^{t} \hat{f}(t, \sigma(s)) \Delta s = \int_{t_0}^{t} f(s) \Delta s, \quad t \in \mathbb{T}, \quad t \geq t_0.$$

Example 1.32. Let $\mathbb{T} = \mathbb{Z}$, $t_0 = 0$, $f(t) = t^2$, $g(t) = t$, $t \in \mathbb{T}$. We will find

$$(f \star g)^{\Delta}(t), \quad t \in \mathbb{T}, \quad t \geq t_0.$$

Here,

$$\sigma(t) = t + 1, \quad t \in \mathbb{T}.$$

By Example 1.29, we have

$$(f \star g)(t) = \frac{1}{12}t(t-1)^2(t-2)$$

$$= \frac{1}{12}t\left(t^2 - 2t + 1\right)(t-2)$$

$$= \frac{1}{12}\left(t^3 - 2t^2 + t\right)(t-2)$$

$$= \frac{1}{12}\left(t^4 - 2t^3 - 2t^3 + 4t^2 + t^2 - 2t\right)$$

$$= \frac{1}{12}\left(t^4 - 4t^3 + 5t^2 - 2t\right), \quad t \in \mathbb{T}, \quad t \geq 0.$$

Then

$$(f \star g)^\Delta(t) = \frac{1}{12}\Big((\sigma(t))^3 + t\,(\sigma(t))^2 + t^2\sigma(t) + t^3$$

$$- 4\left((\sigma(t))^2 + t\sigma(t) + t^2\right) + 5\,(\sigma(t) + t) - 2\Big)$$

$$= \frac{1}{12}\Big((t+1)^3 + t(t+1)^2 + t^2(t+1) + t^3$$

$$- 4\left((t+1)^2 + t(t+1) + t^2\right) + 5(t+1+1) - 2\Big)$$

$$= \frac{1}{12}\Big(t^3 + 3t^2 + 3t + 1 + t\left(t^2 + 2t + 1\right) + t^3 + t^2 + t^3$$

$$- 4\left(t^2 + 2t + 1 + t^2 + t + t^2\right) + 5(2t+1) - 2\Big)$$

$$= \frac{1}{12}\Big(t^3 + 3t^2 + 3t + 1 + t^3 + 2t^2 + t + 2t^3 + t^2$$

$$- 4\left(3t^2 + 3t + 1\right) + 10t + 5 - 2\Big)$$

$$= \frac{1}{12}\left(4t^3 + 6t^2 + 4t + 1 - 12t^2 - 12t - 4 + 10t + 3\right)$$

$$= \frac{1}{12}\left(4t^3 - 6t^2 + 2t\right)$$

$$= \frac{1}{6}\left(2t^3 - 3t^2 + t\right), \quad t \in \mathbb{T}, \quad t \geq 0.$$

Now, we will compute $(f \star g)^{\Delta}(t)$ using Theorem 1.24. We have

$$f^{\Delta}(t) = \sigma(t) + t$$
$$= t + 1 + t$$
$$= 2t + 1,$$
$$f^{\Delta}(0) = 1,$$
$$f^{\Delta^2}(t) = 2,$$
$$f^{\Delta^2}(0) = 2,$$
$$f^{\Delta^k}(t) = 0, \quad k \in \mathbb{N}, \quad k \geq 3.$$

Therefore,

$$f^{\Delta}(t) = h_0(t,0) + 2h_1(t,0), \quad t \in \mathbb{T}, \quad t \geq 0.$$

Hence, from Theorem 1.20, we obtain

$$\widehat{f^{\Delta}}(t,s) = h_0(t,s) + 2h_1(t,s)$$
$$= 1 + 2(t - s)$$
$$= 2t - 2s + 1,$$
$$\widehat{f^{\Delta}}(t,\sigma(s)) = 2t - 2\sigma(s) + 1$$
$$= 2t - 2(s + 1) + 1$$
$$= 2t - 2s - 2 + 1$$
$$= 2t - 2s - 1, \quad t, s \in \mathbb{T}, \quad t \geq s \geq 0.$$

Let

$$g(s) = (2t - 1)\left(\frac{1}{2}s^2 - \frac{1}{2}s\right) - \frac{2}{3}s^3 + s^2 - \frac{1}{3}s, \quad s \in \mathbb{T},$$

for some $t \in \mathbb{T}$. Then

$$g^{\Delta}(s) = (2t - 1)\left(\frac{1}{2}(\sigma(s) + s) - \frac{1}{2}\right) - \frac{2}{3}\left((\sigma(s))^2 + s\sigma(s) + s^2\right)$$
$$+ \sigma(s) + s - \frac{1}{3}$$

$$= (2t - 1)\left(\frac{1}{2}(s + 1 + s) - \frac{1}{2}\right) - \frac{2}{3}\left((s + 1)^2 + s(s + 1) + s^2\right)$$

$$+ s + 1 + s - \frac{1}{3}$$

$$= (2t - 1)s - \frac{2}{3}\left(s^2 + 2s + 1 + s^2 + s + s^2\right) + 2s + \frac{2}{3}$$

$$= (2t - 1)s - \frac{2}{3}\left(3s^2 + 3s + 1\right) + 2s + \frac{2}{3}$$

$$= (2t - 1)s - 2s^2 - 2s - \frac{2}{3} + 2s + \frac{2}{3}$$

$$= (2t - 1)s - 2s^2$$

and

$$\left(f^\Delta \star g\right)(t) = \int_0^t f^\Delta(t, \sigma(s))g(s)\Delta s$$

$$= \int_0^t (2t - 2s - 1)s\Delta s$$

$$= \int_0^t \left((2t - 1)s - 2s^2\right)\Delta s$$

$$= \int_0^t g^\Delta(s)\Delta s$$

$$= g(s)\Big|_{s=0}^{s=t}$$

$$= \left((2t - 1)\left(\frac{1}{2}s^2 - \frac{1}{2}s\right) - \frac{2}{3}s^3 + s^2 - \frac{1}{3}s\right)\Big|_{s=0}^{s=t}$$

$$= (2t - 1)\left(\frac{1}{2}t^2 - \frac{1}{2}t\right) - \frac{2}{3}t^3 + t^2 - \frac{1}{3}t$$

$$= \frac{1}{2}(2t - 1)\left(t^2 - t\right) - \frac{2}{3}t^3 + t^2 - \frac{1}{3}t$$

$$= \frac{1}{2}\left(2t^3 - 2t^2 - t^2 + t\right) - \frac{2}{3}t^3 + t^2 - \frac{1}{3}t$$

$$= \frac{1}{2}\left(2t^3 - 3t^2 + t\right) - \frac{2}{3}t^3 + t^2 - \frac{1}{3}t$$

$$= t^3 - \frac{3}{2}t^2 + \frac{1}{2}t - \frac{2}{3}t^3 + t^2 - \frac{1}{3}t$$

$$= \frac{1}{3}t^3 - \frac{1}{2}t^2 + \frac{1}{6}t$$

$$= \frac{1}{6}\left(2t^3 - 3t^2 + t\right), \quad t \in \mathbb{T}, \quad t \geq 0.$$

Hence, using $f(0) = 0$ and (1.14), we obtain

$$(f \star g)^\Delta (t) = \frac{1}{6}\left(2t^3 - 3t^2 + t\right), \quad t \in \mathbb{T}, \quad t \geq 0.$$

Example 1.33. Let $\mathbb{T} = 2^{\mathbb{N}_0}$, $f(t) = 1$, $g(t) = t^2$, $t \in \mathbb{T}$, $t_0 = 1$. We will find

$$(f \star g)^\Delta (t), \quad t \in \mathbb{T}, \quad t \geq 1.$$

Here,

$$\sigma(t) = 2t, \quad t \in \mathbb{T}.$$

Then

$$g^\Delta(t) = \sigma(t) + t$$

$$= 2t + t$$

$$= 3t,$$

$$f(t) = h_0(t, 1), \quad t \in \mathbb{T}, \quad t \geq 1.$$

Hence, from Theorem 1.20, we obtain

$$\hat{f}(t, s) = h_0(t, s), \quad t, s \in \mathbb{T}, \quad t \geq s \geq 1.$$

Therefore,

$$(f \star g^\Delta)(t) = \int_1^t f(t, \sigma(s))g^\Delta(s)\Delta s$$

$$= \int_1^t g^\Delta(s)\Delta s$$

$$= g(s)\Big|_{s=1}^{s=t}$$

$$= s^2\Big|_{s=1}^{s=t}$$

$$= t^2 - 1,$$

$$f(t)g(1) = 1, \quad t \in \mathbb{T}, \quad t \geq 1.$$

Hence, from (1.15), we get

$$(f \star g)^\Delta (t) = t^2 - 1 + 1$$
$$= t^2, \quad t \in \mathbb{T}, \quad t \geq 1.$$

Example 1.34. Let $\mathbb{T} = 3^{\mathbb{N}_0}$, $t_0 = 1$, $f(t) = t$, $g(t) = t^3$, $t \in \mathbb{T}$. We will find

$$(f \star g)^\Delta (t), \quad t \in \mathbb{T}, \quad t \geq 1.$$

Here,

$$\sigma(t) = 3t, \quad t \in \mathbb{T}, \quad t \geq 1.$$

We have

$$g^\Delta(t) = (\sigma(t))^2 + t\sigma(t) + t^2$$
$$= (3t)^2 + t(3t) + t^2$$
$$= 9t^2 + 3t^2 + t^2$$
$$= 13t^2,$$
$$f(1) = 1,$$
$$f^\Delta(t) = 1,$$
$$f^\Delta(1) = 1, \quad t \in \mathbb{T}, \quad t \geq 1.$$

Then

$$f(t) = h_0(t,1) + h_1(t,1), \quad t \in \mathbb{T}.$$

Hence, from Theorem 1.20, we get

$$\hat{f}(t, s) = h_0(t, s) + h_1(t, s)$$
$$= 1 + t - s,$$
$$\hat{f}(t, \sigma(s)) = t - \sigma(s) + 1$$
$$= t - 3s + 1, \quad t, s \in \mathbb{T}, \quad t \geq s \geq 1.$$

Let

$$l(s) = \frac{1}{13}(1 + t)s^3 - \frac{3}{40}s^4, \quad s \in \mathbb{T},$$

for some $t \in \mathbb{T}$. Then

$$l^\Delta(s) = \frac{1}{13}(1 + t)\left((\sigma(s))^2 + s\sigma(s) + s^2\right)$$
$$- \frac{3}{40}\left((\sigma(s))^3 + s(\sigma(s))^2 + s^2\sigma(s) + s^3\right)$$
$$= \frac{1}{13}(1 + t)\left((3s)^2 + s(3s) + s^2\right)$$
$$- \frac{3}{40}\left((3s)^3 + s(3s)^2 + s^2(3s) + s^3\right)$$
$$= \frac{1}{13}(1 + t)\left(9s^2 + 3s^2 + s^2\right)$$
$$- \frac{3}{40}\left(27s^3 + 9s^3 + 3s^3 + s^3\right)$$
$$= (1 + t)s^2 - 3s^3, \quad s \in \mathbb{T},$$

for some $t \in \mathbb{T}$. Therefore,

$$\left(f \star g^\Delta\right)(t) = \int_1^t \hat{f}(t, \sigma(s))g^\Delta(s)\Delta s$$
$$= 13\int_1^t (1 + t - 3s)s^2 \Delta s$$
$$= 13\int_1^t l^\Delta(s)\Delta s$$
$$= 13l(s)\Big|_{s=1}^{s=t}$$

$$= 13 \left(\frac{1}{13}(1+t)s^3 - \frac{3}{40}s^4 \right) \Big|_{s=1}^{s=t}$$

$$= 13 \left(\frac{(1+t)t^3}{13} - \frac{3}{40}t^4 \right)$$

$$- 13 \left(\frac{1}{13}(1+t) - \frac{3}{40} \right)$$

$$= \frac{t^3(40 + 40t - 39t)}{40} - \frac{40 + 40t - 39}{40}$$

$$= \frac{t^3(40 + t) - 1 - 40t}{40}$$

$$= \frac{t^4 + 40t^3 - 40t - 1}{40}, \quad t \in \mathbb{T}, \quad t \geq 1.$$

Hence, from (1.15), we obtain

$$(f \star g)^\Delta (t) = \frac{t^4 + 40t^3 - 40t - 1}{40} + t$$

$$= \frac{t^4 + 40t^3 - 1}{40}, \quad t \in \mathbb{T}, \quad t \geq 1.$$

Exercise 1.11. Let $\mathbb{T} = 4^{\mathbb{N}_0}$, $f(t) = 1$, $g(t) = 3t + 7$, $t \in \mathbb{T}$, $t_0 = 1$. Find

$$(f \star g)^\Delta (t), \quad t \in \mathbb{T}, \quad t \geq 1.$$

Answer: $3t + 7$, $t \in \mathbb{T}$, $t \geq 1$.

Example 1.35. Let $\mathbb{T} = \mathbb{Z}$, $t_0 = 0$, $f(t) = t^2 - 3t$, $t \in \mathbb{T}$. We will compute

$$\int_0^t \hat{f}(t, \sigma(s))\Delta s, \quad t \in \mathbb{Z}, \quad t \geq 0.$$

Here,

$$\sigma(t) = t + 1, \quad t \in \mathbb{T}.$$

Let

$$g(t) = \frac{1}{3}t^3 - 2t^2 + \frac{5}{3}t, \quad t \in \mathbb{T}.$$

Then

$$g^{\Delta}(t) = \frac{1}{3}\left((\sigma(t))^2 + t\sigma(t) + t^2\right)$$

$$- 2\left(\sigma(t) + t\right) + \frac{5}{3}$$

$$= \frac{1}{3}\left((t+1)^2 + t(t+1) + t^2\right)$$

$$- 2(t+1+t) + \frac{5}{3}$$

$$= \frac{1}{3}\left(t^2 + 2t + 1 + t^2 + t + t^2\right)$$

$$- 2(2t+1) + \frac{5}{3}$$

$$= \frac{1}{3}\left(3t^2 + 3t + 1\right) - 4t - 2 + \frac{5}{3}$$

$$= t^2 + t + \frac{1}{3} - 4t - \frac{1}{3}$$

$$= t^2 - 3t$$

$$= f(t), \quad t \in \mathbb{T}, \quad t \geq 0.$$

Hence,

$$\int_0^t f(s)\Delta s = \int_0^t g^{\Delta}(s)\Delta s$$

$$= g(s)\Big|_{s=0}^{s=t}$$

$$= \left(\frac{1}{3}s^3 - 2s^2 + \frac{5}{3}s\right)\Big|_{s=0}^{s=t}$$

$$= \frac{1}{3}t^3 - 2t^2 + \frac{5}{3}t, \quad t \in \mathbb{T}, \quad t \geq 0.$$

From here and from Corollary 1.2, we get that

$$\int_0^t \hat{f}(t, \sigma(s))\Delta s = \frac{1}{3}t^3 - 2t^2 + \frac{5}{3}t, \quad t \in \mathbb{T}, \quad t \geq 0.$$

Example 1.36. Let $\mathbb{T} = 2^{\mathbb{N}_0}$, $t_0 = 1$,

$$f(t) = \frac{-2t^2 - 3t + 2}{2\left(t^2 + 2\right)\left(2t^2 + 1\right)}, \quad t \in \mathbb{T}.$$

We will find

$$\int_1^t \hat{f}(t, \sigma(s))\Delta s, \quad t \in \mathbb{T}, \quad t \geq 1.$$

Here,

$$\sigma(t) = 2t, \quad t \in \mathbb{T}.$$

Let

$$g(t) = \frac{t+1}{t^2 + 2}, \quad t \in \mathbb{T}, \quad t \geq 1.$$

Then

$$g^{\Delta}(t) = \frac{t^2 + 2 - (t+1)(\sigma(t) + t)}{(t^2 + 2)\left((\sigma(t))^2 + 2\right)}$$

$$= \frac{t^2 + 2 - (t+1)(2t + t)}{(t^2 + 2)\left((2t)^2 + 2\right)}$$

$$= \frac{t^2 + 2 - 3t(t+1)}{(t^2 + 2)\left(4t^2 + 2\right)}$$

$$= \frac{t^2 + 2 - 3t^2 - 3t}{2\left(t^2 + 2\right)\left(2t^2 + 1\right)}$$

$$= \frac{-2t^2 - 3t + 2}{2\left(t^2 + 2\right)\left(2t^2 + 1\right)}$$

$$= f(t), \quad t \in \mathbb{T}.$$

Hence,

$$\int_1^t f(s)\Delta s = \int_1^t g^{\Delta}(s)\Delta s$$

$$= g(s)\Big|_{s=1}^{s=t}$$

$$= \frac{s+1}{s^2+2}\Big|_{s=1}^{s=t}$$

$$= \frac{t+1}{t^2+2} - \frac{2}{3}$$

$$= \frac{3t+3-2t^2-4}{3\,(t^2+2)}$$

$$= \frac{-2t^2+3t-1}{3\,(t^2+2)}, \quad t \in \mathbb{T}, \quad t > 1.$$

Consequently, using Corollary 1.2, we get

$$\int_1^t \hat{f}(t,\sigma(s))\Delta s = \frac{-2t^2+3t-1}{3\,(t^2+2)}, \quad t \in \mathbb{T}, \quad t > 1.$$

Example 1.37. Let $\mathbb{T} = 3^{\mathbb{N}_0}$, $t_0 = 1$,

$$f(t) = \frac{1}{(t+1)(3t+2)} + \sin_1(t,1) + 3t\cos_1(t,1), \quad t \in \mathbb{T}, \quad t > 1.$$

We will find

$$\int_1^t \hat{f}(t,\sigma(s))\Delta s, \quad t \in \mathbb{T}, \quad t > 1.$$

Here,

$$\sigma(t) = 3t, \quad t \in \mathbb{T}.$$

Let

$$g(t) = \frac{t+1}{t+2} + t\sin_1(t,1), \quad t \in \mathbb{T}.$$

We have

$$g^\Delta(t) = \frac{t+2-(t+1)}{(t+2)(\sigma(t)+2)} + \sin_1(t,1) + \sigma(t)\cos_1(t,1)$$

$$= \frac{1}{(t+2)(3t+2)} + \sin_1(t,1) + 3t\cos_1(t,1)$$

$$= f(t), \quad t \in \mathbb{T}, \quad t > 1.$$

Then

$$\int_1^t f(s)\Delta s = \int_1^t g^\Delta(s)\Delta s$$

$$= g(s)\Big|_{s=1}^{s=t}$$

$$= \left(\frac{s+1}{s+2} + s\sin_1(s,1)\right)\Big|_{s=1}^{s=t}$$

$$= \frac{t+1}{t+2} + t\sin_1(t,1) - \frac{2}{3}$$

$$= \frac{3t+3-2t-4}{3(t+2)} + t\sin_1(t,1)$$

$$= \frac{t-1}{3(t+2)} + t\sin_1(t,1), \quad t\in\mathbb{T}, \quad t>1.$$

Consequently, using Corollary 1.2, we get

$$\int_1^t \hat{f}(t,\sigma(s))\Delta s = \frac{t-1}{3(t+2)} + t\sin_1(t,1), \quad t\in\mathbb{T}, \quad t>1.$$

Exercise 1.12. Let $\mathbb{T} = 2^{\mathbb{N}_0}$, $t_0 = 1$. Let also

$$f(t) = \frac{t^2 - 3t - 1}{t^2(t+1)(2t+1)}, \quad t\in\mathbb{T}.$$

Find

$$\int_1^t \hat{f}(t,\sigma(s))\Delta s, \quad t\in\mathbb{T}, \quad t>1.$$

Hint: *Use the function*

$$g(t) = \frac{t^2+2}{t^2+t}, \quad t\in\mathbb{T}.$$

Answer:

$$\frac{-t^2-3t+4}{2t(t+1)}, \quad t\in\mathbb{T}, \quad t>1.$$

Theorem 1.25 ([8]). *If f and g are infinitely often Δ-differentiable, then, for all $k \in \mathbb{N}_0$, we have*

$$(f \star g)^{\Delta^k} = f^{\Delta^k} \star g + \sum_{\nu=0}^{k-1} f^{\Delta^\nu}(t_0) g^{\Delta^{k-1-\nu}}$$

$$= f \star g^{\Delta^k} + \sum_{\nu=0}^{k-1} f^{\Delta^\nu} g^{\Delta^{k-1-\nu}}(t_0), \qquad (1.16)$$

$$(f \star g)^{\Delta^k}(t_0) = \sum_{\nu=0}^{k-1} f^{\Delta^\nu}(t_0) g^{\Delta^{k-1-\nu}}(t_0).$$

Example 1.38. Let $\mathbb{T} = 2\mathbb{Z}$, $t_0 = 0$,

$$f(t) = t^2 + 3t + 1, \quad g(t) = t^2, \quad t \in \mathbb{T}.$$

We will find

$$(f \star g)^{\Delta^4}(t), \quad t \in \mathbb{T}, \quad t > 0.$$

Here,

$$\sigma(t) = t + 2, \quad t \in \mathbb{T}.$$

We have

$$\begin{aligned}
f(0) &= 1, \\
f^{\Delta}(t) &= \sigma(t) + t + 3 \\
&= t + 2 + t + 3 \\
&= 2t + 5, \\
f^{\Delta}(0) &= 5, \\
f^{\Delta^2}(t) &= 2, \\
f^{\Delta^2}(0) &= 2, \\
f^{\Delta^k}(t) &= 0, \\
f^{\Delta^k}(0) &= 0, \quad k \in \mathbb{N}, \quad k \geq 3, \\
g^{\Delta}(t) &= \sigma(t) + t
\end{aligned}$$

$$= t + 2 + t$$
$$= 2t + 2,$$
$$g^{\Delta^2}(t) = 2,$$
$$g^{\Delta^k}(t) = 0, \quad k \in \mathbb{N}, \quad k \geq 3.$$

Then, using (1.16), we get

$$(f \star g)^{\Delta^4}(t) = \left(f^{\Delta^4} \star g\right)(t) + \sum_{\nu=0}^{3} f^{\Delta^\nu}(0) g^{\Delta^{3-\nu}}(t)$$

$$= f(0) f^{\Delta^3}(t) + f^{\Delta}(0) g^{\Delta^2}(t) + f^{\Delta^2}(0) g^{\Delta}(t)$$
$$\quad + f^{\Delta^3}(0) g(t)$$
$$= 10 + 2(2t + 2)$$
$$= 10 + 4t + 4$$
$$= 4t + 14, \quad t \in \mathbb{T}, \quad t > 0.$$

Example 1.39. Let $\mathbb{T} = 2^{\mathbb{N}_0}$, $t_0 = 1$,

$$f(t) = t^3 + t^2 + t, \quad g(t) = t^2 - 2t + 1, \quad t \in \mathbb{T}.$$

We will find

$$(f \star g)^{\Delta^3}(t), \quad t \in \mathbb{T}, \quad t > 1.$$

Here,

$$\sigma(t) = 2t, \quad t \in \mathbb{T}.$$

We have

$$f(1) = 3,$$
$$f^{\Delta}(t) = (\sigma(t))^2 + t\sigma(t) + t^2 + \sigma(t) + t + 1$$
$$= (2t)^2 + t(2t) + t^2 + 2t + t + 1$$
$$= 4t^2 + 2t^2 + t^2 + 3t + 1$$
$$= 7t^2 + 3t + 1,$$
$$f^{\Delta}(1) = 11,$$

$$f^{\Delta^2}(t) = 7(\sigma(t) + t) + 3$$
$$= 7(2t + t) + 3$$
$$= 21t + 3,$$
$$f^{\Delta^2}(1) = 24,$$
$$f^{\Delta^3}(t) = 21,$$
$$f^{\Delta^3}(1) = 21,$$
$$g^{\Delta}(t) = \sigma(t) + t - 2$$
$$= 2t + t - 2$$
$$= 3t - 2,$$
$$g^{\Delta^2}(t) = 3,$$
$$g^{\Delta^3}(t) = 0, \quad t \in \mathbb{T}.$$

Also,

$$f^{\Delta^3}(t) = 21h_0(t, 1), \quad t \in \mathbb{T}.$$

Hence, from Theorem 1.20, we obtain

$$\widehat{f^{\Delta^3}}(t, s) = 21h_0(t, s)$$
$$= 21, \quad t, s \in \mathbb{T}, \quad t \geq s \geq 1.$$

Let

$$l(t) = \frac{1}{7}t^3 - \frac{2}{3}t^2 + t, \quad t \in \mathbb{T}.$$

Then

$$l^{\Delta}(t) = \frac{1}{7}\left((\sigma(t))^2 + t\sigma(t) + t^2\right) - \frac{2}{3}(\sigma(t) + t) + 1$$
$$= \frac{1}{7}\left((2t)^2 + t(2t) + t^2\right) - \frac{2}{3}(2t + t) + 1$$
$$= \frac{1}{7}\left(4t^2 + 2t^2 + t^2\right) - \frac{2}{3}(3t) + 1$$

$$= \frac{1}{7}(7t^2) - 2t + 1$$

$$= t^2 - 2t + 1, \quad t \in \mathbb{T}.$$

Therefore,

$$\left(f^{\Delta^3} \star g\right)(t) = \int_1^t \widehat{f^{\Delta^3}}(t, \sigma(s))g(s)\Delta s$$

$$= 21 \int_1^t g(s)\Delta s$$

$$= 21 \int_1^t l^{\Delta}(s)\Delta s$$

$$= 21 l(s)\Big|_{s=1}^{s=t}$$

$$= 21 \left(\frac{1}{7}s^3 - \frac{2}{3}s^2 + s\right)\Big|_{s=1}^{s=t}$$

$$= 21 \left(\frac{1}{7}t^3 - \frac{2}{3}t^2 + t - \frac{1}{7} + \frac{2}{3} - 1\right)$$

$$= 21 \left(\frac{1}{7}t^3 - \frac{2}{3}t^2 + t - \frac{10}{21}\right)$$

$$= 3t^3 - 14t^2 + 21t - 10, \quad t \in \mathbb{T}, \quad t > 1.$$

Hence, from (1.16), we get

$$(f \star g)^{\Delta^3}(t) = \left(f^{\Delta^3} \star g\right)(t) + \sum_{\nu=0}^{2} f^{\Delta^\nu}(1)g^{\Delta^{2-\nu}}(t)$$

$$= 3t^3 - 14t^2 + 21t - 10$$

$$+ f(1)g^{\Delta^2}(t) + f^{\Delta}(1)g^{\Delta}(t) + f^{\Delta^2}(1)g(t)$$

$$= 3t^3 - 14t^2 + 21t - 10 + 9$$

$$+ 11(3t - 2) + 24\left(t^2 - 2t + 1\right)$$

$$= 3t^3 - 14t^2 + 21t - 10 + 9 + 33t - 22$$

$$+ 24t^2 - 48t + 24$$

$$= 3t^3 + 10t^2 + 6t + 1, \quad t \in \mathbb{T}, \quad t > 1.$$

Example 1.40. Let $\mathbb{T} = 3^{\mathbb{N}_0}$, $t_0 = 1$,

$$f(t) = t^2, \quad g(t) = \frac{1}{3t^2}, \quad t \in \mathbb{T}.$$

We will find

$$(f \star g)^{\Delta^2}(t), \quad t \in \mathbb{T}, \quad t > 1.$$

Here,

$$\sigma(t) = 3t, \quad t \in \mathbb{T}.$$

We have

$$\begin{aligned}
f(1) &= 1, \\
f^\Delta(t) &= \sigma(t) + t \\
&= 3t + t \\
&= 4t, \\
f^\Delta(1) &= 4, \\
f^{\Delta^2}(t) &= 4, \\
f^{\Delta^2}(1) &= 4, \\
g^\Delta(t) &= -\frac{\sigma(t) + t}{3t^2 (\sigma(t))^2} \\
&= -\frac{4t}{3t^2 (3t)^2} \\
&= -\frac{4}{3t(9t^2)} \\
&= -\frac{4}{27t^3}, \\
g^{\Delta^2}(t) &= \frac{4}{27} \frac{(\sigma(t))^2 + t\sigma(t) + t^2}{t^3 (\sigma(t))^3} \\
&= \frac{4}{27} \frac{(3t)^2 + 3t^2 + t^2}{t^3 27t^3}
\end{aligned}$$

$$= \frac{4}{729} \frac{9t^2 + 4t^2}{t^6}$$

$$= \frac{52}{729} \frac{1}{t^4}, \quad t \in \mathbb{T}, \quad t > 1.$$

Also,

$$f^{\Delta^2}(t) = 4h_0(t, 1), \quad t \in \mathbb{T}.$$

Hence, from Theorem 1.20, we get

$$\widehat{f^{\Delta^2}}(t, s) = 4h_0(t, s), \quad t, s \in \mathbb{T}, \quad t \geq s \geq 1.$$

Therefore,

$$\left(f^{\Delta^2} \star g\right)(t) = \int_1^t \widehat{f^{\Delta^2}}(t, \sigma(s))g(s)\Delta s$$

$$= 4 \int_1^t g(s)\Delta s$$

$$= 4 \left(-\frac{1}{s}\right) \Big|_{s=1}^{s=t}$$

$$= -4 \left(\frac{1}{t} - 1\right)$$

$$= -4 \frac{1 - t}{t}$$

$$= \frac{4(t - 1)}{t}, \quad t \in \mathbb{T}, \quad t > 1,$$

$$(f \star g)^{\Delta^2}(t) = \left(f^{\Delta^2} \star g\right)(t) + \sum_{\nu=0}^{1} f^{\Delta^\nu}(1)g^{\Delta^{1-\nu}}(t)$$

$$= \frac{4(t - 1)}{t} + f(1)g^\Delta(t) + f^\Delta(1)g(t)$$

$$= \frac{4(t - 1)}{t} + \left(-\frac{4}{27t^3}\right) + \frac{4}{3t^2}$$

$$= 4 \frac{27t^3 - 27t^2 - 1 + 9t}{27t^3}, \quad t \in \mathbb{T}, \quad t > 1.$$

Exercise 1.13. Let $\mathbb{T} = 2^{\mathbb{N}_0}$, $t_0 = 1$,

$$f(t) = t^2 + t + 3, \quad g(t) = t^3 - 3t^2 + t + 1, \quad t \in \mathbb{T}, \quad t > 1.$$

Find

$$(f \star g)^{\Delta^{11}}(t), \quad t \in \mathbb{T}, \quad t > 1.$$

Answer: 0.

Theorem 1.26 ([8]). *If $\hat{f}$ has partial Δ-derivatives of all orders, then*

$$\hat{f}^{\Delta_t^k}(t, t) = f^{\Delta^k}(t_0) \tag{1.17}$$

for all $k \in \mathbb{N}_0$, where $\hat{f}^{\Delta_t}$ indicates the Δ-derivatives of $\hat{f}$ with respect to its first variable.

Example 1.41. Let $\mathbb{T} = \mathbb{Z}$, $t_0 = 0$,

$$f(t) = t^3 + t^2 - 2t + 1, \quad t \in \mathbb{T}.$$

We will find

$$\hat{f}^{\Delta_t^2}(t, t), \quad t \in \mathbb{T}.$$

Here,

$$\sigma(t) = t + 1, \quad t \in \mathbb{T}.$$

We have

$$\begin{aligned}
f^{\Delta}(t) &= (\sigma(t))^2 + t\sigma(t) + t^2 + \sigma(t) + t - 2 \\
&= (t+1)^2 + t(t+1) + t^2 + t + 1 + t - 2 \\
&= t^2 + 2t + 1 + t^2 + t + t^2 + 2t - 1 \\
&= 3t^2 + 5t, \\
f^{\Delta^2}(t) &= 3(\sigma(t) + t) + 5 \\
&= 3(t + 1 + t) + 5
\end{aligned}$$

$$= 3(2t + 1) + 5$$
$$= 6t + 8,$$
$$f^{\Delta^2}(0) = 8, \quad t \in \mathbb{T}.$$

Hence, from (1.17), we get

$$\hat{f}^{\Delta_t^2}(t, t) = 8, \quad t \in \mathbb{T}.$$

Example 1.42. Let $\mathbb{T} = 2^{\mathbb{N}_0}$, $t_0 = 1$,

$$f(t) = \frac{t + 1}{t + 2}, \quad t \in \mathbb{T}.$$

We will find

$$\hat{f}^{\Delta_t^3}(t, t), \quad t \in \mathbb{T}.$$

Here,

$$\sigma(t) = 2t, \quad t \in \mathbb{T}.$$

We have

$$f^{\Delta}(t) = \frac{t + 2 - (t + 1)}{(t + 2)(\sigma(t) + 2)}$$
$$= \frac{t + 2 - t - 1}{(t + 2)(2t + 2)}$$
$$= \frac{1}{2(t + 1)(t + 2)}$$
$$= \frac{1}{2\left(t^2 + 3t + 2\right)},$$

$$f^{\Delta^2}(t) = -\frac{\sigma(t) + t + 3}{2\left(t^2 + 3t + 2\right)\left((\sigma(t))^2 + 3\sigma(t) + 2\right)}$$
$$= -\frac{2t + t + 3}{2(t + 1)(t + 2)\left(4t^2 + 6t + 2\right)}$$

$$= -\frac{3(t+1)}{4(t+1)(t+2)(t+1)\left(t+\frac{1}{2}\right)}$$

$$= -\frac{3}{2(t+1)(t+2)(2t+1)}$$

$$= -\frac{3}{2\left(2t^3 + 7t^2 + 7t + 2\right)},$$

$$f^{\Delta^3}(t) = \frac{3}{2} \frac{2\left((\sigma(t))^2 + t\sigma(t) + t^2\right) + 7\left(\sigma(t) + t\right) + 7}{(t+1)(t+2)(2t+1)\left(\sigma(t)+1\right)\left(\sigma(t)+2\right)\left(2\sigma(t)+1\right)}$$

$$= \frac{3}{2} \frac{2\left(4t^2 + 2t^2 + t^2\right) + 7(3t) + 7}{(t+1)(t+2)(2t+1)(2t+1)(2t+2)(4t+1)}$$

$$= \frac{3}{4} \frac{14t^2 + 21t + 7}{(t+1)^2(t+2)(2t+1)^2(4t+1)},$$

$$f^{\Delta^3}(1) = \frac{7}{120}, \quad t \in \mathbb{T}.$$

Hence, from (1.17), we get

$$\hat{f}^{\Delta_t^3}(t,t) = \frac{7}{120}, \quad t \in \mathbb{T}.$$

Example 1.43. Let $\mathbb{T} = 3^{\mathbb{N}_0}$, $t_0 = 1$,

$$f(t) = \frac{t+2}{t^2 + 2t + 1}, \quad t \in \mathbb{T}.$$

We will find

$$\hat{f}^{\Delta_t}(t,t), \quad t \in \mathbb{T}.$$

Here,

$$\sigma(t) = 3t, \quad t \in \mathbb{T}.$$

We have

$$f^{\Delta}(t) = \frac{t^2 + 2t + 1 - (t+2)\left(\sigma(t) + t + 2\right)}{(t^2 + 2t + 1)\left((\sigma(t))^2 + 2\sigma(t) + 1\right)}$$

$$= \frac{t^2 + 2t + 1 - (t+2)(3t + t + 2)}{(t^2 + 2t + 1)\left(9t^2 + 6t + 1\right)}$$

$$= \frac{t^2 + 2t + 1 - (t+2)(4t+2)}{(t+1)^2(3t+1)^2}, \quad t \in \mathbb{T},$$

$$f^{\Delta}(1) = -\frac{7}{32}.$$

Hence, from (1.17), we get

$$\hat{f}^{\Delta_t}(t,t) = -\frac{7}{32}, \quad t \in \mathbb{T}.$$

Exercise 1.14. Let $\mathbb{T} = \mathbb{Z}$, $t_0 = 0$,

$$f(t) = t^2 - 11t + 12, \quad t \in \mathbb{T}.$$

Find

$$\hat{f}^{\Delta_t^2}(t,t), \quad t \in \mathbb{T}.$$

Answer: 2.

1.9   The Convolution Theorem

We start with the following useful theorem.

Theorem 1.27 ([8]). *Let* $f : \mathbb{T} \to \mathbb{C}$ *and*

$$\psi(s) = \int_s^\infty \frac{\hat{f}(t,s)}{e_z(\sigma(t),s)} \Delta t, \quad z \in \mathcal{D}\{f\}, \quad s \in \mathbb{T}.$$

Then ψ *is a constant.*

Corollary 1.3 ([8]). *Let* f *and* ψ *be as in Theorem 1.27. Then*

$$\psi(t_0) = \mathcal{L}(f)(z), \quad z \in \mathcal{D}\{f\}.$$

Theorem 1.28 ([8] (Convolution Theorem)). *Suppose* $f, g :$ $\mathbb{T} \to \mathbb{C}$ *are locally* Δ-*integrable functions on* $\mathbb{T}$. *Then*

$$\mathcal{L}\left(f \star g\right)(z) = \mathcal{L}(f)(z)\mathcal{L}(g)(z), \quad z \in \mathcal{D}\{f\}\bigcap \mathcal{D}\{g\}.$$

Theorem 1.29 ([8]). *Let* $s, t_0 \in \mathbb{T}$, $s \geq t_0$, *and*

$$u_s(t) = \begin{cases} 0 & if \quad t < s, \\ 1 & if \quad s \geq t. \end{cases}$$

Then

$$\mathcal{L}\left(u_s\hat{f}(\cdot,s)\right)(z) = e_{\ominus z}(s,t_0)\mathcal{L}(f)(z), \quad z \in \mathcal{D}\{f\}.$$

1.10   Advanced Practical Problems

Problem 1.1. Let $\mathbb{T} = \{-4, -\frac{7}{2}, -3, -2, -1, 0, \frac{1}{8}, \frac{1}{4}, 3, 5, 9\}$. Find $\sigma(t)$, $\rho(t)$, $t \in \mathbb{T}$.

Problem 1.2. Let $\mathbb{T} = 3^{\mathbb{N}_0}$. Find $\sigma(t)$, $\rho(t)$, $t \in \mathbb{T}$.

Problem 1.3. Classify the points of $\mathbb{T} = \{0\} \cup \left\{\frac{3n}{3n+4}\right\}_{n\in\mathbb{N}_0} \cup \{1\} \cup \left\{\frac{2}{n}\right\}_{n\in\mathbb{N}}$.

Problem 1.4. Let $\mathbb{T} = 2^{\mathbb{N}_0}$, $f, g : \mathbb{T} \to \mathbb{R}$ be given by

$$f(t) = 1 - t^2, \quad g(t) = \frac{1+t}{1+t+t^2}, \quad t \in \mathbb{T}.$$

Find $f^\sigma(t)$, $f^\rho(t)$, $g^\sigma(t)$, $g^\rho(t)$, $f^\sigma(t)g^\rho(t)$, $t \in \mathbb{T}$.

Problem 1.5. Let $\mathbb{T} = 3^{\mathbb{N}_0}$ and $f : \mathbb{T} \to \mathbb{R}$ be defined by $f(t) = 1 + 3t + t^4$, $t \in \mathbb{T}$. Find $f^\Delta(t)$, $f^{\Delta^2}(t)$, $f^{\Delta\sigma}(t)$, $f^{\Delta^2\rho}(t)$, $f^\nabla(t)$, $f^{\nabla^2}(t)$, $f^{\nabla\rho}(t)$, $f^{\nabla^2\sigma}(t)$, $t \in \mathbb{T}$.

Problem 1.6. Let $\mathbb{T} = 2^{\mathbb{N}_0}$ and $f : \mathbb{T} \to \mathbb{R}$ be defined by

$$f(t) = \frac{1}{(1+t^2)(1+4t^2)} + t, \quad t \in \mathbb{T}.$$

Find

$$\int f(t)\Delta t, \quad \int f^\sigma(t)\Delta t, \quad \int f^\rho(t)\Delta t, \quad \int_8^{64} f(t)\nabla t.$$

Problem 1.7. Let $\mathbb{T} = \mathbb{N}_0$, $t_0 = 0$,

$$f(t) = \frac{t+1}{t+2}, \quad t \in \mathbb{T}.$$

Find the shift of f.

Answer:

$$\hat{f}(t,s) = \frac{1}{2}h_0(t,s) + \frac{1}{6}h_1(t,s) - \frac{1}{12}h_2(t,s) + \cdots, \quad t,s \in \mathbb{T}, \quad t \geq s \geq 0.$$

Problem 1.8. Let $\mathbb{T} = 2^{\mathbb{N}_0}$, $t_0 = 1$,

$$f(t) = \sum_{k=0}^{\infty} \frac{k}{2k+1} h_k(t,1), \quad g(t) = \sum_{k=0}^{\infty} \frac{k}{k+2} h_k(t,1), \quad t \in \mathbb{T}, \quad t \geq 1.$$

Find $(f \star g)(t)$, $t \in \mathbb{T}$, $t \geq 1$.

Answer:

$$(f \star g)(t) = \sum_{k=0}^{\infty} \sum_{l=0}^{k} \left(\frac{k-l}{2(k-l)+1} \right) \left(\frac{l}{l+2} \right) h_{k+1}(t,1), \quad t \in \mathbb{T}, \quad t \geq 1.$$

Problem 1.9. Let $\mathbb{T} = 2^{\mathbb{N}_0}$, $f(t) = t^7$, $g(t) = 1$, $t \in \mathbb{T}$, $t_0 = 1$. Find

$$(f \star g)^{\Delta}(t), \quad t \in \mathbb{T}, \quad t \geq 1.$$

Answer: t^7, $t \in \mathbb{T}$, $t \geq 1$.

Problem 1.10. Let $\mathbb{T} = 2^{\mathbb{N}_0}$, $t_0 = 1$,

$$f(t) = \frac{-4t^2 - 3t + 4}{2(t^2+2)(2t^2+1)}, \quad t \in \mathbb{T}, \quad t > 1.$$

Find

$$\int_{t_0}^{t} \hat{f}(t, \sigma(s)) \, \Delta s, \quad t \in \mathbb{T}, \quad t > t_0.$$

Answer:

$$-\frac{(t-1)^2}{t^2+2}, \quad t \in \mathbb{T}, \quad t > 1.$$

Problem 1.11. Let $\mathbb{T} = 4^{\mathbb{N}_0}$, $t_0 = 1$,

$$f(t) = t^4 + 4t^3 + 2t^2 + t - 1, \quad g(t) = t^5 + t^4 + t^3 - 3t^2, \quad t \in \mathbb{T}, \quad t > 1.$$

Find

$$(f \star g)^{\Delta^{40}}(t), \quad t \in \mathbb{T}, \quad t > 1.$$

Answer: 0.

Problem 1.12. Let $\mathbb{T} = 2^{\mathbb{N}_0}$, $t_0 = 1$,

$$f(t) = t^7 + 12t^5 + t^3 + 3t^2 + t + 15, \quad t \in \mathbb{T}.$$

Find

$$\hat{f}^{\Delta_t^{20}}(t, t), \quad t \in \mathbb{T}.$$

Answer: 0.

Chapter 2

Elements of Fractional Dynamic Calculus on Time Scales

In this chapter, we introduce the basic definitions and results of the fractional dynamic calculus on time scales. For more details regarding the fractional dynamic calculus on time scales and the proofs of the basic results in this chapter, we refer the reader to Ref. [8].

In this chapter, we suppose that $\mathbb{T}$ is a time scale with forward jump operator and delta differentiation operator σ and Δ, respectively, such that

$$\mathbb{T} = \{t_n : n \in \mathbb{N}_0\},$$

$$\lim_{n \to \infty} t_n = \infty,$$

$$\sigma(t_n) = t_{n+1}, \quad n \in \mathbb{N}_0,$$

$$w = \inf_{n \in \mathbb{N}_0} \mu(t_n) > 0.$$

2.1 The Δ-Power Function

Suppose that $\alpha \in \mathbb{R}$.

Definition 2.1 ([8] (The Δ-Power Function)). We define generalized Δ-power function $h_\alpha(t, t_0)$ on $\mathbb{T}$ as follows:

$$h_\alpha(t, t_0) = \mathcal{L}^{-1}\left(\frac{1}{z^{\alpha+1}}\right)(t), \quad t \geq t_0, \tag{2.1}$$

for those $z \in \mathbb{C} \backslash \{0\}$ such that $\mathcal{L}^{-1}$ exist, $t \geq t_0$. The fractional generalized Δ-power function $h_\alpha(t, s)$ on $\mathbb{T}$, $t \geq s \geq t_0$, is defined as the shift of $h_\alpha(t, t_0)$, i.e.,

$$h_\alpha(t, s) = \widehat{h_\alpha(\cdot, t_0)}(t, s), \quad t, s \in \mathbb{T}, \quad t \geq s \geq t_0. \qquad (2.2)$$

Let

$$\gamma = \left\{ z \in \mathbb{C} : |z| = \frac{1}{w}, \ \sqrt{1} = 1 \right\}.$$

We can rewrite (2.1) as follows:

$$h_\alpha(t_n, t_0) = \frac{1}{2\pi i} \int_\gamma \frac{1}{z^{\alpha+1}} \prod_{k=0}^{n-1} (1 + \mu(t_k)z) \, dz, \quad n \in \mathbb{N}.$$

We have

$$h_\alpha(t_0, t_0) = \lim_{z \to \infty} \frac{z}{z^{\alpha+1}}$$

$$= \lim_{z \to \infty} \frac{1}{z^\alpha}$$

$$= \begin{cases} 0 & \text{if } \alpha > 0, \\ 1 & \text{if } \alpha = 0, \\ \infty & \text{if } \alpha < 0, \end{cases}$$

$$\lim_{t \to \infty} h_\alpha(t, t_0) = \lim_{z \to 0} \frac{z}{z^{\alpha+1}}$$

$$= \lim_{z \to 0} \frac{1}{z^\alpha}$$

$$= \begin{cases} \infty & \text{if } \alpha > 0, \\ 1 & \text{if } \alpha = 0, \\ 0 & \text{if } \alpha < 0. \end{cases}$$

Theorem 2.1 ([8]). *Let $\alpha, \beta \in \mathbb{R}$. Then*

$$(h_\alpha(\cdot, t_0) \star h_\beta(\cdot, t_0))(t) = h_{\alpha+\beta+1}(t, t_0), \quad t \in \mathbb{T}. \qquad (2.3)$$

By (2.3), we get

$$\int_{t_0}^t \widehat{h_\alpha(\cdot, t_0)}(t, \sigma(u)) h_\beta(u, t_0) \Delta u = h_{\alpha+\beta+1}(t, t_0), \quad t \in \mathbb{T}.$$

Hence, for $\alpha = 0$, we get

$$\int_{t_0}^t h_\beta(u, t_0) \Delta u = h_{\beta+1}(t, t_0), \quad t \in \mathbb{T},$$

and

$$h_{\beta+1}^\Delta(t, t_0) = h_\beta(t, t_0), \quad t \in \mathbb{T}.$$

2.2 Definition for the Riemann–Liouville Fractional Δ-Integral and the Riemann–Lioville Fractional Δ-Derivative

Suppose that $\alpha \geq 0$, Ω is a finite interval on the time scale $\mathbb{T}$. With $-\overline{[-\alpha]}$, we will denote the integer part of $-\alpha$.

Definition 2.2 ([8]). For a function $f : \mathbb{T} \to \mathbb{R}$, the Riemann–Liouville fractional Δ-integral of order α is defined by

$$I_{\Delta, t_0}^0 f(t) = f(t),$$

$$\left(I_{\Delta, t_0}^\alpha f\right)(t) = (h_{\alpha-1}(\cdot, t_0) \star f)(t)$$

$$= \int_{t_0}^t \widehat{h_\alpha(\cdot, t_0)}(t, \sigma(u)) f(u) \Delta u$$

$$= \int_{t_0}^t h_{\alpha-1}(t, \sigma(u)) f(u) \Delta u$$

for $\alpha > 0$, $t > t_0$.

Example 2.1. Let $\mathbb{T} = \mathbb{R}$. Then

$$h_{\alpha-1}(t, t_0) = \frac{(t - t_0)^{\alpha-1}}{\Gamma(\alpha)}, \quad t \in \mathbb{T}, \quad t > t_0,$$

and

$$\left(I_{t_0+}^\alpha f\right)(t) = \frac{1}{\Gamma(\alpha)} \int_{t_0}^t (t - \tau)^{\alpha-1} f(\tau) d\tau, \quad t \in \mathbb{T}, \quad t > t_0.$$

Example 2.2. Let $\mathbb{T} = \mathbb{Z}$. Define the factorial polynomial

$$t^{(\alpha)} = \frac{\Gamma(t+1)}{\Gamma(t+1-\alpha)}, \quad t \in \mathbb{T}, \quad \alpha \in \mathbb{R}.$$

The factorial polynomial has the following properties:

1. $\Delta t^{(\alpha)} = \alpha t^{(\alpha-1)}$, $t \in \mathbb{T}$, $\alpha \in \mathbb{R}$, where Δ is the forward difference operator,
2. $(t - \alpha)t^{(\alpha)} = t^{(\alpha+1)}$, $t \in \mathbb{T}$, $\alpha \in \mathbb{R}$,
3. $\alpha^{(\alpha)} = \Gamma(\alpha+1)$, $\alpha \in \mathbb{R}$,
4. $t^{(\alpha+\beta)} = (t - \beta)^{(\alpha)}t^{(\beta)}$, $t \in \mathbb{T}$, $\alpha, \beta \in \mathbb{R}$.

Using the above properties, we conclude that

$$h_{\alpha-1}(t,s) = \frac{(t-s)^{(\alpha-1)}}{\Gamma(\alpha)}, \quad t \geq s \geq t_0 - \overline{[-\alpha]},$$

and then

$$\left(I^{\alpha}_{\Delta,t_0}f\right)(t) = \int_{t_0}^{t} h_{\alpha-1}(t,\sigma(s))f(s)\Delta s$$

$$= \frac{1}{\Gamma(\alpha)} \int_{t_0}^{t} (t - \sigma(s))^{(\alpha-1)} f(s)\Delta s$$

$$= \frac{1}{\Gamma(\alpha)} \sum_{s=t_0}^{t-1} (t - s - 1)^{(\alpha-1)} f(s), \quad t \geq t_0 - \overline{[-\alpha]}.$$

Example 2.3. Let $\mathbb{T} = \mathbb{Z}$, $f(t) = t$, $t \in \mathbb{T}$, $t_0 = 0$, $\alpha = \frac{1}{3}$. Then

$$\overline{[-\alpha]} = \overline{\left[-\frac{1}{3}\right]}$$

$$= 0,$$

$$\left(I^{\frac{1}{3}}_{\Delta,0}f\right)(t) = \frac{1}{\Gamma\left(\frac{1}{3}\right)} \sum_{s=0}^{t-1}(t - s - 1)^{\left(-\frac{2}{3}\right)}s$$

$$= \frac{1}{\Gamma\left(\frac{1}{3}\right)} \sum_{s=0}^{t-1} \frac{\Gamma(t-s)}{\Gamma\left(t - s + \frac{2}{3}\right)}s, \quad t \in \mathbb{Z}, \quad t \geq 1.$$

Example 2.4. Let $\mathbb{T} = q^{\mathbb{N}_0}$, $q > 1$. Let also $\alpha \in [0, \infty)\backslash\mathbb{N}_0$. We define the q-factorial function

$$(t - s)_q^{\alpha} = t^{\alpha} \prod_{n=0}^{\infty} \frac{1 - \frac{s}{t}q^n}{1 - \frac{s}{t}q^{\alpha+n}}$$

$$= t^{\alpha} \prod_{n=0}^{\infty} \frac{t - sq^n}{t - sq^{\alpha+n}}, \qquad t, s \in \mathbb{T}, \quad t \geq s. \qquad (2.4)$$

Note that if $t, s \in \mathbb{T}$, $t \geq s$, then there exists $l \in \mathbb{N}_0$ such that $\frac{t}{s} = q^l$. Hence, $\alpha + n \neq l$ for any $n \in \mathbb{N}_0$, and

$$q^l \neq q^{\alpha+n} \quad \text{for any } n \in \mathbb{N}_0$$

or

$$\frac{t}{s} \neq q^{\alpha+n} \quad \text{for any } n \in \mathbb{N}_0$$

or

$$t - sq^{\alpha+n} \neq 0 \quad \text{for any } n \in \mathbb{N}_0.$$

Consequently, (2.4) is well defined. We will give some properties of the q-factorial function:

1. Let $\alpha, \beta \in [0, \infty)\backslash\mathbb{N}_0$. Then

$$(t - s)_q^{\alpha+\beta} = (t - s)_q^{\alpha} (t - q^{\alpha}s)_q^{\beta}.$$

2. Let $\alpha \in [0, \infty)\backslash\mathbb{N}_0$, $a \in \mathbb{R}$. Then

$$(at - as)_q^{\alpha} = a^{\alpha}(t - s)_q^{\alpha}, \qquad t, s \in \mathbb{T}, \quad t \geq s \geq t_0.$$

3. Let $\alpha \in [0, \infty)\backslash\mathbb{N}_0$. Then

$$\left((t - s)_q^{\alpha}\right)^{\Delta_t} = \frac{q^{\alpha} - 1}{q - 1}(t - s)_q^{\alpha-1}, \qquad t, s \in \mathbb{T}, \quad t \geq s \geq 1. \qquad (2.5)$$

4. Let $\alpha \in [0, \infty) \backslash \mathbb{N}_0$. Then

$$\left((t - s)_q^\alpha \right)^{\Delta_s} = -\frac{q^\alpha - 1}{q - 1} (t - \sigma(s))_q^{\alpha - 1}, \quad t, s \in \mathbb{T}, \quad t \geq s \geq 1.$$

(2.6)

Define the q-Gamma function $\Gamma_q : \mathbb{R} \backslash \mathbb{Z} \to \mathbb{R}$ as follows:

$$\Gamma_q \left(\frac{1}{2} \right) = 1, \quad \Gamma_q(\alpha) \frac{q^\alpha - 1}{q - 1} = \Gamma_q(\alpha - 1), \quad \alpha \in \mathbb{R} \backslash \mathbb{Z}.$$

Let

$$h_\alpha(t, s) = \Gamma_q(\alpha)(t - s)_q^\alpha, \quad t, s \in \mathbb{T}, \quad t \geq s \geq t_0.$$

By (2.5), (2.6), it follows that

$$h_\alpha^{\Delta_t}(t, s) = h_{\alpha - 1}(t, s),$$
$$h_\alpha^{\Delta_t}(t, \sigma(s)) = -h_\alpha^{\Delta_s}(t, s), \quad t, s \in \mathbb{T}, \quad t \geq s \geq 1,$$

for any $\alpha \in [0, \infty) \backslash \mathbb{N}_0$. Then

$$\left(I_{\Delta, t}^\alpha f \right)(t) = \int_1^t h_{\alpha - 1}(t, \sigma(s)) f(s) \Delta s$$

$$= \Gamma_q(\alpha - 1) \int_1^t (t - qs)_q^{\alpha - 1} f(s) \Delta s, \quad t, s \in \mathbb{T}, \quad t \geq s \geq 1.$$

If $t = q^l$ for some $l \in \mathbb{N}$, then

$$\left(I_{\Delta, t}^\alpha f \right) \left(q^l \right) = \Gamma_q(\alpha - 1) q^{(\alpha - 1)l} \sum_{r=0}^{l-1} \mu(q^r) \prod_{n=0}^\infty \frac{q^l - q^r q^n}{q^l - q^r q^{\alpha + n}} f(q^r)$$

$$= \Gamma_q(\alpha - 1) q^{(\alpha - 1)l} \sum_{r=0}^{l-1} (q - 1) q^r \prod_{n=0}^\infty \frac{q^l - q^{r+n}}{q^l - q^{\alpha + r + n}} f(q^r).$$

This completes the example.

Suppose that $\alpha \in [0,\infty)\backslash\mathbb{N}_0$. Let $s \in \{0,\dots,n-1\}$ and

$$a_{n-s}(t_n,t_s) = 1,$$

$$a_{n-1-s}(t_n,t_s) = \frac{1}{\mu(t_s)} + \cdots + \frac{1}{\mu(t_{n-1})},$$

$$a_{n-2-s}(t_n,t_s) = \frac{1}{\mu(t_s)\mu(t_{s+1})} + \cdots + \frac{1}{\mu(t_s)\mu(t_{n-1})} \qquad (2.7)$$

$$+ \cdots + \frac{1}{\mu(t_{n-2})\mu(t_{n-1})},$$

$$\vdots$$

$$a_0(t_n,t_s) = \frac{1}{\mu(t_s)\cdots\mu(t_{n-1})}.$$

Then

$$\prod_{k=s}^{n-1}(1+\mu(t_k)z) = \prod_{k=s}^{n-1}\frac{1}{\mu(t_k)}\left(z+\frac{1}{\mu(t_k)}\right)$$

$$= a_0(t_n,t_s)\prod_{k=s}^{n-1}\left(z+\frac{1}{\mu(t_k)}\right)$$

$$= a_0(t_n,t_s)\sum_{k=s}^{n-1}a_{n-k}(t_n,t_s)z^{n-k}$$

and

$$\mathcal{L}^{-1}\left(\frac{1}{z^{\alpha+1}}\right) = \frac{1}{2\pi i}\int_\gamma \frac{1}{z^{\alpha+1}}a_0(t_n,t_s)\sum_{k=s}^{n-1}a_{n-k}(t_n,t_s)z^{n-k}dz$$

$$= \frac{1}{2\pi i}a_0(t_n,t_s)\sum_{k=s}^{n-1}a_{n-k}(t_n,t_s)\int_\gamma z^{n-k-\alpha-1}dz$$

$$= \frac{1}{2\pi i}a_0(t_n,t_s)\sum_{k=s}^{n-1}a_{n-k}(t_n,t_s)\left(\frac{1}{w}\right)^{n-k-\alpha}$$

$$\times \int_0^{2\pi}e^{i(n-k-\alpha)\theta}d(i\theta)$$

$$= \frac{1}{2\pi i} a_0(t_n, t_s) \sum_{k=s}^{n-1} a_{n-k}(t_n, t_s) \left(\frac{1}{w}\right)^{n-k-\alpha}$$

$$\times \frac{1}{n-k-\alpha} \left(e^{2\pi(n-k-\alpha)i} - 1\right),$$

i.e.,

$$h_\alpha(t_n, t_s) = \frac{1}{2\pi i} a_0(t_n, t_s) \sum_{k=s}^{n-1} a_{n-k}(t_n, t_s) \left(\frac{1}{w}\right)^{n-k-\alpha}$$

$$\times \frac{1}{n-k-\alpha} \left(e^{2\pi(n-k-\alpha)i} - 1\right).$$

Therefore,

$$\left(I_{\Delta,t_0}^\alpha f\right)(t_n) = \int_{t_0}^{t_n} h_{\alpha-1}(t_n, \sigma(u)) f(u) \Delta u$$

$$= \int_{t_0}^{t_n} h_{\alpha-1}(t_n, \sigma(t_s)) f(t_s) \Delta t_s$$

$$= \int_{t_0}^{t_n} h_{\alpha-1}(t_n, t_{s+1}) f(t_s) \Delta t_s$$

$$= \sum_{r=0}^{n-1} \mu(t_r) h_{\alpha-1}(t_n, t_{r+1}) f(t_r)$$

$$= \sum_{r=0}^{n-2} \mu(t_r) f(t_r) \frac{1}{2\pi i} a_0(t_n, t_{r+1}) \sum_{k=r+1}^{n-1} a_{n-k}(t_n, t_{r+1})$$

$$\times \left(\frac{1}{w}\right)^{n-k-\alpha} \frac{1}{n-k-\alpha} \left(e^{2\pi(n-k-\alpha)i} - 1\right).$$

For a function $f : \mathbb{T} \to \mathbb{R}$, we will denote

$$f^{\Delta^n} = D_\Delta^n f, \quad n \in \mathbb{N}_0.$$

Definition 2.3 ([8]). Let $\alpha \geq 0$, $m = -\overline{[-\alpha]}$, $f : \mathbb{T} \to \mathbb{R}$. For $s, t \in \mathbb{T}^{\kappa^m}$, $s < t$, the Riemann–Liouville fractional Δ-derivative of order α is defined by the expression

$$D_{\Delta,s}^{\alpha} f(t) = D_{\Delta}^{m} I_{\Delta,s}^{m-\alpha} f(t), \quad t \in \mathbb{T},$$

if it exists. For $\alpha < 0$, we define

$$D_{\Delta,s}^{\alpha} f(t) = I_{\Delta,s}^{-\alpha} f(t), \quad t, s \in \mathbb{T}, \quad t > s,$$

$$I_{\Delta,s}^{\alpha} f(t) = D_{\Delta,s}^{-\alpha} f(t), \quad t, s \in \mathbb{T}^{\kappa^r}, \quad t > s, \quad r = \overline{[-\alpha]} + 1.$$

Suppose that $\alpha \in [0, \infty) \backslash \mathbb{N}_0$, $m = -\overline{[-\alpha]}$. Then, using the definition, we have

$$D_{\Delta,s}^{\alpha} f(t) = D_{\Delta}^{m} I_{\Delta,s}^{m-\alpha} f(t)$$

$$= D_{\Delta}^{m} \left(\int_{t_0}^{t} h_{m-\alpha-1}(t, \sigma(u)) f(u) \Delta u \right), \quad t \in \mathbb{T}, \quad t \geq t_0.$$

If $m = 0$, then

$$D_{\Delta,s}^{\alpha} f(t) = \int_{t_0}^{t} h_{-\alpha-1}(t, \sigma(u)) f(u) \Delta u, \quad t \in \mathbb{T}, \quad t \geq t_0.$$

If $m = 1$, then

$$D_{\Delta,s}^{\alpha} f(t) = D_{\Delta} \left(\int_{t_0}^{t} h_{-\alpha}(t, \sigma(u)) f(u) \Delta u \right)$$

$$= \int_{t_0}^{t} h_{-\alpha}^{\Delta_t}(t, \sigma(u)) f(u) \Delta u + h_{-\alpha}(\sigma(t), \sigma(t)) f(t)$$

$$= \int_{t_0}^{t} h_{-\alpha-1}(t, \sigma(u)) f(u) \Delta u + h_{-\alpha}(\sigma(t), \sigma(t)) f(t),$$

$$t \in \mathbb{T}, \quad t \geq t_0.$$

Example 2.5. Let $\mathbb{T} = \mathbb{Z}$, $t_0 \in \mathbb{T}$, $\alpha \in [0,\infty)\backslash\mathbb{N}_0$. Using Example 2.2, we get

$$I_{\Delta,t_0}^{m-\alpha} f(t) = \frac{1}{\Gamma(\alpha)} \sum_{s=t_0}^{t-1} (t-s-1)^{(m-\alpha-1)} f(s)$$

$$= \frac{1}{\Gamma(\alpha)} \sum_{s=t_0}^{t-1} \frac{\Gamma(t-s)}{\Gamma(t-s-1-m+\alpha+1)} f(s)$$

$$= \frac{1}{\Gamma(\alpha)} \sum_{s=t_0}^{t-1} \frac{\Gamma(t-s)}{\Gamma(t-s-m+\alpha+1)} f(s), \quad t \in \mathbb{T}, \quad t \ge t_0.$$

Then

$$D_{\Delta,s}^{\alpha} f(t)$$

$$= D_{\Delta}^m \left(I_{\Delta,t_0}^{m-\alpha} f(t) \right) = D_{\Delta}^m \left(\frac{1}{\Gamma(\alpha)} \sum_{s=t_0}^{t-1} \frac{\Gamma(t-s)}{\Gamma(t-s-m+\alpha+1)} f(s) \right)$$

$$= \sum_{k=0}^{m} \binom{m}{k} (-1)^k \left(\frac{1}{\Gamma(\alpha)} \sum_{s=t_0}^{t+m-k-1} \frac{\Gamma(t+m-k-s)}{\Gamma(t+m-k-s-m+\alpha+1)} f(s) \right)$$

$$= \sum_{k=0}^{m} \binom{m}{k} (-1)^k \left(\frac{1}{\Gamma(\alpha)} \sum_{s=t_0}^{t+m-k-1} \frac{\Gamma(t+m-k-s)}{\Gamma(t-s-k+\alpha+1)} f(s) \right),$$

$$t \in \mathbb{T}, \ t > t_0.$$

Definition 2.4 ([8]). The Δ-Mittag-Leffler function is defined by

$$_\Delta F_{\alpha,\beta}(\lambda, t, t_0) = \sum_{j=0}^{\infty} \lambda^j h_{j\alpha+\beta-1}(t, t_0), \tag{2.8}$$

provided that the right-hand side is convergent, where $\alpha, \beta > 0$, $\lambda \in \mathbb{R}$.

Remark 2.1. When $0 < \lambda < 1$, $\alpha > 0$, $\beta > 0$, and

$$|h_{j\alpha+\beta-1}(t, t_0)| < M, \quad t \in \mathbb{T}, \quad t > t_0,$$

for any $j \in \mathbb{N}_0$, then the series (2.8) is convergent.

Remark 2.2. When $\mathbb{T} = \mathbb{R}$, $\alpha > 0$, $\beta > 0$, then

$$\Delta F_{\alpha,\beta}(\lambda, t, t_0) = \sum_{j=0}^{\infty} \lambda^j \frac{(t-t_0)^{j\alpha+\beta-1}}{\Gamma(j\alpha+\beta)},$$

which is a convergent series for any $t \geq t_0$, i.e., $\Delta F_{\alpha,\beta}(\lambda, t, t_0)$ is defined as $t \geq t_0$.

Example 2.6. Let $\mathbb{T} = \mathbb{Z}$, $\alpha > 0$, $\beta > 0$. Then

$$h_{j\alpha+\beta-1}(t, t_0) = \frac{(t-t_0)^{(j\alpha+\beta-1)}}{\Gamma(j\alpha+\beta)}$$

$$= \frac{\Gamma(t-t_0+1)}{\Gamma(t-t_0+1-j\alpha-\beta)\Gamma(j\alpha+\beta)}$$

$$= \frac{\Gamma(t-t_0+1)}{\Gamma(t-t_0-j\alpha-\beta+2)\Gamma(j\alpha+\beta)}, \quad j \in \mathbb{N}_0.$$

Hence,

$$\Delta F_{\alpha,\beta}(\lambda, t, t_0) = \sum_{j=0}^{\infty} \lambda^j \frac{\Gamma(t-t_0+1)}{\Gamma(t-t_0-j\alpha-\beta+2)\Gamma(j\alpha+\beta)}.$$

Example 2.7. Let $\mathbb{T} = q^{\mathbb{N}_0}$, $q > 1$, $\alpha > 0$, $\beta > 0$. Then

$$h_{j\alpha+\beta-1}(t, t_0) = \Gamma_q(j\alpha+\beta-1)t^{j\alpha+\beta-1} \prod_{n=0}^{\infty} \frac{t - t_0 q^n}{t - t_0 q^{j\alpha+\beta-1+n}}, \quad j \in \mathbb{N}_0,$$

and

$$\Delta F_{\alpha,\beta}(\lambda, t, t_0) = \sum_{j=0}^{\infty} \lambda^j h_{j\alpha+\beta-1}(t, t_0)$$

$$= \sum_{j=0}^{\infty} \lambda^j \Gamma_q(j\alpha+\beta-1)t^{j\alpha+\beta-1}$$

$$\times \prod_{n=0}^{\infty} \frac{t - t_0 q^n}{t - t_0 q^{j\alpha+\beta-1+n}}, \quad t, t_0 \in \mathbb{T}, \quad t > t_0.$$

Theorem 2.2 ([8]). *We have*

$$\mathcal{L}\left(\Delta F_{\alpha,\beta}(\lambda, t, t_0)\right)(z, t_0) = \frac{z^{\alpha-\beta}}{z^\alpha - \lambda}, \quad |\lambda| < |z|^\alpha, \quad \alpha, \beta > 0. \quad (2.9)$$

Suppose that all conditions of Theorem 2.2 hold. We differentiate k times, $k \in \mathbb{N}$, the equality (2.9) with respect to λ and get

$$\mathcal{L}\left(\frac{\partial^k}{\partial \lambda^k} {}_\Delta F_{\alpha,\beta}(\lambda, t, t_0)\right)(z, t_0) = \frac{k! z^{\alpha-\beta}}{(z^\alpha - \lambda)^{k+1}}, \quad |\lambda| < |z|^\alpha.$$

2.3 Properties of the Riemann–Liouville Fractional Δ-Integral and the Riemann–Liouville Fractional Δ-Derivative on Time Scales

Theorem 2.3 ([8]). *Let $\alpha > 0$, $m = -\overline{[-\alpha]}$, $\beta \in \mathbb{R}$. Then*

$$I^\alpha_{\Delta, t_0} h_{\beta-1}(t, t_0) = h_{\beta+\alpha-1}(t, t_0), \quad t \in \mathbb{T}, \quad t \geq t_0.$$

Example 2.8. Let $\alpha = \beta = \frac{1}{2}$. Then

$$\begin{aligned}
I^{\frac{1}{2}}_{\Delta, t_0} h_{-\frac{1}{2}}(t, t_0) &= h_{\frac{1}{2}+\frac{1}{2}-1}(t, t_0) \\
&= h_0(t, t_0) \\
&= 1, \quad t \in \mathbb{T}, \quad t > t_0.
\end{aligned}$$

Example 2.9. Let $\alpha = \frac{1}{3}$, $\beta = \frac{5}{3}$. Then

$$\begin{aligned}
I^{\frac{1}{3}}_{\Delta, t_0} h_{\frac{5}{3}-1}(t, t_0) &= I^{\frac{1}{3}}_{\Delta, t_0} h_{\frac{2}{3}}(t, t_0) \\
&= h_{\frac{1}{3}+\frac{5}{3}-1}(t, t_0) \\
&= h_1(t, t_0) \\
&= t - t_0, \quad t \in \mathbb{T}, \quad t > t_0.
\end{aligned}$$

Example 2.10. Let $\mathbb{T} = 2^{\mathbb{N}_0}$, $t_0 = 1$, $\alpha = \frac{4}{3}$, $\beta = \frac{5}{3}$. Then

$$\begin{aligned}
I^{\frac{4}{3}}_{\Delta, 1} h_{\frac{2}{3}}(t, 1) &= h_{\frac{4}{3}+\frac{5}{3}-1}(t, 1) \\
&= h_2(t, 1) \\
&= \frac{(t-1)_{2}^{2}}{[2]!}
\end{aligned}$$

$$= \frac{(t-1)(t-2)}{\frac{2^2-1}{2-1}}$$

$$= \frac{(t-1)(t-2)}{3}, \quad t \in \mathbb{T}, \quad t > 1.$$

Exercise 2.1. Let $\mathbb{T} = \mathbb{Z}$, $t_0 = 0$, $\alpha = \frac{1}{2}$, $\beta = \frac{5}{2}$. Find

$$I_{\Delta,0}^{\alpha} h_{\beta-1}(t,0), \quad t \in \mathbb{T}, \quad t > 0.$$

Answer: $\frac{t(t-1)}{2}$, $t \in \mathbb{T}$, $t > 0$.   □

Theorem 2.4 ([8]). *Let $\alpha > 0$, $m = -\overline{[-\alpha]}$, $\beta \in \mathbb{R}$. Then*

$$D_{\Delta,t_0}^{\alpha} h_{\beta-1}(t,t_0) = h_{\beta-\alpha-1}(t,t_0), \quad t \in \mathbb{T}, \quad t \geq t_0.$$

Example 2.11. Let $\alpha = \frac{1}{2}$, $\beta = \frac{3}{2}$. Then

$$D_{\Delta,t_0}^{\frac{1}{2}} h_{\frac{3}{2}-1}(t,t_0) = D_{\Delta,t_0}^{\frac{1}{2}} h_{\frac{1}{2}}(t,t_0)$$

$$= h_{\frac{3}{2}-\frac{1}{2}-1}(t,t_0)$$

$$= h_0(t,t_0)$$

$$= 1, \quad t \in \mathbb{T}, \quad t > 1.$$

Example 2.12. Let $\alpha = \frac{1}{3}$, $\beta = \frac{7}{3}$. Then

$$D_{\Delta,t_0}^{\frac{1}{3}} h_{\frac{7}{3}-1}(t,t_0) = D_{\Delta,t_0}^{\frac{1}{3}} h_{\frac{4}{3}}(t,t_0)$$

$$= h_{\frac{7}{3}-\frac{1}{3}-1}(t,t_0)$$

$$= h_1(t,t_0)$$

$$= t - t_0, \quad t \in \mathbb{T}, \quad t > t_0.$$

Example 2.13. Let $\mathbb{T} = 4^{\mathbb{N}_0}$, $\alpha = \frac{1}{4}$, $\beta = \frac{13}{4}$. Then

$$D_{\Delta,1}^{\frac{1}{4}} h_{\frac{13}{4}-1}(t,1) = D_{\Delta,1}^{\frac{1}{4}} h_{\frac{9}{4}}(t,1)$$

$$= h_{\frac{13}{4}-\frac{1}{4}-1}(t,1)$$

$$= h_2(t, 1)$$

$$= \frac{(t-1)(t-4)}{5}, \quad t \in \mathbb{T}, \quad t > 1.$$

Exercise 2.2. Let $\mathbb{T} = 2\mathbb{Z}$, $t_0 = 0$, $\alpha = \frac{1}{7}$, $\beta = \frac{22}{7}$. Find

$$D^{\frac{1}{7}}_{\Delta,0} h_{\frac{15}{7}}(t, 0), \quad t \in \mathbb{T}, \quad t > 0.$$

Answer: $\frac{1}{2}t^2 - t$, $t \in \mathbb{T}$, $t > 0$.

Example 2.14. Let $\alpha = \frac{1}{3}$, $\beta = \frac{2}{3}$. We will find

$$\mathcal{L}\left(D^{\frac{1}{3}}_{\Delta,t_0} h_{-\frac{1}{3}}(\cdot, t_0)\right)(z, t_0).$$

We have

$$D^{\alpha}_{\Delta,t_0} h_{\beta-1}(t, t_0) = D^{\frac{1}{3}}_{\Delta,t_0} h_{\frac{2}{3}-1}(t, t_0)$$

$$= D^{\frac{1}{3}}_{\Delta,t_0} h_{-\frac{1}{3}}(t, t_0)$$

$$= h_{\frac{2}{3}-\frac{1}{3}-1}(t, t_0)$$

$$= h_{-\frac{2}{3}}(t, t_0), \quad t \in \mathbb{T}, \quad t > t_0.$$

Hence,

$$\mathcal{L}\left(D^{\frac{1}{3}}_{\Delta,t_0} h_{-\frac{1}{3}}(\cdot, t_0)\right)(z, t_0) = \mathcal{L}\left(h_{-\frac{2}{3}}(\cdot, t_0)\right)(z, t_0)$$

$$= \frac{1}{z^{1-\frac{2}{3}}}$$

$$= \frac{1}{z^{\frac{1}{3}}}.$$

Example 2.15. Let $\alpha = \frac{1}{7}$, $\beta = \frac{2}{3}$. We will find

$$\mathcal{L}\left(I^{\frac{1}{7}}_{\Delta,t_0} h_{-\frac{1}{3}}(\cdot, t_0)\right)(z, t_0).$$

We have

$$I^{\alpha}_{\Delta,t_0} h_{\beta-1}(t, t_0) = I^{\frac{1}{7}}_{\Delta,t_0 9} h_{\frac{2}{3}-1}(t, t_0)$$

$$= I^{\frac{1}{7}}_{\Delta,t_0} h_{-\frac{1}{3}}(t, t_0)$$

$$= h_{\frac{1}{7}+\frac{2}{3}-1}(t, t_0)$$

$$= h_{-\frac{4}{21}}(t, t_0), \quad t \in \mathbb{T}, \quad t > t_0.$$

Therefore,

$$\mathcal{L}\left(I^{\alpha}_{\Delta,t_0} h_{\beta-1}(\cdot, t_0)\right)(z, t_0) = \mathcal{L}\left(h_{-\frac{4}{21}}(\cdot, t_0)\right)(z, t_0)$$

$$= \frac{1}{z^{-\frac{4}{21}+1}}$$

$$= \frac{1}{z^{\frac{17}{21}}}.$$

Example 2.16. Let $\alpha = \frac{1}{2}$, $\beta = \frac{3}{2}$. We will find

$$\mathcal{L}\left(I^{\frac{1}{2}}_{\Delta,t_0} h_{\frac{1}{2}}(\cdot, t_0) + h_4(\cdot, t_0) \star D^{\frac{1}{2}}_{\Delta,t_0} h_{\frac{1}{2}}(\cdot, t_0)\right)(z, t_0).$$

We have

$$I^{\alpha}_{\Delta,t_0} h_{\beta-1}(t, t_0) = I^{\frac{1}{2}}_{\Delta,t_0} h_{\frac{3}{2}-1}(t, t_0)$$

$$= I^{\frac{1}{2}}_{\Delta,t_0} h_{\frac{1}{2}}(t, t_0)$$

$$= h_{\frac{1}{2}+\frac{3}{2}-1}(t, t_0)$$

$$= h_1(t, t_0), \quad t \in \mathbb{T}, \quad t > t_0,$$

$$\mathcal{L}\left(I^{\frac{1}{2}}_{\Delta,t_0} h_{\frac{1}{2}}(\cdot, t_0)\right)(z, t_0) = \mathcal{L}\left(h_1(\cdot, t_0)\right)(z, t_0)$$

$$= \frac{1}{z^2},$$

$$D^{\alpha}_{\Delta,t_0} h_{\beta-1}(t, t_0) = D^{\frac{1}{2}}_{\Delta,t_0} h_{\frac{3}{2}-1}(t, t_0)$$

$$= D^{\frac{1}{2}}_{\Delta,t_0} h_{\frac{1}{2}}(t, t_0)$$

$$= h_{\frac{3}{2}-\frac{1}{2}-1}(t, t_0)$$

$$= h_0(t, t_0),$$

$$\left(h_4\left(\cdot, t_0\right) \star D^{\frac{1}{2}}_{\Delta, t_0} h_{\frac{1}{2}}\left(\cdot, t_0\right) \right) = \left(h_4(\cdot, t_0) \star h_0(\cdot, t_0) \right)(t, t_0)$$

$$= h_{4+0+1}(t, t_0)$$

$$= h_5(t, t_0), \quad t \in \mathbb{T}, \quad t > t_0,$$

$$\mathcal{L}\left(h_4(\cdot, t_0) \star D^{\frac{1}{2}}_{\Delta, t_0} h_{\frac{1}{2}}(\cdot, t_0) \right)(z, t_0) = \mathcal{L}\left(h_5(\cdot, t_0) \right)(z, t_0)$$

$$= \frac{1}{z^6}.$$

Consequently,

$$\mathcal{L}\left(I^{\frac{1}{2}}_{\Delta, t_0} h_{\frac{1}{2}}(\cdot, t_0) + h_4(\cdot, t_0) \star D^{\frac{1}{2}}_{\Delta, t_0} h_{\frac{1}{2}}(\cdot, t_0) \right)(z, t_0)$$

$$= \mathcal{L}\left(I^{\frac{1}{2}}_{\Delta, t_0} h_{\frac{1}{2}}(\cdot, t_0) \right)(z, t_0)$$

$$+ \mathcal{L}\left(h_4(\cdot, t_0) \star D^{\frac{1}{2}}_{\Delta, t_0} h_{\frac{1}{2}}(\cdot, t_0) \right)(z, t_0)$$

$$= \frac{1}{z^2} + \frac{1}{z^6}$$

$$= \frac{z^4 + 1}{z^6}.$$

Exercise 2.3. Let $\alpha = \frac{1}{3}$, $\beta = \frac{5}{3}$. Find

$$\mathcal{L}\left(h_{\frac{1}{2}}(\cdot, t_0) \star I^{\frac{1}{3}}_{\Delta, t_0} h_{\frac{2}{3}}(\cdot, t_0) - 2h_2(\cdot, t_0) \star D^{\frac{1}{3}}_{\Delta, t_0} h_{\frac{2}{3}}(\cdot, t_0) \right)(z, t_0).$$

Answer: $\dfrac{1}{z^{\frac{7}{2}}} - \dfrac{2}{z^{\frac{13}{3}}}.$

Corollary 2.1 ([8]). *Let* $0 < \alpha < 1$. *Then*

$$D^{\alpha}_{\Delta, t_0} 1 = h_{-\alpha}(t, t_0), \quad t \in \mathbb{T}, \quad t \geq t_0. \tag{2.10}$$

Example 2.17. Let $\alpha = \frac{1}{3}$, $\beta = \frac{2}{3}$. We will find

$$f(z) = \mathcal{L}\left(h_1(\cdot, t_0) \star \left(D_{\Delta,t_0}^{\frac{1}{3}}\left(I_{\Delta,t_0}^{\frac{1}{3}}h_{-\frac{1}{3}}\right)\right)(\cdot, t_0)\right)$$

$$- 4h_2(\cdot, t_0) \star h_{\frac{1}{3}}(\cdot, t_0)\right)(z, t_0).$$

We have

$$I_{\Delta,t_0}^{\alpha}h_{\beta-1}(t, t_0) = I_{\Delta,t_0}^{\frac{1}{3}}h_{\frac{2}{3}-1}(t, t_0)$$

$$= I_{\Delta,t_0}^{\frac{1}{3}}h_{-\frac{1}{3}}(t, t_0)$$

$$= h_{\frac{1}{3}+\frac{2}{3}-1}(t, t_0)$$

$$= h_0(t, t_0) = 1,$$

$$D_{t,t_0}^{\frac{1}{3}}\left(I_{\Delta,t_0}^{\frac{1}{3}}h_{-\frac{1}{3}}(t, t_0)\right) = D_{\Delta,t_0}^{\frac{1}{3}}1$$

$$= h_{\frac{5}{3}}(t, t_0), \quad t \in \mathbb{T}, \quad t > t_0,$$

$$h_1(\cdot, t_0) \star \left(D_{\Delta,t_0}^{\frac{1}{3}}\left(I_{\Delta,t_0}^{\frac{1}{3}}h_{-\frac{1}{3}}\right)\right)(\cdot, t_0) = h_1(\cdot, t_0) \star h_{-\frac{1}{3}}(\cdot, t_0)$$

$$= h_{-\frac{1}{3}}(\cdot, t_0),$$

$$\mathcal{L}\left(h_1(\cdot, t_0) \star \left(D_{\Delta,t_0}^{\frac{1}{3}}\left(I_{\Delta,t_0}^{\frac{1}{3}}h_{-\frac{1}{3}}\right)\right)(\cdot, t_0)\right)(z, t_0) = \mathcal{L}\left(h_{\frac{5}{3}}(\cdot, t_0)\right)(z, t_0)$$

$$= \frac{1}{z^{\frac{5}{3}+1}} = \frac{1}{z^{\frac{8}{3}}},$$

$$h_2(\cdot, t_0) \star h_{\frac{1}{3}}(\cdot, t_0) = h_{2+\frac{1}{3}+1}(\cdot, t_0)$$

$$= h_{\frac{10}{3}}(\cdot, t_0),$$

$$\mathcal{L}\left(h_2(\cdot, t_0) \star h_{\frac{1}{3}}(\cdot, t_0)\right)(z, t_0) = \mathcal{L}\left(h_{\frac{10}{3}}(\cdot, t_0)\right)(z, t_0)$$

$$= \frac{1}{z^{\frac{10}{3}+1}} = \frac{1}{z^{\frac{13}{3}}}.$$

Consequently,

$$f(z) = \mathcal{L}\left(h_1(\cdot,t_0) \star \left(D_{\Delta,t_0}^{\frac{1}{3}}\left(I_{\Delta,t_0}^{\frac{1}{3}}h_{-\frac{1}{3}}\right)\right)(\cdot,t_0)\right)(z,t_0)$$

$$- 4\mathcal{L}\left(h_2(\cdot,t_0) \star h_{\frac{1}{3}}(\cdot,t_0)\right)(z,t_0)$$

$$= \frac{1}{z^{\frac{8}{3}}} - \frac{4}{z^{\frac{13}{3}}}.$$

Example 2.18. Let $\alpha \in (0,1)$, $\beta = 1 - \alpha$. We will prove that

$$h_{\alpha-1}(\cdot,t_0) \star \left(D_{\Delta,t_0}^\alpha\left(I_{\Delta,t_0}^\alpha h_{\beta-1}\right)\right)(\cdot,t_0) = 1.$$

Really, we have

$$I_{\Delta,t_0}^\alpha h_{\beta-1}(t,t_0) = h_{\beta+\alpha-1}(t,t_0)$$
$$= h_{1-\alpha+\alpha-1}(t,t_0)$$
$$= h_0(t,t_0)$$
$$= 1,$$
$$D_{\Delta,t_0}^\alpha\left(I_{\Delta,t_0}^\alpha h_{\beta-1}(t,t_0)\right) = D_{\Delta,t_0}^\alpha 1$$
$$= h_{-\alpha}(t,t_0),$$
$$h_{\alpha-1}(\cdot,t_0) \star D_{\Delta,t_0}^\alpha\left(I_{\Delta,t_0}^\alpha h_{\beta-1}(t,t_0)\right)(\cdot,t_0) = h_{\alpha-1}(\cdot,t_0) \star h_{-\alpha}(\cdot,t_0)$$
$$= h_{\alpha+1-\alpha-1}(\cdot,t_0)$$
$$= h_0(\cdot,t_0)$$
$$= 1.$$

Corollary 2.2 ([8]). *Let* $\alpha > 0$, $m = -\overline{[-\alpha]}$. *Then*

$$D_{\Delta,t_0}^\alpha h_{\alpha-j}(t,s) = 0, \quad t,s \in \mathbb{T}, \quad t \geq s \geq t_0,$$

for any $j \in \{1,\ldots,m\}$.

Corollary 2.3 ([8]). *Let* $\mathbb{T} = \mathbb{R}$, $\alpha \geq 0$, $\beta > 0$, $t_0 \in \mathbb{R}$. *Then*

$$\left(I_{t_0+}^{\alpha}(t-t_0)^{\beta-1}\right)(x) = \frac{\Gamma(\beta)}{\Gamma(\beta+\alpha)}(x-t_0)^{\beta+\alpha-1},$$

$$\alpha > 0, \quad x \in \mathbb{R}, \quad x \geq t_0,$$

$$\left(D_{t_0+}^{\alpha}(t-t_0)^{\beta-1}\right)(x) = \frac{\Gamma(\beta)}{\Gamma(\beta-\alpha)}(x-t_0)^{\beta-\alpha-1},$$

$$\alpha \geq 0, \quad x \in \mathbb{R}, \quad x \geq t_0.$$

In particular, if $\beta = 1$ *and* $\alpha \geq 0$, *then the Riemann–Liouville fractional derivatives of a constant are, in general, not equal to zero,*

$$\left(D_{t_0+}^{\alpha}1\right)(x) = \frac{(x-t_0)^{-\alpha}}{\Gamma(1-\alpha)}, \quad 0 < \alpha < 1, \quad x \in \mathbb{T}, \quad x \geq t_0.$$

On the other hand, for $j \in \{1,\ldots,-\overline{[-\alpha]}\}$,

$$\left(D_{t_0^+}^{\alpha}(t-t_0)^{\alpha-j}\right)(x) = 0, \quad x \in \mathbb{T}, \quad x \geq t_0.$$

Corollary 2.4 ([8]). *Let* $\mathbb{T} = \mathbb{Z}$, $\alpha \geq 0$, $\beta \in \mathbb{R}\backslash\{\ldots,-2,-1\}$. *Then*

$$D_{\Delta,t_0}^{-\alpha}(t-t_0)^{(\beta)} = \beta^{(-\alpha)}(t-t_0)^{(\alpha+\beta)}, \quad t \geq t_0 - \overline{[-\beta]}.$$

Theorem 2.5 ([8]). *Let* $\alpha > 0$, $m = -\overline{[-\alpha]}$, f *is a function which is* m-*times* Δ-*differentiable on* $\mathbb{T}^{\kappa^m}$ *with* f^{Δ^m} *rd-continuous over* $\mathbb{T}$. *Then*

$$I_{\Delta,t_0}^{\alpha}f(t) = \sum_{k=0}^{m-1}h_{k+\alpha}(t,t_0)f^{\Delta^k}(t_0) + \left(h_{m+\alpha-1}(\cdot,t_0)\star f^{\Delta^m}\right)(t), \quad t \in \mathbb{T}^{\kappa^m}.$$

Example 2.19. Let $\alpha = \frac{1}{2}$, $f(t) = h_2(t,t_0)$, $t \in \mathbb{T}$, $t > t_0$. Then $m = 1$ and

$$I_{\Delta,t_0}^{\frac{1}{2}}h_2(t,t_0) = \sum_{k=0}^{0}h_{k+\frac{1}{2}}(t,t_0)h_2^{\Delta^k}(t_0,t_0)$$

$$+ \left(h_{\frac{1}{2}}(\cdot,t_0)\star h_2^{\Delta}(\cdot,t_0)\right)(t,t_0)$$

$$= h_{\frac{1}{2}}(t, t_0) h_2(t_0, t_0)$$

$$+ \left(h_{\frac{1}{2}}(\cdot, t_0) \star h_1(\cdot, t_0) \right)(t, t_0)$$

$$= h_{\frac{5}{2}}(t, t_0), \quad t \in \mathbb{T}, \quad t > t_0.$$

Example 2.20. Let $\alpha = \frac{9}{2}$, $f(t) = h_3(t, t_0)$, $t \in \mathbb{T}$, $t > t_0$. Then

$$m = \left[\frac{9}{2} \right] + 1$$

$$= 4 + 1 = 5$$

and

$$I_{\Delta, t_0}^{\frac{9}{2}} h_3(t, t_0) = \sum_{k=0}^{4} h_{k+\frac{9}{2}}(t, t_0) h_3^{\Delta^k}(t_0, t_0)$$

$$+ \left(h_{5+\frac{9}{2}-1}(\cdot, t_0) \star h_3^{\Delta^5}(\cdot, t_0) \right)(t)$$

$$= h_{\frac{9}{2}}(t, t_0) h_3(t_0, t_0)$$

$$+ h_{\frac{11}{2}}(t, t_0) h_3^{\Delta}(t_0, t_0)$$

$$+ h_{\frac{13}{2}}(t, t_0) h_3^{\Delta^2}(t_0, t_0)$$

$$+ h_{\frac{15}{2}}(t, t_0) h_3^{\Delta^3}(t_0, t_0)$$

$$+ h_{\frac{17}{2}}(t_0, t_0) h_3^{\Delta^4}(t_0, t_0)$$

$$= h_{\frac{15}{2}}(t, t_0), \quad t \in \mathbb{T}, \quad t > t_0.$$

Example 2.21. Let $\alpha = \frac{5}{3}$, $f(t) = e_a(t, t_0)$, $t \in \mathbb{T}$, $t \geq t_0$, $a \in \mathbb{C}$ be a given constant. We will find

$$\mathcal{L}\left(I_{\Delta, t_0}^{\frac{5}{3}} e_a(\cdot, t_0) \right)(z, t_0).$$

Note that

$$m = \left\lceil \frac{5}{3} \right\rceil + 1$$

$$= 1 + 1 = 2$$

and

$$e_a(t_0, t_0) = 1,$$
$$e_a^{\Delta}(t, t_0) = a e_a(t, t_0),$$
$$e_a^{\Delta}(t_0, t_0) = a,$$
$$e_a^{\Delta^2}(t, t_0) = a^2 e_a(t, t_0),$$
$$e_a^{\Delta^2}(t_0, t_0) = a^2.$$

We have

$$I_{\Delta, t_0}^{\frac{5}{3}} e_a(t, t_0) = \sum_{k=0}^{1} h_{k+\frac{5}{3}}(t, t_0) e_a^{\Delta^k}(t_0, t_0)$$

$$+ \left(h_{2+\frac{5}{3}-1}(\cdot, t_0) \star e_a^{\Delta^2}(\cdot, t_0) \right)(t, t_0)$$

$$= h_{\frac{5}{3}}(t, t_0) e_a(t_0, t_0)$$

$$+ h_{\frac{8}{3}}(t, t_0) e_a^{\Delta}(t_0, t_0)$$

$$+ \left(h_{\frac{8}{3}}(\cdot, t_0) \star \left(a^2 e_a(\cdot, t_0) \right) \right)(t, t_0)$$

$$= h_{\frac{5}{3}}(t, t_0) + a h_{\frac{8}{3}}(t, t_0)$$

$$+ a^2 \left(h_{\frac{8}{3}}(\cdot, t_0) \star e_a(\cdot, t_0) \right)(t), \quad t \in \mathbb{T}, \quad t > t_0.$$

Therefore,

$$\mathcal{L} \left(I_{\Delta, t_0}^{\frac{5}{3}} e_a(t, t_0) \right)(z, t_0) = \mathcal{L} \left(h_{\frac{5}{3}}(\cdot, t_0) \right)(z, t_0)$$

$$+ a \mathcal{L} \left(h_{\frac{8}{3}}(\cdot, t_0) \right)(z, t_0)$$

$$+ a^2 \mathcal{L} \left(h_{\frac{8}{3}}(\cdot, t_0) \star e_a(\cdot, t_0) \right)(z, t_0)$$

$$= \frac{1}{z^{\frac{5}{3}+1}} + a \frac{1}{z^{\frac{8}{3}+1}}$$

$$+ a^2 \mathcal{L}\left(h_{\frac{8}{3}}(\cdot, t_0)\right)(z, t_0) \mathcal{L}\left(e_a(\cdot, t_0)\right)(z, t_0)$$

$$= \frac{1}{z^{\frac{8}{3}}} + \frac{a}{z^{\frac{11}{3}}} + \frac{a^2}{z^{\frac{8}{3}+1}} \frac{1}{z - a}$$

$$= \frac{1}{z^{\frac{8}{3}}} + \frac{a}{z^{\frac{11}{3}}} + \frac{a^2}{(z - a)z^{\frac{11}{3}}}.$$

Exercise 2.4. Let $\alpha = \frac{3}{2}$, $f(t) = \cos_2(t, t_0)$, $t \in \mathbb{T}$, $t \geq t_0$. Find

$$\mathcal{L}\left(I_{\Delta, t_0}^{\frac{3}{2}} \cos_2(\cdot, t_0)\right)(z, t_0).$$

Answer: $\frac{1}{z^{\frac{1}{2}}(z^2 + 4)}$.

Theorem 2.6 ([8]). *Let $\alpha > 0$, $m = -\overline{[-\alpha]}$, f be a function which is m-times Δ-differentiable on $\mathbb{T}^{\kappa^m}$ with f^{Δ^m} rd-continuous over $\mathbb{T}$, and $D_{\Delta, t_0}^{\alpha} f$ exists almost on $\mathbb{T}$. Then*

$$D_{\Delta, t_0}^{\alpha} f(t) = \sum_{k=0}^{m-1} h_{k-\alpha}(t, t_0) f^{\Delta^k}(t_0) + \left(h_{m-\alpha-1}(\cdot, t_0) \star f^{\Delta^m}\right)(t),$$

$$t \in \mathbb{T}^{\kappa^m}.$$

Example 2.22. Let $\alpha > 0$ be arbitrarily chosen, $m = -\overline{[-\alpha]}$. If $r \in \mathbb{N}$, $r < m$, we get

$$D_{\Delta, t_0}^{\alpha} h_r(t, t_0) = \sum_{k=0}^{m-1} h_{k-\alpha}(t, t_0) h_r^{\Delta^k}(t_0, t_0)$$

$$+ \left(h_{m-\alpha-1}(\cdot, t_0) \star h_r^{\Delta^m}(\cdot, t_0)\right)(t, t_0)$$

$$= h_{r-\alpha}(t, t_0), \quad t \in \mathbb{T}, \quad t > t_0.$$

If $r \in \mathbb{N}$, $r \geq m$, then

$$D_{\Delta, t_0}^{\alpha} h_r(t, t_0) = \sum_{k=0}^{m-1} h_{k-\alpha}(t, t_0) h_r^{\Delta^k}(t_0, t_0)$$

$$+ \left(h_{m-\alpha-1}(\cdot, t_0) \star h_r^{\Delta^m}(\cdot, t_0)\right)(t, t_0)$$

$$= (h_{m-\alpha-1}(\cdot, t_0) \star h_{r-m}(\cdot, t_0)) (t, t_0)$$
$$= h_{r-\alpha}(t, t_0), \quad t \in \mathbb{T}, \quad t > t_0.$$

Example 2.23. Let $\alpha > 0$ be arbitrarily chosen, $m = -\overline{[-\alpha]}$, $a \in \mathbb{C}$. We will find

$$\mathcal{L} \left(D_{\Delta, t_0}^{\alpha} e_a(\cdot, t_0) \right) (z, t_0).$$

We have

$$e_a^{\Delta^k}(t, t_0) = a^k e_a(t, t_0),$$
$$e_a^{\Delta^k}(t_0, t_0) = a^k, \quad t \in \mathbb{T}, \quad t \geq t_0.$$

Then

$$D_{\Delta, t_0}^{\alpha} e_a(t, t_0)$$

$$= \sum_{k=0}^{m-1} h_{k-\alpha}(t, t_0) e_a^{\Delta^k}(t_0, t_0) + \left(h_{m-\alpha-1}(\cdot, t_0) \star e_a^{\Delta^m}(\cdot, t_0) \right) (t, t_0)$$

$$= \sum_{k=0}^{m-1} a^k h_{k-\alpha}(t, t_0) + a^m \left(h_{m-\alpha-1}(\cdot, t_0) \star e_a(\cdot, t_0) \right) (t, t_0),$$

$$t \in \mathbb{T}, \quad t \geq t_0.$$

Hence,

$$\mathcal{L} \left(D_{\Delta, t_0}^{\alpha} e_a(\cdot, t_0) \right) (z, t_0)$$

$$= \sum_{k=0}^{m-1} a^k \mathcal{L} \left(h_{k-\alpha}(\cdot, t_0) \right) (z, t_0)$$

$$+ a^m \mathcal{L} \left(h_{m-\alpha-1}(\cdot, t_0) \star e_a(\cdot, t_0) \right) (z, t_0)$$

$$= \sum_{k=0}^{m-1} \frac{a^k}{z^{k-\alpha+1}} + a^m \mathcal{L} \left(h_{m-\alpha-1}(\cdot, t_0) \right) (z, t_0) \mathcal{L} \left(e_a(\cdot, t_0) \right) (z, t_0)$$

$$= \sum_{k=0}^{m-1} \frac{a^k}{z^{k-\alpha+1}} + \frac{a^m}{z^{m-\alpha}(z - a)}.$$

Example 2.24. Let $\alpha = \frac{5}{4}$, $f(t) = \sin_1(t, t_0)$, $t \in \mathbb{T}$, $t \geq t_0$. We will find

$$\mathcal{L}\left(D_{\Delta, t_0}^{\frac{5}{4}} \sin(\cdot, t_0)\right)(z, t_0).$$

We have

$$m = \left\lceil \frac{5}{4} \right\rceil + 1$$
$$= 1 + 1 = 2$$

and

$$D_{\Delta, t_0}^{\frac{5}{4}} \sin_1(t, t_0) = \sum_{k=0}^{1} h_{k-\frac{5}{4}}(t, t_0) \sin_1^{\Delta^k}(t_0, t_0)$$

$$+ \left(h_{2-\frac{5}{4}-1}(\cdot, t_0) \star \sin_1^{\Delta^2}(\cdot, t_0)\right)(z, t_0)$$

$$= h_{-\frac{5}{4}}(t, t_0) \sin_1(t_0, t_0) + h_{1-\frac{5}{4}}(t, t_0) \cos_1(t_0, t_0)$$

$$- \left(h_{-\frac{1}{4}}(\cdot, t_0) \star \sin_1(\cdot, t_0)\right)(t, t_0)$$

$$= h_{-\frac{1}{4}}(t, t_0) - \left(h_{-\frac{1}{4}}(\cdot, t_0) \star \sin_1(\cdot, t_0)\right)(t, t_0),$$

$$t \in \mathbb{T}, \quad t \geq t_0.$$

Hence,

$$\mathcal{L}\left(D_{\Delta, t_0}^{\frac{5}{4}} \sin_1(t, t_0)\right)(z, t_0) = \mathcal{L}\left(h_{-\frac{1}{4}}(\cdot, t_0)\right)(z, t_0)$$

$$- \mathcal{L}\left(h_{-\frac{1}{4}}(\cdot, t_0) \star \sin_1(\cdot, t_0)\right)(z, t_0)$$

$$= \frac{1}{z^{-\frac{1}{4}+1}} - \mathcal{L}\left(h_{-\frac{1}{4}}(\cdot, t_0)\right)(z, t_0)$$

$$\times \mathcal{L}\left(\sin_1(\cdot, t_0)\right)(z, t_0)$$

$$= \frac{1}{z^{\frac{3}{4}}} - \frac{1}{z^{-\frac{1}{4}+1}} \frac{1}{z^2 + 1}$$

$$= \frac{z^2 + 1 - 1}{z^{\frac{3}{4}} (z^2 + 1)}$$

$$= \frac{z^{\frac{5}{4}}}{z^2 + 1}.$$

Exercise 2.5. Let $\alpha = \frac{1}{2}$, $f(t) = \sin_1(t, t_0)$, $t \in \mathbb{T}$, $t \geq t_0$. Find

$$\mathcal{L} \left(D^{\frac{1}{2}}_{\Delta, t_0} \sin_1(\cdot, t_0) \right) (z, t_0).$$

Answer: $\frac{z^{\frac{1}{2}}}{z^2 + 1}$.

Theorem 2.7 ([8]). *Let $\alpha > 0$, $\beta > 0$. Then*

$$\left(I^\alpha_{\Delta, t_0} I^\beta_{\Delta, t_0} f \right) (t) = I^{\alpha + \beta}_{\Delta, t_0} f(t), \quad t \in \mathbb{T}, \quad t > t_0.$$

Theorem 2.8 ([8]). *Let $\alpha > 0$, $m = -\overline{[-\alpha]}$, $n \in \mathbb{N}$, f be m-times Δ-differentiable and f^{Δ^m} be rd-continuous on $\mathbb{T}^{\kappa^m}$. Then*

$$D^n_\Delta D^\alpha_{\Delta, t_0} f(t) = D^{n+\alpha}_{\Delta, t_0} f(t), \quad t \in \mathbb{T}^{\kappa^m}.$$

Theorem 2.9 ([8]). *Let $\alpha > 0$, $m = -\overline{[-\alpha]}$, $n \in \mathbb{N}$, f be m-times Δ-differentiable and f^{Δ^m} be rd-continuous on $\mathbb{T}^{\kappa^m}$. Then*

$$D^n_\Delta I^\alpha_{\Delta, t_0} f(t) = I^{\alpha - n}_{\Delta, t_0} f(t), \quad t \in \mathbb{T}^{\kappa^m}.$$

Theorem 2.10 ([8]). *Let $\alpha > 0$, $m = -\overline{[-\alpha]}$, $n \in \mathbb{N}$, f be $(n + m)$-times Δ-differentiable and $f^{\Delta^{n+m}}$ be rd-continuous over $\mathbb{T}$. Then*

$$D^{n+\alpha}_{\Delta, t_0} f(t) = D^\alpha_{\Delta, t_0} D^n_\Delta f(t) + \sum_{k=0}^{n-1} h_{k-\alpha-n}(t, t_0) f^{\Delta^k}(t_0), \quad t \in \mathbb{T}^{\kappa^{n+m}}.$$

Theorem 2.11 ([8]). *Let $\alpha > 0$, $m = -\overline{[-\alpha]}$, $n \in \mathbb{N}$, f be $(n + m)$-times Δ-differentiable and $f^{\Delta^{n+m}}$ be rd-continuous over $\mathbb{T}$. Then*

$$D^n_\Delta I^\alpha_{\Delta, t_0} f(t) = I^\alpha_{\Delta, t_0} D^n_\Delta f(t) + \sum_{k=0}^{n-1} h_{k+\alpha-n}(t, t_0) f^{\Delta^k}(t_0), \quad t \in \mathbb{T}^{\kappa^{n+m}}.$$

Theorem 2.12 ([8]). *Let* $\alpha > 0$, $\beta > 0$, $M = -\overline{[-\beta]}$, f *be* Δ*-differentiable and* f^{Δ^M} *be rd-continuous over* $\mathbb{T}$. *Then*

$$\left(I_{\Delta,t_0}^\alpha D_{\Delta,t_0}^\beta f\right)(t) = D_{\Delta,t_0}^{\beta-\alpha} f(t) - \sum_{k=1}^M h_{\alpha-k}(t,t_0) D_{\Delta,t_0}^{\beta-k} f(t_0), \quad t \in \mathbb{T}^{\kappa^M}.$$

Theorem 2.13 ([8]). *Let* $\alpha > 0$, $\beta > 0$, $M = -\overline{[-\beta]}$, f *be* Δ*-differentiable and* f^{Δ^M} *be rd-continuous over* $\mathbb{T}$. *Then*

$$\left(D_{\Delta,t_0}^\beta I_{\Delta,t_0}^\alpha f\right)(t) = I_{\Delta,t_0}^{\alpha-\beta} f(t), \quad t \in \mathbb{T}^{\kappa^M}.$$

Theorem 2.14 ([8]). *Let* $\alpha > 0$, $f : \mathbb{T} \to \mathbb{R}$ *be locally* Δ*-integrable. For* $s,t \in \mathbb{T}$, $t \geq s \geq t_0$, *one has*

$$\mathcal{L}\left(I_{\Delta,t_0}^\alpha f(t)\right)(z,t_0) = \frac{1}{z^\alpha} \mathcal{L}\left(f(t)\right)(z,t_0).$$

Example 2.25. Let $\alpha > 0$ and $\beta \in \mathbb{R}$ be arbitrarily chosen. Then

$$\mathcal{L}\left(I_{\Delta,t_0}^\alpha h_\beta(\cdot,t_0)\right)(z,t_0) = \frac{1}{z^\alpha} \mathcal{L}\left(h_\beta(\cdot,t_0)\right)(z,t_0)$$

$$= \frac{1}{z^\alpha} \frac{1}{z^{\beta+1}}$$

$$= \frac{1}{z^{\alpha+\beta+1}}.$$

Example 2.26. Let $\alpha > 0$ and $a \in \mathbb{C}$ be arbitrarily chosen. Then

$$\mathcal{L}\left(I_{\Delta,t_0}^\alpha e_a(\cdot,t_0)\right)(z,t_0) = \frac{1}{z^\alpha} \mathcal{L}\left(e_a(\cdot,t_0)\right)(z,t_0)$$

$$= \frac{1}{z^\alpha(z-a)}.$$

Example 2.27. Let $\alpha > 0$ and $a \in \mathbb{C}$ be arbitrarily chosen. Then

$$\mathcal{L}\left(I_{\Delta,t_0}^\alpha \cos_a(\cdot,t_0)\right)(z,t_0) = \frac{1}{z^\alpha} \mathcal{L}\left(\cos_a(\cdot,t_0)\right)(z,t_0)$$

$$= \frac{z}{z^\alpha(z^2+a^2)}$$

$$= \frac{z^{1-\alpha}}{z^2+a^2}.$$

Exercise 2.6. Let $\alpha > 0$ and $a \in \mathbb{C}$ be arbitrarily chosen. Find

$$\mathcal{L}\left(I^\alpha_{\Delta,t_0}\sin_a(\cdot,t_0)\right)(z,t_0).$$

Answer: $\frac{a}{z^\alpha(z^2+a^2)}$.

Theorem 2.15 ([8]). *Let* $\alpha > 0$, $m = -\overline{[-\alpha]}$, $f : \mathbb{T} \to \mathbb{R}$ *be locally* Δ-*integrable. For* $s,t \in \mathbb{T}$, $t \geq s \geq t_0$, *one has*

$$\mathcal{L}\left(D^\alpha_{\Delta,t_0}f(t)\right)(z,t_0) = z^\alpha \mathcal{L}(f(t))(z,t_0) - \sum_{j=1}^{m} z^{j-1} D^{\alpha-j}_{\Delta,t_0} f(t_0).$$

2.4 Definition for the Caputo Fractional Δ-Derivative: Examples

Definition 2.5 ([8]). Let $t \in \mathbb{T}$. The Caputo fractional Δ-derivative of order $\alpha \geq 0$ is defined via the Riemann–Liouville fractional Δ-derivative as follows:

$$^C D^\alpha_{\Delta,t_0} f(t) = D^\alpha_{\Delta,t_0}\left(f(t) - \sum_{k=0}^{m-1} h_k(t,t_0)f^{\Delta^k}(t_0)\right), \quad t > t_0,$$

(2.11)

where $m = \overline{[\alpha]} + 1$ if $\alpha \notin \mathbb{N}$, $m = \overline{[\alpha]}$ if $\alpha \in \mathbb{N}$.

Example 2.28. Let $\alpha \in (0,1)$. Then $m = 1$ and

$$^C D^\alpha_{\Delta,t_0} f(t) = D^\alpha_{\Delta,t_0}\left(f(t) - \sum_{k=0}^{0} h_k(t,t_0)f^{\Delta^k}(t_0)\right)$$

$$= D^\alpha_{\Delta,t_0}\left(f(t) - h_0(t,t_0)f(t_0)\right)$$

$$= D^\alpha_{\Delta,t_0}\left(f(t) - f(t_0)\right), \quad t \in \mathbb{T}, \quad t > t_0.$$

If $f(t_0) = 0$, then the Caputo fractional Δ-derivative coincides with the Riemann–Liouville fractional Δ-derivative.

Example 2.29. Let $\alpha = m \in \mathbb{N}$. Then

$$^C D^m_{\Delta,t_0} f(t) = D^m_{\Delta,t_0}\left(f(t) - \sum_{k=0}^{m-1} h_k(t,t_0)f^{\Delta^k}(t_0)\right)$$

$$= D_\Delta^m \left(f(t) - \sum_{k=0}^{m-1} h_k(t, t_0) f^{\Delta^k}(t_0) \right)$$

$$= D_\Delta^m f(t) - \sum_{k=0}^{m-1} D_\Delta^m h_k(t, t_0) f^{\Delta^k}(t_0)$$

$$= f^{\Delta^m}(t), \quad t \in \mathbb{T}, \quad t > t_0,$$

i.e., when $\alpha = m \in \mathbb{N}$, then the Caputo fractional Δ-derivative coincides with the delta derivative.

We can rewrite (2.11) in the following way:

$$^C D_{\Delta, t_0}^\alpha f(t)$$

$$= D_\Delta^m I_{\Delta, t_0}^{m-\alpha} \left(f(t) - \sum_{k=0}^{m-1} h_k(t, t_0) f^{\Delta^k}(t_0) \right)$$

$$= D_\Delta^m \left(\int_{t_0}^t h_{m-\alpha-1}(t, \sigma(u)) \left(f(u) - \sum_{k=0}^{m-1} h_k(u, t_0) f^{\Delta^k}(t_0) \right) \Delta u \right)$$

$$= D_\Delta^m \left(\int_{t_0}^t h_{m-\alpha-1}(t, \sigma(u)) f(u) \Delta u \right)$$

$$- \sum_{k=0}^{m-1} f^{\Delta^k}(t_0) D_\Delta^m \left(\int_{t_0}^t h_{m-\alpha-1}(t, \sigma(u)) h_k(u, t_0) \Delta u \right)$$

$$= D_\Delta^m \left(\int_{t_0}^t h_{m-\alpha-1}(t, \sigma(u)) f(u) \Delta u \right)$$

$$- \sum_{k=0}^{m-1} f^{\Delta^k}(t_0) D_\Delta^m h_{m+k-\alpha}(t, t_0)$$

$$= D_\Delta^m (h_{m-\alpha-1} \star f)(t, t_0)$$

$$- \sum_{k=0}^{m-1} f^{\Delta^k}(t_0) h_{k-\alpha}(t, t_0), \quad t \in \mathbb{T}, \quad t > t_0. \tag{2.12}$$

Example 2.30. We will find

$$^C D_{\Delta, t_0}^\alpha h_\alpha(t, t_0), \quad t \in \mathbb{T}, \quad t > t_0.$$

We have, using (9.1),

$$^{C}D_{\Delta,t_0}^{\alpha} h_{\alpha}(t,t_0) = D_{\Delta}^{m}\left(h_{m-\alpha-1} \star h_{\alpha}\right)(t,t_0)$$

$$- \sum_{k=0}^{m-1} h_{\alpha}^{\Delta^{k}}(t_0,t_0) h_{k-\alpha}(t,t_0)$$

$$= D_{\Delta}^{m} h_m(t,t_0)$$

$$= 1, \quad t \in \mathbb{T}, \quad t > t_0.$$

Example 2.31. Let $\beta > \alpha$. We will find

$$^{C}D_{\Delta,t_0}^{\alpha} h_{\beta}(t,t_0), \quad t \in \mathbb{T}, \quad t > t_0.$$

Using (9.1), we obtain

$$^{C}D_{\Delta,t_0}^{\alpha} h_{\beta}(t,t_0) = D_{\Delta}^{m}\left(h_{m-\alpha-1} \star h_{\beta}\right)(t,t_0)$$

$$- \sum_{k=0}^{m-1} h_{\beta}^{\Delta^{k}}(t_0,t_0) h_{k-\alpha}(t,t_0)$$

$$= D_{\Delta}^{m} h_{m+\beta-\alpha}(t,t_0)$$

$$= h_{\beta-\alpha}(t,t_0), \quad t \in \mathbb{T}, \quad t > t_0.$$

Exercise 2.7. Find

$$^{C}D_{\Delta,t_0}^{\frac{1}{2}}\left(h_{\frac{4}{3}} \star h_{\frac{1}{4}}\right)(t,t_0), \quad t \in \mathbb{T}, \quad t > t_0.$$

Answer: $h_{\frac{25}{12}}(t,t_0), \quad t \in \mathbb{T}, \quad t > t_0.$

2.5   Properties of the Caputo Fractional Δ-Derivative

Theorem 2.16 ([8]). *Let* $\alpha \geq 0$, $m = [\alpha] + 1$ *if* $\alpha \notin \mathbb{N}$, $m = \alpha$ *if* $\alpha \in \mathbb{N}$.

1. *If* $\alpha \notin \mathbb{N}$, *then*

$$^{C}D_{\Delta,t_0}^{\alpha} f(t) = \left(h_{m-\alpha-1}(\cdot,t_0) \star f^{\Delta^{m}}\right)(t)$$

$$= I_{\Delta,t_0}^{m-\alpha} D_{\Delta}^{m} f(t), \quad t \in \mathbb{T}, \quad t > t_0.$$

2. *If $\alpha = m \in \mathbb{N}$, then*

$$^{C}D^{\alpha}_{\Delta,t_0} f(t) = f^{\Delta^m}(t), \quad t \in \mathbb{T}, \quad t > t_0.$$

Theorem 2.17 ([8]). *Let $\alpha > 0$. Then*

$$^{C}D^{\alpha}_{\Delta,t_0} I^{\alpha}_{\Delta,t_0} f(t) = f(t), \quad t \in \mathbb{T}, \quad t > t_0.$$

Theorem 2.18 ([8]). *Let $\alpha > 0$, $m = -\overline{[-\alpha]}$. Then*

$$I^{\alpha}_{\Delta,t_0} {}^{C}D^{\alpha}_{\Delta,t_0} f(t) = f(t) - \sum_{k=0}^{m-1} h_k(t,t_0) D^{k}_{\Delta,t_0} f(t_0), \quad t \in \mathbb{T}, \quad t > t_0.$$

Remark 2.3. When $\alpha \in (0,1)$, then

$$I^{\alpha}_{\Delta,t_0} {}^{C}D^{\alpha}_{\Delta,t_0} f(t) = f(t) - f(t_0), \quad t \in \mathbb{T}, \quad t > t_0.$$

Example 2.32. We have

$$I^{\frac{1}{2}}_{\Delta,t_0} {}^{C}D^{\frac{1}{2}}_{\Delta,t_0} e_1(t,t_0) = e_1(t,t_0) - e_1(t_0,t_0)$$
$$= e_1(t,t_0) - 1, \quad t \in \mathbb{T}, \quad t > t_0.$$

Exercise 2.8. Find

$$I^{\frac{1}{3}}_{\Delta,t_0} {}^{C}D^{\frac{1}{3}}_{\Delta,t_0} \left(-3e_2(t,t_0) + \sin_1(t,t_0) + h_2(t,t_0) \right), \quad t \in \mathbb{T}, \quad t > t_0.$$

Answer:

$$-3e_2(t,t_0) + \sin_1(t,t_0) + h_2(t,t_0) + 3, \quad t \in \mathbb{T}, \quad t > t_0.$$

Theorem 2.19 ([8]). *Let $m - 1 < \beta < \alpha < m$, $m \in \mathbb{N}$. Then, for all $k \in \{1, \ldots, m-1\}$, we have*

$$^{C}D^{\alpha-m+k}_{\Delta,t_0} D^{m-k}_{\Delta} f(t) = {}^{C}D^{\alpha}_{\Delta,t_0} f(t), \quad t \in \mathbb{T}, \quad t > t_0, \quad (2.13)$$
$$^{C}D^{\alpha-\beta}_{\Delta,t_0} {}^{C}D^{\beta}_{\Delta,t_0} f(t) = {}^{C}D^{\alpha}_{\Delta,t_0} f(t), \quad t \in \mathbb{T}, \quad t > t_0. \quad (2.14)$$

Exercise 2.9. Let $m \in \mathbb{N}$ and $m - 1 < \beta < \alpha < m$. Prove that

$$I_{\Delta,t_0}^\alpha{}^C D_{\Delta,t_0}^{\alpha-m+k} D_{\Delta,t_0}^{m-k} f(t)$$

$$= f(t) - \sum_{k=0}^{m-1} h_k(t,t_0) D_\Delta^k f(t_0), \quad t \in \mathbb{T}, \quad t > t_0.$$

Theorem 2.20 ([8]). *Let* $\alpha > 0$, $m - 1 < \alpha \le m$, $m \in \mathbb{N}$. *Then*

$$\mathcal{L}\left({}^C D_{\Delta,t_0}^\alpha f(t)\right)(z) = z^\alpha \mathcal{L}\left(f(t)\right)(z) - \sum_{k=0}^{m-1} z^{\alpha-k-1} f^{\Delta^k}(t_0)$$

for those $z \in \mathbb{C}$, *for which*

$$\lim_{t\to\infty} \left(f^{\Delta^k}(t) e_{\ominus z}(t,t_0)\right) = 0, \quad k \in \{0,\ldots,m-1\}. \tag{2.15}$$

Example 2.33. Let $\alpha > 0$, $m - 1 < \alpha \le m$, $m \in \mathbb{N}$, $\beta \in \mathbb{C}$. Then

$$\mathcal{L}\left({}^C D_{\Delta,t_0}^\alpha e_\beta(t,t_0)\right)(z) = z^\alpha \mathcal{L}\left(e_\beta(t,t_0)\right)(z) - \sum_{k=0}^{m-1} z^{\alpha-k-1} e_\beta^{\Delta^k}(t_0,t_0)$$

$$= \frac{z^\alpha}{z-\beta} - \sum_{k=0}^{m-1} \beta^k z^{\alpha-k-1},$$

provided that

$$\lim_{t\to\infty} \left(\beta^k e_\beta(t,t_0) e_{\ominus z}(t,t_0)\right) = 0, \quad k \in \{0,\ldots,m-1\}.$$

Example 2.34. Let $\alpha > 0$, $m - 1 < \alpha \le m$, $m \in \mathbb{N}$, $l \in \mathbb{N}$. Then

$$\mathcal{L}\left({}^C D_{\Delta,t_0}^\alpha h_l(t,t_0)\right)(z) = z^\alpha \mathcal{L}\left(h_l(t,t_0)\right)(z)$$

$$- \sum_{k=0}^{m-1} z^{\alpha-k-1} h_l^{\Delta^k}(t_0,t_0)$$

$$= \begin{cases} \dfrac{z^\alpha}{z^{l+1}} & \text{if } l > m-1 \\ \dfrac{z^\alpha}{z^{l+1}} - z^{\alpha-l-1} & \text{if } l \le m-1 \end{cases}$$

$$= \begin{cases} z^{\alpha-l-1} & \text{if } l > m-1, \\ 0 & \text{if } l \le m-1, \end{cases}$$

provided that

$$\lim_{t\to\infty} \left(h_l^{\Delta^k}(t,t_0) e_{\ominus z}(t,t_0) \right) = 0, \quad k \in \{0, \ldots, m-1\}.$$

Example 2.35. We have

$$\mathcal{L}\left({}^C D_{\Delta,t_0}^{\frac{1}{2}} \sin_1(t,t_0) \right) = z^{\frac{1}{2}} \mathcal{L}\left(\sin_1(t,t_0) \right)(z) - z^{-\frac{1}{2}} \sin_1(t_0,t_0)$$

$$= \frac{z^{\frac{1}{2}}}{z^2+1},$$

provided that

$$\lim_{t\to\infty} \left(\sin_1(t,t_0) e_{\ominus z}(t,t_0) \right) = 0.$$

Exercise 2.10. Find

$$\mathcal{L}\left({}^C D_{\Delta,t_0}^{\frac{4}{3}} e_2(t,t_0) - 2 {}^C D_{\Delta,t_0}^{\frac{1}{2}} \cos_1(t,t_0) \right)(z).$$

Answer:

$$\frac{z^{\frac{4}{3}}}{z-2} - z^{\frac{1}{3}} - 2z^{-\frac{2}{3}} - 2\frac{z^{\frac{3}{2}}}{z^2+1} + 2z^{-\frac{1}{2}}.$$

2.6　Advanced Practical Problems

Problem 2.1. Let $\mathbb{T} = 3^{\mathbb{N}_0}$, $t_0 = 1$, $\alpha = \frac{1}{4}$, $\beta = \frac{11}{4}$. Find

$$I_{\Delta,1}^\alpha h_{\beta-1}(t,1), \quad t \in \mathbb{T}, \ t > 1.$$

Answer: $\frac{(t-1)(t-3)}{4}$, $t \in \mathbb{T}, t > 1$.

Problem 2.2. Let $\mathbb{T} = 3^{\mathbb{N}_0}$, $t_0 = 1$, $\alpha = \frac{1}{2}$, $\beta = -\frac{9}{2}$. Find

$$D_{\Delta,1}^{\frac{1}{2}} h_{\frac{7}{2}}(t,1), \quad t \in \mathbb{T}, \quad t > 1.$$

Answer:

$$\frac{(t-1)(t-3)(t-9)}{52}, \quad t \in \mathbb{T}, \quad t > 1.$$

Problem 2.3. Let $\alpha = \frac{1}{5}$, $\beta = \frac{7}{5}$. Find

$$\mathcal{L}\left(h_{\frac{3}{5}}(\cdot, t_0) \star I_{\Delta, t_0}^{\frac{1}{2}} h_{\frac{2}{5}}(\cdot, t_0) - 3D_{\Delta, t_0}^{\frac{1}{5}} h_{\frac{2}{5}}(\cdot, t_0) \right)(z, t_0).$$

Answer: $-\frac{3}{z^{\frac{6}{5}}} + \frac{1}{z^{\frac{5}{2}}}$.

Problem 2.4. Let $\alpha \in (0, 1)$, $\beta = 1 + \alpha$. Prove that

$$\left(h_{\beta-2}(\cdot, t_0) \star D_{\Delta, t_0}^{\alpha}\left(D_{\Delta, t_0}^{\alpha} h_{\beta-1} \right)(\cdot, t_0) \right)(z, t_0) = 1.$$

Problem 2.5. Let $\alpha = \frac{4}{3}$, $f(t) = \sin_3(t, t_0)$, $t \in \mathbb{T}$, $t \geq t_0$. Find

$$\mathcal{L}\left(I_{\Delta, t_0}^{\frac{4}{3}} \sin_3(\cdot, t_0) \right)(z, t_0).$$

Answer: $\dfrac{3}{z^{\frac{4}{3}}(z^2+9)}$.

Problem 2.6. Let $\alpha = \frac{1}{3}$, $f(t) = \sin_3(t, t_0)$, $t \in \mathbb{T}$, $t \geq t_0$. Find

$$\mathcal{L}\left(D_{\Delta, t_0}^{\frac{1}{3}} \sin_3(\cdot, t_0) \right)(z, t_0).$$

Answer: $3\dfrac{z^{\frac{1}{3}}}{z^2+9} - I_{\Delta, t_0}^{\frac{2}{3}} \sin_3(t_0, t_0)$.

Problem 2.7. Let $\alpha > 0$ be arbitrarily chosen and $f : \mathbb{T} \to \mathbb{R}$ be a continuous function on $\mathbb{T}$. Prove that

$$\mathcal{L}\left(I_{\Delta, t_0}^{\alpha} \int_{t_0}^t f(y)\Delta y \right)(z, t_0) = \frac{1}{z^{\alpha+1}}\mathcal{L}(f)(z).$$

Problem 2.8. Find

1. $^C D_{\Delta, t_0}^{\frac{3}{2}}\left(h_1 \star h_{\frac{1}{2}} \right)(t, t_0)$, $t \in \mathbb{T}$, $t > t_0$,

2. $^C D_{\Delta, t_0}^{\frac{4}{3}}\left(h_3 \star h_{\frac{5}{6}} \right)(t, t_0)$, $t \in \mathbb{T}$, $t > t_0$,

3. $^C D_{\Delta, t_0}^{\frac{4}{3}}\left(-4h_2 - 7\left(h_1 \star h_{\frac{5}{6}} \right) \right)(t, t_0)$, $t \in \mathbb{T}$, $t > t_0$.

Answer:

1. $h_1(t, t_0), \quad t \in \mathbb{T}, \quad t > t_0,$
2. $h_{\frac{21}{6}}(t, t_0), \quad t \in \mathbb{T}, \quad t > t_0,$
3. $-4h_{\frac{2}{3}}(t, t_0) - 7h_{\frac{3}{2}}(t, t_0), \quad t \in \mathbb{T}, \quad t > t_0.$

Problem 2.9. Let $\alpha \in (0, 1)$. Prove that

$$I^\alpha_{\Delta, t_0}\, {}^C D^\alpha_{\Delta, t_0} a = 0,$$

where a is a constant.

Problem 2.10. Find

$$I^{\frac{1}{3}}_{\Delta, t_0}\, {}^C D^{\frac{1}{3}}_{\Delta, t_0} \left(\cos_2(t, t_0) - 4h_{10}(t, t_0) + 5 \right), \quad t \in \mathbb{T}, \quad t > t_0.$$

Answer:

$$\cos_2(t, t_0) - 4h_{10}(t, t_0) - 1, \quad t \in \mathbb{T}, \quad t > t_0.$$

Problem 2.11. Let $m \in \mathbb{N}$ and $m - 1 < \beta < \alpha < m$. Prove that

$$I^\alpha_{\Delta, t_0}\, {}^C D^{\alpha-\beta}_{\Delta, t_0}\, {}^C D^\beta_{\Delta, t_0} f(t) = f(t) - \sum_{k=0}^{m-1} h_k(t, t_0) D^k_\Delta f(t_0), \quad t \in \mathbb{T}, \quad t > t_0.$$

Problem 2.12. Find

$$\mathcal{L}\left({}^C D^{\frac{3}{2}}_{\Delta, t_0} e_1(t, t_0) + {}^C D^{\frac{1}{3}}_{\Delta, t_0} \sin_1(t, t_0) \right)(z).$$

Answer:

$$\frac{z^{\frac{3}{2}}}{z - 1} - z^{\frac{1}{2}} - z^{-\frac{1}{2}} + \frac{z^{\frac{1}{3}}}{z^2 + 1}.$$

Chapter 3

Linear Inequalities for Riemann–Liouville Fractional Delta Integral Operator

This chapter is devoted to the Riemann–Liouville fractional delta integral inequalities. The fractional admissible triple is defined, and using this, some Gronwall-type Riemann–Liouville fractional inequalities are deducted. Some classes of Volterra-type fractional inequalities as well as simultaneous Riemann–Liouville fractional inequalities are investigated. Then some analogues of the Pachpatte inequalities are deducted. As applications, we investigate the Cauchy problem for Riemann–Liouville fractional dynamic equations on arbitrary time scales and the dependency of its solution on the initial data. Some of the results in this chapter can be found in Ref. [11].

Suppose that $\mathbb{T}$ is a time scale with forward jump operator and delta differentiation operator σ and Δ, respectively. Let $t_0, a \in \mathbb{T}$, $a < t_0$.

3.1 The Gronwall-Type Fractional Inequalities

Definition 3.1 (α-Fractional Admissible Triple). Let $A \subseteq \mathbb{T}$, $\alpha > 0$, u and v be functions defined on A and Δ-integrable on A. We say that (u, v, A) is an α-fractional admissible triple if

$$\int_A |v(s)|^k h_{k\alpha-1}(t, \sigma(s))|u(s)|\Delta s$$

exists for any $t \in A$ and any $k \in \mathbb{N}$,

$$\sum_{k=1}^{\infty} \int_A |v(s)|^k h_{k\alpha-1}(t, \sigma(s))|u(s)| \Delta s < \infty$$

uniformly on A.

Such α-fractional admissible triple exists. Really, let $A \subseteq \mathbb{T}$ be such that $\mu(A) < \infty$, $|h_{k\alpha-1}(t, s)| \leq M$, $t, s \in A$ for any $k \in \mathbb{N}$ and for some positive constant M, u and v be Δ-integrable functions on A so that $|v(t)| \leq B$, $t \in A$, for some positive constant $B < 1$, $|v(t)| \leq C$, $t \in A$, for some positive constant C. Then

$$\sum_{k=1}^{\infty} \int_A |v(t)|^k h_{k\alpha-1}(t, \sigma(s))|u(s)| \Delta s \leq CM\mu(A) \sum_{k=1}^{\infty} B^k < \infty$$

uniformly on A. Here, $\mu(A)$ is the Δ-Lebesgue measure of the set A.

Theorem 3.1 (The Gronwall-Type Inequality). *Let $\alpha > 0$, $y, u, v : [t_0, a) \to [0, \infty)$ be Δ-integrable, $v(t) \leq B$, $t \in [t_0, a)$, for some positive constant B, v is a non-decreasing function on $[t_0, a)$. Let also $(y, B, [t_0, a))$ be an α-fractional admissible triple. If*

$$y(t) \leq u(t) + v(t) I_{\Delta, t_0}^{\alpha} y(t), \quad t \in [t_0, a),$$

then

$$y(t) \leq u(t) + \sum_{k=1}^{\infty} (v(t))^k I_{\Delta, t_0}^{k\alpha} u(t), \quad t \in [t_0, a).$$

Proof.   Define

$$Qy(t) = v(t) I_{\Delta, t_0}^{\alpha} y(t), \quad t \in [t_0, a).$$

Then

$$y(t) \leq u(t) + Qy(t), \quad t \in [t_0, a).$$

Taking iteration in the above inequality, we get

$$\begin{aligned}
y(t) &\leq u(t) + Qy(t) \\
&\leq u(t) + Q\left(u(t) + Qy(t)\right) \\
&= u(t) + Qu(t) + Q^2 y(t)
\end{aligned}$$

$$\leq u(t) + Qu(t) + Q^2\left(u(t) + Qy(t)\right)$$

$$\leq \cdots$$

$$\leq \sum_{k=0}^{n-1} Q^k u(t) + Q^n y(t), \quad t \in [t_0, a), \quad n \in \mathbb{N},$$

i.e.,

$$y(t) \leq \sum_{k=0}^{n-1} Q^k u(t) + Q^n y(t), \quad t \in [t_0, a), \tag{3.1}$$

for any $n \in \mathbb{N}$. We will prove that

$$Q^k u(t) \leq (v(t))^k I_{\Delta,t_0}^{k\alpha} u(t), \quad t \in [t_0, a), \tag{3.2}$$

for any $k \in \mathbb{N}_0$. For $k = 1$, the inequality (3.2) is obvious. Assume that the inequality (3.2) holds for some $k \in \mathbb{N}$. We will prove the inequality (3.2) for $k+1$. We have, using v as a non-decreasing function on $[t_0, a)$,

$$Q^{k+1} u(t) = Q\left(Q^k u(t)\right)$$

$$\leq Q\left((v(t))^k I_{\Delta,t_0}^{k\alpha} u(t)\right)$$

$$\leq (v(t))^{k+1} I_{\Delta,t_0}^{\alpha}\left(I_{\Delta,t_0}^{k\alpha} u(t)\right)$$

$$= (v(t))^{k+1} I_{\Delta,t_0}^{(k+1)\alpha} u(t), \quad t \in [t_0, a).$$

Thus, the inequality (3.2) holds for any $k \in \mathbb{N}$. Now, applying the inequality (3.2) to the inequality (3.1), we arrive at the inequality

$$y(t) \leq \sum_{k=0}^{n-1} (v(t))^k I_{\Delta,t_0}^{k\alpha} u(t) + Q^n y(t), \quad t \in [t_0, a), \tag{3.3}$$

for any $n \in \mathbb{N}$. Next,

$$Q^n y(t) \leq (v(t))^n I_{\Delta,t_0}^{n\alpha} y(t)$$

$$= (v(t))^n \int_{t_0}^{t} h_{n\alpha-1}(t, \sigma(\tau)) y(\tau) \Delta\tau$$

$$\leq B^n \int_{t_0}^t h_{n\alpha-1}(t, \sigma(\tau)) y(\tau) \Delta\tau$$

$$\leq B^n \int_{t_0}^a h_{n\alpha-1}(t, \sigma(\tau)) y(\tau) \Delta\tau, \quad t \in [t_0, a).$$

Since $(y, B, [t_0, a))$ is an α-fractional admissible triple, we have that

$$\sum_{n=0}^\infty B^n \int_{t_0}^a h_{n\alpha-1}(t, \sigma(\tau)) y(\tau) \Delta\tau < \infty$$

uniformly on $[t_0, a)$. Hence,

$$Q^n y(t) \to 0, \quad \text{as } n \to \infty, \quad t \in [t_0, a).$$

From here and from (3.3), we find

$$y(t) \leq u(t) + \sum_{k=1}^\infty (v(t))^k I_{\Delta,t_0}^{k\alpha} u(t), \quad t \in [t_0, a).$$

This completes the proof.   □

Corollary 3.1. *Let* $\alpha > 0$, $y, u, v : [t_0, a) \to [0, \infty)$ *be* Δ-*integrable,* $v(t) \leq B$, $t \in [t_0, a)$, *for some positive constant* B, v *be a non-decreasing function on* $[t_0, a)$. *Let also* $(y, B, [t_0, a))$ *be an* α-*fractional admissible triple. If*

$$y(t) \leq u(t) + I_{\Delta,t_0}^\alpha (vy)(t), \quad t \in [t_0, a), \tag{3.4}$$

then

$$y(t) \leq u(t) + \sum_{k=1}^\infty (v(t))^k I_{\Delta,t_0}^{k\alpha} u(t), \quad t \in [t_0, a). \tag{3.5}$$

Proof.   Since v is a non-decreasing function on $[t_0, a)$, by the inequality (3.4), we find

$$y(t) \leq u(t) + v(t) I_{\Delta,t_0}^\alpha y(t), \quad t \in [t_0, a).$$

Now, applying Theorem 3.1, we get the desired inequality (3.5). This completes the proof.   □

Corollary 3.2. *Let $\alpha > 0$, $B \geq 0$, $y, u, v : [t_0, a) \to [0, \infty)$ be Δ-integrable, $v(t) \leq B$, $t \in [t_0, a)$. Let also $(y, B, [t_0, a))$ be an α-fractional admissible triple. If*

$$y(t) \leq u(t) + v(t) I^\alpha_{\Delta, t_0} y(t), \quad t \in [t_0, a),$$

then

$$y(t) \leq u(t) + \sum_{k=1}^{\infty} B^k I^{k\alpha}_{\Delta, t_0} u(t), \quad t \in [t_0, a).$$

Proof. Since $v(t) \leq B$, $t \in [t_0, a)$, using Theorem 3.1, we find

$$y(t) \leq u(t) + \sum_{k=1}^{\infty} (v(t))^k I^{k\alpha}_{\Delta, t_0} u(t)$$

$$\leq u(t) + \sum_{k=1}^{\infty} B^k I^{k\alpha}_{\Delta, t_0} u(t), \quad t \in [t_0, a).$$

This completes the proof. $\qquad \square$

Theorem 3.2 (The Gronwall-Type Inequality). *Let $\alpha > 0$, $y, u, v : [t_0, a) \to [0, \infty)$ be Δ-integrable and $\sum_{n=1}^{\infty} (I^\alpha_{\Delta, t_0} v(t))^n$ be uniformly convergent on $[t_0, a)$ and $I^\alpha_{\Delta, t_0}(vy)(t) < \infty$, $I^\alpha_{\Delta, t_0}(vu)(t) < \infty$, $t \in [t_0, a)$. If*

$$y(t) \leq u(t) + I^\alpha_{\Delta, t_0}(vy)(t), \quad t \in [t_0, a),$$

then

$$y(t) \leq u(t) + \sum_{k=1}^{\infty} \left(I^\alpha_{\Delta, t_0} v(t) \right)^{k-1} I^\alpha_{\Delta, t_0}(uv)(t), \quad t \in [t_0, a).$$

Proof. Let

$$Qy(t) = I^\alpha_{\Delta, t_0}(vy)(t), \quad t \in [t_0, a).$$

Then

$$y(t) \leq u(t) + Qy(t), \quad t \in [t_0, a).$$

As in the proof of Theorem 3.1, we have

$$y(t) \le \sum_{k=0}^{n-1} Q^k u(t) + Q^n y(t), \quad t \in [t_0, a), \qquad (3.6)$$

for any $n \in \mathbb{N}$. Note that

$$Q^2 u(t) = Q(Qu(t))$$
$$= I_{\Delta,t_0}^\alpha (v(t) Q u(t))$$
$$\le I_{\Delta,t_0}^\alpha (uv)(t) I_{\Delta,t_0}^\alpha v(t), \quad t \in [t_0, a).$$

Assume that

$$Q^k u(t) \le I_{\Delta,t_0}^\alpha (vu)(t) \left(I_{\Delta,t_0}^\alpha v(t) \right)^{k-1}, \quad t \in [t_0, a), \qquad (3.7)$$

for some $k \in \mathbb{N}$. Then

$$Q^{k+1} u(t) = Q(Q^k u)(t)$$
$$= I_{\Delta,t_0}^\alpha \left(v Q^k u \right)(t)$$
$$\le \left(I_{\Delta,t_0}^\alpha v(t) \right)^k I_{\Delta,t_0}^\alpha (vu)(t), \quad t \in [t_0, a).$$

Thus, (3.7) holds for any $k \in \mathbb{N}$. As above,

$$Q^n y(t) \le \left(I_{\Delta,t_0}^\alpha v(t) \right)^{n-1} I_{\Delta,t_0}^\alpha (vy)(t), \quad t \in [t_0, a).$$

Since the series $\sum_{n=1}^{\infty} (I_{\Delta,t_0}^\alpha v(t))^n$ is uniformly convergent on $[t_0, a)$, we get

$$Q^n y(t) \to 0, \quad \text{as } n \to \infty, \quad t \in [t_0, a).$$

Hence, applying (3.6) and (3.7), we arrive at

$$y(t) \le u(t) + \sum_{k=1}^{\infty} \left(I_{\Delta,t_0}^\alpha v(t) \right)^{k-1} I_{\Delta,t_0}^\alpha (vu)(t) + \lim_{n \to \infty} Q^n y(t)$$

$$= u(t) + I_{\Delta,t_0}^\alpha (vu)(t) \sum_{k=1}^{\infty} \left(I_{\Delta,t_0}^\alpha v(t) \right)^{k-1}, \quad t \in [t_0, a).$$

This completes the proof. $\qquad \qquad \square$

Theorem 3.3 (The Gronwall-Type Inequality). *Let* $\alpha > 0$, $y, u, v, w : [t_0, a) \to [0, \infty)$ *be* Δ-*integrable*, $I^{\alpha}_{\Delta, t_0}(wy)(t) < \infty$, $I^{\alpha}_{\Delta, t_0}(wu)(t) < \infty$, $I^{\alpha}_{\Delta, t_0}(wv)(t) < \infty$, $t \in [t_0, a)$. *Let also* $\sum_{n=1}^{\infty} (I^{\alpha}_{\Delta, t_0}(wv)(t))^{n-1}$ *be uniformly convergent on* $[t_0, a)$. *If*

$$y(t) \le u(t) + v(t) I^{\alpha}_{\Delta, t_0}(wy)(t), \quad t \in [t_0, a),$$

then

$$y(t) \le u(t) + v(t) I^{\alpha}_{\Delta, t_0}(wu)(t) \sum_{k=1}^{\infty} \left(I^{\alpha}_{\Delta, t_0}(wv)(t) \right)^{k-1}, \quad t \in [t_0, a).$$

Proof. Let

$$Qy(t) = v(t) I^{\alpha}_{\Delta, t_0}(wy)(t), \quad t \in [t_0, a).$$

Then

$$y(t) \le u(t) + Qy(t), \quad t \in [t_0, a).$$

As in the proof of Theorem 3.1, we get

$$y(t) \le \sum_{k=1}^{n-1} Q^k u(t) + Q^n y(t), \quad t \in [t_0, a), \tag{3.8}$$

for any $n \in \mathbb{N}$. Observe that

$$\begin{aligned}
Q^2 u(t) &= Q(Qu)(t) \\
&= Q \left(v I^{\alpha}_{\Delta, t_0}(wu) \right)(t) \\
&= v(t) I^{\alpha}_{\Delta, t_0} \left(wv I^{\alpha}_{\Delta, t_0}(wu) \right)(t) \\
&\le v(t) I^{\alpha}_{\Delta, t_0}(wu)(t) I^{\alpha}_{\Delta, t_0}(wv)(t), \quad t \in [t_0, a),
\end{aligned}$$

and

$$\begin{aligned}
Q^3 u(t) &= Q \left(Q^2 u \right)(t) \\
&\le Q \left(v I^{\alpha}_{\Delta, t_0}(wv) I^{\alpha}_{\Delta, t_0}(wu) \right)(t) \\
&= v(t) I^{\alpha}_{\Delta, t_0} \left(wv I^{\alpha}_{\Delta, t_0}(wv) I^{\alpha}_{\Delta, t_0}(wu) \right)(t) \\
&\le v(t) I^{\alpha}_{\Delta, t_0}(wu)(t) \left(I^{\alpha}_{\Delta, t_0}(wv)(t) \right)^2, \quad t \in [t_0, a).
\end{aligned}$$

Assume that

$$Q^k u(t) \le v(t) I_{\Delta,t_0}^\alpha (wu)(t) \left(I_{\Delta,t_0}^\alpha (wv)(t) \right)^{k-1}, \quad t \in [t_0, a), \quad (3.9)$$

for some $k \in \mathbb{N}$. Then

$$
\begin{aligned}
Q^{k+1} u(t) &= Q \left(Q^k u \right)(t) \\
&\le Q \left(v I_{\Delta,t_0}^\alpha (wu) \left(I_{\Delta,t_0}^\alpha (wv) \right)^{k-1} \right)(t) \\
&= v(t) I_{\Delta,t_0}^\alpha \left(wv I_{\Delta,t_0}^\alpha (wu) \left(I_{\Delta,t_0}^\alpha (wv) \right)^{k-1} \right)(t) \\
&\le v(t) I_{\Delta,t_0}^\alpha (wu)(t) \left(I_{\Delta,t_0}^\alpha (wv)(t) \right)^{k}, \quad t \in [t_0, a).
\end{aligned}
$$

Therefore, the inequality (3.9) holds for any $k \in \mathbb{N}$. As above,

$$Q^n y(t) \le v(t) I_{\Delta,t_0}^\alpha (wy)(t) \left(I_{\Delta,t_0}^\alpha (wv)(t) \right)^{n-1}, \quad t \in [t_0, a), \quad n \in \mathbb{N}.$$

Since $\sum_{n=1}^\infty (I_{\Delta,t_0}^\alpha (wv)(t))^{n-1}$ is uniformly convergent on $[t_0, a)$, we have that

$$\left(I_{\Delta,t_0}^\alpha (wv)(t) \right)^{n-1} \to 0, \quad \text{as } n \to \infty,$$

uniformly on $[t_0, a)$. Therefore,

$$Q^n y(t) \to 0, \quad \text{as } n \to \infty,$$

uniformly on $[t_0, a)$. Now, applying the inequality (3.9) to the inequality (3.8), we get

$$y(t) \le u(t) + v(t) I_{\Delta,t_0}^\alpha (wu)(t) \sum_{k=1}^\infty \left(I_{\Delta,t_0}^\alpha (wv)(t) \right)^{k-1} + \lim_{n\to\infty} Q^n y(t)$$

$$= u(t) + v(t) I_{\Delta,t_0}^\alpha (wu)(t) \sum_{k=1}^\infty \left(I_{\Delta,t_0}^\alpha (wv)(t) \right)^{k-1}, \quad t \in [t_0, a).$$

This completes the proof. $\qquad\qquad\qquad\qquad\qquad\qquad\qquad\qquad \square$

Theorem 3.4 (The Gronwall-Type Inequality). *Let $\alpha > 0$, $u :$ $[t_0, a) \to (0, \infty)$ be Δ-integrable and non-decreasing, $y, v : [t_0, a) \to$ $[0, \infty)$ be Δ-integrable and $I^\alpha_{\Delta, t_0}\left(\frac{v}{u}y\right)(t) < \infty$, $I^\alpha_{\Delta, t_0}v(t) < \infty$, $t \in$ $[t_0, a)$. Let also $\sum_{n=1}^{\infty} \left(I^\alpha_{\Delta, t_0}v(t)\right)^n < \infty$ uniformly on $[t_0, a)$. If*

$$y(t) \le u(t) + I^\alpha_{\Delta, t_0}(vy)(t), \quad t \in [t_0, a), \tag{3.10}$$

then

$$y(t) \le u(t) \left(1 + \sum_{k=1}^{\infty} \left(I^\alpha_{\Delta, t_0}v(t)\right)^k\right), \quad t \in [t_0, a).$$

Proof. Let

$$w(t) = \frac{y(t)}{u(t)}, \quad t \in [t_0, a).$$

Since u is non-decreasing on $[t_0, a)$, by the inequality (3.10), we get

$$\frac{y(t)}{u(t)} \le 1 + \frac{1}{u(t)} I^\alpha_{\Delta, t_0}(vy)(t)$$

$$\le 1 + I^\alpha_{\Delta, t_0}\left(\frac{v}{u}y\right)(t), \quad t \in [t_0, a),$$

whereupon

$$w(t) \le 1 + I^\alpha_{\Delta, t_0}(vw)(t), \quad t \in [t_0, a).$$

Now, we apply Theorem 3.2 and find

$$w(t) \le 1 + \sum_{k=1}^{\infty} \left(I^\alpha_{\Delta, t_0}v(t)\right)^{k-1} I^\alpha_{\Delta, t_0}v(t)$$

$$= 1 + \sum_{k=1}^{\infty} \left(I^\alpha_{\Delta, t_0}v(t)\right)^k, \quad t \in [t_0, a).$$

Hence,

$$y(t) \le u(t) \left(1 + \sum_{k=1}^{\infty} \left(I^\alpha_{\Delta, t_0}v(t)\right)^k\right), \quad t \in [t_0, a).$$

This completes the proof. □

Theorem 3.5 (The Gronwall-Type Inequality). *Let $\alpha > 0$, $m = -\overline{[-\alpha]}$, $y, u, v : [t_0, a) \to [0, \infty)$ be Δ-integrable, $\sum_{n=1}^{\infty} (I_{\Delta,t_0}^{\alpha} v(t))^n$ be uniformly convergent on $[t_0, a)$, $I_{\Delta,t_0}^{\alpha}(vy)(t) < \infty$, $t \in [t_0, a)$,*

$$I_{\Delta,t_0}^{\alpha}\left(v(\cdot)\left(\sum_{k=1}^{m} h_{\alpha-k}(\cdot, t_0) D_{\Delta,t_0}^{\alpha-k} y(t_0) + I_{\Delta,t_0}^{\alpha} u(\cdot)\right)\right)(t) < \infty,$$

$t \in [t_0, a)$.

If

$$D_{\Delta,t_0}^{\alpha} y(t) \leq u(t) + v(t) y(t), \quad t \in [t_0, a), \tag{3.11}$$

then

$$y(t) \leq \sum_{k=1}^{\infty} h_{\alpha-k}(t, t_0) D_{\Delta,t_0}^{\alpha-k} y(t_0) + I_{\Delta,t_0}^{\alpha} u(t) + \sum_{k=1}^{\infty} \left(I_{\Delta,t_0}^{\alpha} v(t)\right)^{k-1} I_{\Delta,t_0}^{\alpha}$$

$$\times \left(v(\cdot)\left(\sum_{k=1}^{m} h_{\alpha-k}(\cdot, t_0) D_{\Delta,t_0}^{\alpha-k} y(t_0) + I_{\Delta,t_0}^{\alpha} u(\cdot)\right)\right)(t),$$

$t \in [t_0, a)$.

Here, with $\overline{[-\alpha]}$, we denote the integer part of $-\alpha$.

Proof. We take I_{Δ,t_0}^{α} of both sides of the inequality (3.11) and find

$$y(t) - \sum_{k=1}^{m} h_{\alpha-k}(t, t_0) D_{\Delta,t_0}^{\alpha-k} y(t_0) \leq I_{\Delta,t_0}^{\alpha} u(t) + I_{\Delta,t_0}^{\alpha}(vy)(t), \quad t \in [t_0, a),$$

or

$$y(t) \leq \sum_{k=1}^{m} h_{\alpha-k}(t, t_0) D_{\Delta,t_0}^{\alpha-k} y(t_0) + I_{\Delta,t_0}^{\alpha} u(t) + I_{\Delta,t_0}^{\alpha}(vy)(t), \quad t \in [t_0, a).$$

Now, we apply Theorem 3.2 and get the desired inequality. This completes the proof. $\quad\square$

3.2 Volterra-Type Fractional Inequalities

In this section, we deduct some fractional analogues of the Volterra-type integral inequalities. We start with the Fubini theorem needed for our investigations.

Theorem 3.6 (The Fubini Theorem). *Let* $g : [t_0, a) \times [t_0, a) \to \mathbb{R}$ *be* Δ*-integrable and the integrals*

$$\int_{t_0}^{t} \int_{t_0}^{s} g(s, y) \Delta s \Delta y, \quad \int_{t_0}^{t} \int_{\sigma(s)}^{t} g(s, y) \Delta y \Delta s, \quad t \in [t_0, a),$$

exist. Then

$$\int_{t_0}^{t} \int_{t_0}^{s} g(s, y) \Delta s \Delta y = \int_{t_0}^{t} \int_{\sigma(s)}^{t} g(s, y) \Delta y \Delta s, \quad t \in [t_0, a).$$

Proof.   Set

$$q(t) = \int_{t_0}^{t} \int_{t_0}^{s} g(s, y) \Delta s \Delta y - \int_{t_0}^{t} \int_{\sigma(s)}^{t} g(s, y) \Delta y \Delta s, \quad t \in [t_0, a).$$

Then

$$q^{\Delta}(t) = \int_{t_0}^{t} g(s, t) \Delta s - \int_{\sigma(t)}^{\sigma(t)} g(t, y) \Delta y - \int_{t_0}^{t} g(s, t) \Delta s$$
$$= 0, \quad t \in [t_0, a).$$

Therefore, q is a constant on $[t_0, a)$ and

$$g(t) = g(t_0)$$
$$= 0, \quad t \in [t_0, a).$$

This completes the proof. $\square$

Theorem 3.7. *Let* $y, f : [t_0, a) \to [0, \infty)$ *be* Δ-*integrable,* $k : [t_0, a) \times [t_0, a) \to [0, \infty)$ *be* Δ-*integrable and*

$$k_1(t, s) = k(t, s),$$
$$k_n(t, s) = I^{\alpha}_{\Delta, \sigma(s)} (k_{n-1}(t, \cdot)k(\cdot, s)) (t), \quad (t, s) \in [t_0, a) \times [t_0, a).$$

Let also

$$\sum_{n=1}^{\infty} I^{\alpha}_{\Delta, t_0} (k_n(t, \cdot)y(\cdot))$$

be uniformly convergent on $[t_0, a)$. *If*

$$y(t) \le f(t) + I^{\alpha}_{\Delta, t_0}(k(t, \cdot)y(\cdot))(t), \quad t \in [t_0, a), \qquad (3.12)$$

then

$$y(t) \le f(t) + I^{\alpha}_{\Delta, t_0}(h(t, \cdot)f(\cdot))(t), \quad t \in [t_0, a),$$

where

$$h(t, s) = \sum_{r=1}^{\infty} k_r(t, s), \quad (t, s) \in [t_0, a).$$

Proof. By the inequality (3.12), applying the Fubini theorem, we obtain

$$y(t) \le f(t) + I^{\alpha}_{\Delta, t_0}(k(t, \cdot)y(\cdot))(t)$$

$$= f(t) + \int_{t_0}^{t} h_{\alpha-1}(t, \sigma(s))k(t, s)y(s)\Delta s$$

$$\le f(t) + \int_{t_0}^{t} h_{\alpha-1}(t, \sigma(s))k(t, s)$$

$$\times \left(f(s) + \int_{t_0}^{s} h_{\alpha-1}(s, \sigma(z))k(s, z)y(z)\Delta z \right) \Delta s$$

$$= f(t) + \int_{t_0}^{t} h_{\alpha-1}(t, \sigma(s))k(t, s)f(s)\Delta s$$

$$+ \int_{t_0}^{t} \int_{t_0}^{s} h_{\alpha-1}(t, \sigma(s))k(t, s)h_{\alpha-1}(s, \sigma(z))k(s, z)y(z)\Delta z\Delta s$$

$$= f(t) + \int_{t_0}^t h_{\alpha-1}(t, \sigma(s))k(t, s)f(s)\Delta s$$

$$+ \int_{t_0}^t \left(\int_{\sigma(z)}^t h_{\alpha-1}(t, \sigma(s))k(t, s)h_{\alpha-1}(s, \sigma(z))k(s, z)\Delta s \right) y(z)\Delta z$$

$$\leq f(t) + I_{\Delta, t_0}^\alpha(k(t, \cdot)f(\cdot))(t)$$

$$+ \int_{t_0}^t h_{\alpha-1}(t, \sigma(z)) \left(\int_{\sigma(z)}^t h_{\alpha-1}(t, \sigma(s))k(t, s)k(s, z)\Delta s \right) y(z)\Delta z$$

$$= f(t) + I_{\Delta, t_0}^\alpha(k(t, \cdot)f(\cdot))(t)$$

$$+ \int_{t_0}^t h_{\alpha-1}(t, \sigma(z))I_{\Delta, \sigma(z)}^\alpha(k(t, \cdot)k(\cdot, z))(t)y(z)\Delta z$$

$$= f(t) + I_{\Delta, t_0}^\alpha(k(t, \cdot)f(\cdot))(t) + \int_{t_0}^t h_{\alpha-1}(t, \sigma(z))k_2(t, z)y(z)\Delta z$$

$$= f(t) + I_{\Delta, t_0}^\alpha(k_1(t, \cdot)f(\cdot))(t) + I_{\Delta, t_0}^\alpha(k_2(t, \cdot)y(\cdot))(t), \quad t \in [t_0, a).$$

Assume that

$$y(t) \leq f(t) + \sum_{r=1}^l I_{\Delta, t_0}^\alpha(k_r(t, \cdot)f(\cdot))(t) + I_{\Delta, t_0}^\alpha(k_{l+1}(t, \cdot)y(\cdot))(t),$$

$$t \in [t_0, a), \tag{3.13}$$

for some $l \in \mathbb{N}$. We will prove that

$$y(t) \leq f(t) + \sum_{r=1}^{l+1} I_{\Delta, t_0}^\alpha(k_r(t, \cdot)f(\cdot))(t) + I_{\Delta, t_0}^\alpha(k_{l+2}(t, \cdot)y(\cdot))(t),$$

$$t \in [t_0, a).$$

Applying (3.12) and the Fubini theorem, we arrive at

$$y(t) \leq f(t) + \sum_{r=1}^l I_{\Delta, t_0}^\alpha(k_r(t, \cdot)f(\cdot))(t)$$

$$+ \int_{t_0}^t h_{\alpha-1}(t, \sigma(s))k_{l+1}(t, s)y(s)\Delta s$$

$$\leq f(t) + \sum_{r=1}^{l} I_{\Delta,t_0}^{\alpha}(k_r(t,\cdot)f(\cdot))(t) + \int_{t_0}^{t} h_{\alpha-1}(t,\sigma(s))k_{l+1}(t,s)$$

$$\times \left(f(s) + \int_{t_0}^{s} h_{\alpha-1}(s,\sigma(z))k(s,z)y(z)\Delta z \right) \Delta s$$

$$= f(t) + \sum_{r=1}^{l+1} I_{\Delta,t_0}^{\alpha}(k_r(t,\cdot)f(\cdot))(t)$$

$$+ \int_{t_0}^{t} \int_{t_0}^{s} h_{\alpha-1}(t,\sigma(s))h_{\alpha-1}(s,\sigma(z))k_{l+1}(t,s)k(s,z)y(z)\Delta z \Delta s$$

$$= f(t) + \sum_{r=1}^{l+1} I_{\Delta,t_0}^{\alpha}(k_r(t,\cdot)f(\cdot))(t)$$

$$+ \int_{t_0}^{t} \left(\int_{\sigma(z)}^{t} h_{\alpha-1}(t,\sigma(s))h_{\alpha-1}(s,\sigma(z))k_{l+1}(t,s)k(s,z)\Delta s \right) y(z)\Delta z$$

$$\leq f(t) + \sum_{r=1}^{l+1} I_{\Delta,t_0}^{\alpha}(k_r(t,\cdot)f(\cdot))(t)$$

$$+ \int_{t_0}^{t} h_{\alpha-1}(t,\sigma(z)) \left(\int_{\sigma(z)}^{t} h_{\alpha-1}(t,\sigma(s))k_{l+1}(t,s)k(s,z)\Delta s \right) y(z)\Delta z$$

$$= f(t) + \sum_{r=1}^{l+1} I_{\Delta,t_0}^{\alpha}(k_r(t,\cdot)f(\cdot))(t)$$

$$+ \int_{t_0}^{t} h_{\alpha-1}(t,\sigma(z)) I_{\Delta,\sigma(z)}^{\alpha}(k_{l+1}(t,\cdot)k(\cdot,z))(t)y(z)\Delta z$$

$$= f(t) + \sum_{r=1}^{l+1} I_{\Delta,t_0}^{\alpha}(k_r(t,\cdot)f(\cdot))(t)$$

$$+ \int_{t_0}^{t} h_{\alpha-1}(t,\sigma(s))k_{l+2}(t,z)y(z)\Delta z$$

$$= f(t) + \sum_{r=1}^{l+1} I_{\Delta,t_0}^{\alpha}(k_r(t,\cdot)f(\cdot))(t) + I_{\Delta,t_0}^{\alpha}(k_{l+2}(t,\cdot)y(\cdot))(t),$$

$$t \in [t_0, a).$$

Thus, (3.13) holds for any $l \in \mathbb{N}$. Since

$$\sum_{n=1}^{\infty} I_{\Delta,t_0}^{\alpha}(k_n(t,\cdot)y(\cdot))$$

is uniformly convergent on $[t_0, a)$, by (3.13), we find

$$y(t) \le f(t) + \sum_{r=1}^{\infty} I_{\Delta,t_0}^{\alpha}(k_r(t,\cdot)f(\cdot))(t) + \lim_{l\to\infty} I_{\Delta,t_0}^{\alpha}(k_l(t,\cdot)y(\cdot))(t)$$

$$= f(t) + \sum_{r=1}^{\infty} I_{\Delta,t_0}^{\alpha}(k_r(t,\cdot)f(\cdot))$$

$$= f(t) + I_{\Delta,t_0}^{\alpha}\left(\sum_{r=1}^{\infty} k_r(t,\cdot)f(\cdot)\right)(t)$$

$$= f(t) + I_{\Delta,t_0}^{\alpha}(h(t,\cdot)f(\cdot))(t), \quad t \in [t_0, a).$$

This completes the proof. □

Theorem 3.8. *Let* $\alpha > 0$, $m = -\overline{[-\alpha]}$, $B \ge 0$, $y : [t_0, a) \to [0, \infty)$ *be* Δ-*integrable*, $k : [t_0, a) \times [t_0, a) \to [0, \infty)$ *be* Δ-*integrable, bounded and non-decreasing with respect to the first argument function. Let also*

$$\sum_{n=1}^{\infty} \left(\left(I_{\Delta,t_0}^{\alpha}(k(a,\cdot))\right)(t)\right)^n$$

be uniformly convergent on $[t_0, a)$, $I_{\Delta,t_0}^{\alpha}(k(a,\cdot)y(\cdot))(t) < \infty$, $t \in [t_0, a)$, *and*

$$I_{\Delta,t_0}^{\alpha}\left(k(a,\cdot)\left(\sum_{k=1}^{\infty} h_{\alpha-k}(\cdot, t_0)D_{\Delta,t_0}^{\alpha-k}y(t_0) + I_{\Delta,t_0}^{\alpha}(Bh_{-\alpha}(\cdot, t_0))\right)\right)(t) < \infty,$$

$$t \in [t_0, a).$$

If

$$y(t) \le B + I_{\Delta,t_0}^{\alpha}(k(t,\cdot)y(\cdot))(t), \quad t \in [t_0, a), \tag{3.14}$$

then

$$y(t) \le \sum_{r=1}^{\infty} h_{\alpha-r}(t, t_0)D_{\Delta,t_0}^{\alpha-r}\left(B + I_{\Delta,t_0}^{\alpha}(k(a,\cdot)y(\cdot))\right)(t_0)$$

$$+ B + \sum_{r=1}^{m}\left(I_{\Delta,t_0}^{\alpha}k(a,t)\right)^{r-1} I_{\Delta,t_0}^{\alpha}\left(k(a,\cdot)\left(\sum_{r=1}^{\infty} h_{\alpha-r}(t, t_0)D_{\Delta,t_0}^{\alpha-r}\right.\right.$$

$$\left.\left. \times \left(B + I_{\Delta,t_0}^{\alpha}(k(a,\cdot)y(\cdot))\right)(t_0) + B\right)(t)\right), \quad t \in [t_0, a).$$

Proof.   Let

$$z(t) = B + I_{\Delta,t_0}^\alpha (k(t,\cdot)y(\cdot))(t), \quad t \in [t_0, a).$$

Then

$$D_{\Delta,t_0}^\alpha z(t_0) = D_{\Delta,t_0}^\alpha \left(B + I_{\Delta,t_0}^\alpha (k(t,\cdot)y(\cdot)) \right)(t_0)$$

and

$$y(t) \le z(t), \quad t \in [t_0, a).$$

Next, using k as a non-decreasing function with respect to its first argument, we get

$$\begin{aligned}
D_{\Delta,t_0}^\alpha z(t) &= D_{\Delta,t_0}^\alpha B + D_{\Delta,t_0}^\alpha \left(I_{\Delta,t_0}^\alpha (k(t,\cdot)y(\cdot)) \right)(t) \\
&= B h_{-\alpha}(t,t_0) + k(t,t)y(t) \\
&\le B h_{-\alpha}(t,t_0) + k(a,t)z(t), \quad t \in [t_0, a).
\end{aligned}$$

Now, we apply Theorem 3.5 and arrive at

$$\begin{aligned}
y(t) &\le \sum_{r=1}^\infty h_{\alpha-r}(t,t_0) D_{\Delta,t_0}^{\alpha-r} z(t_0) \\
&\quad + I_{\Delta,t_0}^\alpha \left(B h_{-\alpha}(\cdot, t_0) \right)(t) + \sum_{r=1}^\infty \left(I_{\Delta,t_0}^\alpha k(a,t) \right)^{r-1} I_{\Delta,t_0}^\alpha \left(k(a,\cdot) \right. \\
&\quad \left. \times \left(\sum_{r=1}^m h_{\alpha-r}(t,t_0) D_{\Delta,t_0}^{\alpha-r} z(t_0) + I_{\Delta,t_0}^\alpha \left(B h_{-\alpha}(\cdot, t_0) \right)(t) \right) \right) \\
&\le \sum_{r=1}^\infty h_{\alpha-r}(t,t_0) D_{\Delta,t_0}^{\alpha-r} \left(B + I_{\Delta,t_0}^\alpha (k(a,\cdot)y(\cdot)) \right)(t_0) \\
&\quad + B + \sum_{r=1}^m \left(I_{\Delta,t_0}^\alpha k(a,t) \right)^{r-1} I_{\Delta,t_0}^\alpha \left(k(a,\cdot) \left(\sum_{r=1}^\infty h_{\alpha-r}(t,t_0) D_{\Delta,t_0}^{\alpha-r} \right. \right. \\
&\quad \left. \left. \times \left(B + I_{\Delta,t_0}^\alpha (k(a,\cdot)y(\cdot)) \right)(t_0) + B \right)(t) \right), \quad t \in [t_0, a).
\end{aligned}$$

This completes the proof. $\square$

3.3 Simultaneous Fractional Inequalities

In this section, we deduct some simultaneous Riemann–Liouville fractional inequalities.

Theorem 3.9. *Let* $u_1, u_2, v_1, v_2, y_1, y_2 : [t_0, a) \to [0, \infty)$ *be* Δ*-integrable,*

$$\sum_{n=1}^{\infty}(I^{\alpha}_{\Delta,t_0}v_1(t))^n \quad and \quad \sum_{n=1}^{\infty}(I^{\alpha}_{\Delta,t_0}v_2(t))^n$$

be uniformly convergent on $[t_0, a)$*, and*

$$I^{\alpha}_{\Delta,t_0}(v_1y_1)(t) < \infty, \quad I^{\alpha}_{\Delta,t_0}(v_2y_2)(t) < \infty,$$
$$I^{\alpha}_{\Delta,t_0}(v_1(v_1+g_2))(t) < \infty, \quad I^{\alpha}_{\Delta,t_0}(v_2(v_2+g_1))(t) < \infty,$$
$$1 - I^{\alpha}_{\Delta,t_0}v_1(t) - I^{\alpha}_{\Delta,t_0}v_2(t) > 0, \quad t \in [t_0, a).$$

If

$$y_1(t) \le u_1(t) + I^{\alpha}_{\Delta,t_0}(v_1y_1)(t) + I^{\alpha}_{\Delta,t_0}(v_2y_2)(t)$$
$$y_2(t) \le u_2(t) + I^{\alpha}_{\Delta,t_0}(v_1y_1)(t) + I^{\alpha}_{\Delta,t_0}(v_2y_2)(t), \quad t \in [t_0, a),$$
$$(3.15)$$

then

$$y_1(t) \le u_1(t) + g_1(t) + g_2(t)$$
$$y_2(t) \le u_2(t) + g_1(t) + g_2(t), \quad t \in [t_0, a),$$

and

$$y_1(t) \le u_1(t) + g_2(t) + \sum_{k=1}^{\infty}\left(I^{\alpha}_{\Delta,t_0}v_1(t)\right)^{k-1} I^{\alpha}_{\Delta,t_0}((v_1+g_2)v_1)(t)$$

$$y_2(t) \le u_2(t) + g_1(t) + \sum_{k=1}^{\infty}\left(I^{\alpha}_{\Delta,t_0}v_2(t)\right)^{k-1} I^{\alpha}_{\Delta,t_0}(v_2(v_2+g_1))(t),$$

$$t \in [t_0, a),$$

where

$$g_1(t) = \frac{I^\alpha_{\Delta,t_0}(v_1 u_1)(t) - I^\alpha_{\Delta,t_0}(v_1 u_1)(t) I^\alpha_{\Delta,t_0} v_2(t) + I^\alpha_{\Delta,t_0} v_1(t) I^\alpha_{\Delta,t_0}(v_2 u_2)(t)}{1 - I^\alpha_{\Delta,t_0} v_1(t) - I^\alpha_{\Delta,t_0} v_2(t)},$$

$$g_2(t) \leq \frac{I^\alpha_{\Delta,t_0}(v_2 u_2)(t) - I^\alpha_{\Delta,t_0}(v_2 u_2)(t) I^\alpha_{\Delta,t_0} v_1(t) + I^\alpha_{\Delta,t_0} v_2(t) I^\alpha_{\Delta,t_0}(v_1 u_1)(t)}{1 - I^\alpha_{\Delta,t_0} v_1(t) - I^\alpha_{\Delta,t_0} v_2(t)},$$

$t \in [t_0, a)$.

Proof. We multiply the first equation of the system (3.15) with $v_1(t)$ and the second equation of the system (3.15) with $v_2(t)$ and find

$$y_1(t) v_1(t) \leq v_1(t) u_1(t) + v_1(t) I^\alpha_{\Delta,t_0}(v_1 y_1)(t) + v_1(t) I^\alpha_{\Delta,t_0}(v_2 y_2)(t)$$

$$v_2(t) y_2(t) \leq v_2(t) u_2(t) + v_2(t) I^\alpha_{\Delta,t_0}(v_1 y_1)(t) + v_2(t) I^\alpha_{\Delta,t_0}(v_2 y_2)(t),$$

$$t \in [t_0, a).$$

Now, we take both sides of the last system the operator I^α_{Δ,t_0} and find

$$I^\alpha_{\Delta,t_0}(y_1 v_1)(t) \leq I^\alpha_{\Delta,t_0}\left(v_1 u_1 + v_1 I^\alpha_{\Delta,t_0}(v_1 y_1) + v_1 I^\alpha_{\Delta,t_0}(v_2 y_2)\right)(t)$$

$$= I^\alpha_{\Delta,t_0}(v_1 u_1)(t) + I^\alpha_{\Delta,t_0}\left(v_1 I^\alpha_{\Delta,t_0}(v_1 y_1)\right)(t)$$

$$+ I^\alpha_{\Delta,t_0}\left(v_1 I^\alpha_{\Delta,t_0}(v_2 y_2)\right)(t)$$

$$\leq I^\alpha_{\Delta,t_0}(v_1 u_1)(t) + I^\alpha_{\Delta,t_0} v_1(t) I^\alpha_{\Delta,t_0}(v_1 y_1)(t)$$

$$+ I^\alpha_{\Delta,t_0} v_1(t) I^\alpha_{\Delta,t_0}(v_2 y_2)(t), \quad t \in [t_0, a),$$

and

$$I^\alpha_{\Delta,t_0}(y_2 v_2)(t) \leq I^\alpha_{\Delta,t_0}\left(v_2 u_2 + v_2 I^\alpha_{\Delta,t_0}(v_1 y_1) + v_2 I^\alpha_{\Delta,t_0}(v_2 y_2)\right)(t)$$

$$= I^\alpha_{\Delta,t_0}(v_2 u_2)(t) + I^\alpha_{\Delta,t_0}\left(v_2 I^\alpha_{\Delta,t_0}(v_1 y_1)\right)(t)$$

$$+ I^\alpha_{\Delta,t_0}\left(v_2 I^\alpha_{\Delta,t_0}(v_2 y_2)\right)(t)$$

$$\leq I^\alpha_{\Delta,t_0}(v_2 u_2)(t) + I^\alpha_{\Delta,t_0} v_2(t) I^\alpha_{\Delta,t_0}(v_1 y_1)(t)$$

$$+ I^\alpha_{\Delta,t_0} v_2(t) I^\alpha_{\Delta,t_0}(v_2 y_2)(t), \quad t \in [t_0, a).$$

Thus, we get the system

$$I_{\Delta,t_0}^{\alpha}(v_1 y_1)(t) \le I_{\Delta,t_0}^{\alpha}(v_1 u_1)(t) + I_{\Delta,t_0}^{\alpha} v_1(t) I_{\Delta,t_0}^{\alpha}(v_1 y_1)(t)$$
$$+ I_{\Delta,t_0}^{\alpha} v_1(t) I_{\Delta,t_0}^{\alpha}(v_2 y_2)(t),$$
$$I_{\Delta,t_0}^{\alpha}(v_2 y_2)(t) \le I_{\Delta,t_0}^{\alpha}(v_2 u_2)(t) + I_{\Delta,t_0}^{\alpha} v_2(t) I_{\Delta,t_0}^{\alpha}(v_1 y_1)(t)$$
$$+ I_{\Delta,t_0}^{\alpha} v_2(t) I_{\Delta,t_0}^{\alpha}(v_2 y_2)(t), \quad t \in [t_0, a),$$

or

$$\left(1 - I_{\Delta,t_0}^{\alpha} v_1(t)\right) I_{\Delta,t_0}^{\alpha}(y_1 v_1)(t)$$
$$\le I_{\Delta,t_0}^{\alpha}(v_1 u_1)(t) + I_{\Delta,t_0}^{\alpha} v_1(t) I_{\Delta,t_0}^{\alpha}(v_2 y_2)(t),$$
$$\left(1 - I_{\Delta,t_0}^{\alpha} v_2(t)\right) I_{\Delta,t_0}^{\alpha}(y_2 v_2)(t)$$
$$\le I_{\Delta,t_0}^{\alpha}(v_2 u_2)(t) + I_{\Delta,t_0}^{\alpha} v_2(t) I_{\Delta,t_0}^{\alpha}(v_1 y_1)(t), \quad t \in [t_0, a),$$

or

$$I_{\Delta,t_0}^{\alpha}(y_1 v_1)(t) \le \frac{1}{1 - I_{\Delta,t_0}^{\alpha} v_1(t)} \left(I_{\Delta,t_0}^{\alpha}(v_1 u_1)(t) \right.$$
$$\left. + I_{\Delta,t_0}^{\alpha} v_1(t) I_{\Delta,t_0}^{\alpha}(v_2 y_2)(t) \right),$$
$$I_{\Delta,t_0}^{\alpha}(y_2 v_2)(t) \le \frac{1}{1 - I_{\Delta,t_0}^{\alpha} v_2(t)} \left(I_{\Delta,t_0}^{\alpha}(v_2 u_2)(t) \right.$$
$$\left. + I_{\Delta,t_0}^{\alpha} v_2(t) I_{\Delta,t_0}^{\alpha}(v_1 y_1)(t) \right), \quad t \in [t_0, a). \quad (3.16)$$

Hence,

$$I_{\Delta,t_0}^{\alpha}(y_1 v_1)(t)$$
$$\le \frac{I_{\Delta,t_0}^{\alpha}(v_1 u_1)(t)}{1 - I_{\Delta,t_0}^{\alpha} v_1(t)} + \frac{I_{\Delta,t_0}^{\alpha} v_1(t)}{1 - I_{\Delta,t_0}^{\alpha} v_1(t)} I_{\Delta,t_0}^{\alpha}(v_2 y_2)(t)$$
$$\le \frac{I_{\Delta,t_0}^{\alpha}(v_1 u_1)(t)}{1 - I_{\Delta,t_0}^{\alpha} v_1(t)} + \frac{I_{\Delta,t_0}^{\alpha} v_1(t)}{1 - I_{\Delta,t_0}^{\alpha} v_1(t)}$$
$$\times \left(\frac{I_{\Delta,t_0}^{\alpha}(v_2 u_2)(t)}{1 - I_{\Delta,t_0}^{\alpha} v_2(t)} + \frac{I_{\Delta,t_0}^{\alpha} v_2(t)}{1 - I_{\Delta,t_0}^{\alpha} v_2(t)} I_{\Delta,t_0}^{\alpha}(v_1 y_1)(t) \right)$$

$$= \frac{I^\alpha_{\Delta,t_0}(v_1 u_1)(t) - I^\alpha_{\Delta,t_0}(v_1 u_1)(t) I^\alpha_{\Delta,t_0} v_2(t) + I^\alpha_{\Delta,t_0} v_1(t) I^\alpha_{\Delta,t_0}(v_2 u_2)(t)}{\left(I^\alpha_{\Delta,t_0} v_1(t)\right)\left(1 - I^\alpha_{\Delta,t_0} v_2(t)\right)}$$

$$+ \frac{I^\alpha_{\Delta,t_0} v_1(t) I^\alpha_{\Delta,t_0} v_2(t)}{\left(1 - I^\alpha_{\Delta,t_0} v_1(t)\right)\left(1 - I^\alpha_{\Delta,t_0} v_2(t)\right)} I^\alpha_{\Delta,t_0}(v_1 y_1)(t), \quad t \in [t_0, a),$$

whereupon

$$\left(1 - \frac{I^\alpha_{\Delta,t_0} v_1(t) I^\alpha_{\Delta,t_0} v_2(t)}{\left(1 - I^\alpha_{\Delta,t_0} v_1(t)\right)\left(1 - I^\alpha_{\Delta,t_0} v_2(t)\right)}\right) I^\alpha_{\Delta,t_0}(v_1 y_1)(t)$$

$$\leq \frac{I^\alpha_{\Delta,t_0}(v_1 u_1)(t) - I^\alpha_{\Delta,t_0}(v_1 u_1)(t) I^\alpha_{\Delta,t_0} v_2(t) + I^\alpha_{\Delta,t_0} v_1(t) I^\alpha_{\Delta,t_0}(v_2 u_2)(t)}{\left(1 - I^\alpha_{\Delta,t_0} v_1(t)\right)\left(1 - I^\alpha_{\Delta,t_0} v_2(t)\right)},$$

$$t \in [t_0, a),$$

or

$$I^\alpha_{\Delta,t_0}(v_1 y_1)(t)$$

$$\leq \frac{I^\alpha_{\Delta,t_0}(v_1 u_1)(t) - I^\alpha_{\Delta,t_0}(v_1 u_1)(t) I^\alpha_{\Delta,t_0} v_2(t) + I^\alpha_{\Delta,t_0} v_1(t) I^\alpha_{\Delta,t_0}(v_2 u_2)(t)}{1 - I^\alpha_{\Delta,t_0} v_1(t) - I^\alpha_{\Delta,t_0} v_2(t)}$$

$$= g_1(t), \quad t \in [t_0, a).$$

Now, by the second equation of the system (3.16), we find

$$I^\alpha_{\Delta,t_0}(y_2 v_2)(t) \leq \frac{1}{1 - I^\alpha_{\Delta,t_0} v_2(t)} I^\alpha_{\Delta,t_0}(v_2 u_2)(t) + \frac{I^\alpha_{\Delta,t_0} v_2(t)}{1 - I^\alpha_{\Delta,t_0} v_2(t)}$$

$$\times \left(\frac{I^\alpha_{\Delta,t_0}(v_1 u_1)(t)}{1 - I^\alpha_{\Delta,t_0} v_1(t)} + \frac{I^\alpha_{\Delta,t_0} v_1(t)}{1 - I^\alpha_{\Delta,t_0} v_1(t)} I^\alpha_{\Delta,t_0}(v_2 y_2)(t)\right),$$

$$t \in [t_0, a),$$

or

$$\left(1 - \frac{I^\alpha_{\Delta,t_0} v_1(t) I^\alpha_{\Delta,t_0} v_2(t)}{(1 - v_1(t))\left(1 - I^\alpha_{\Delta,t_0} v_2(t)\right)}\right) I^\alpha_{\Delta,t_0}(y_2 v_2)(t)$$

$$\leq \frac{I^\alpha_{\Delta,t_0}(v_2 u_2)(t) - I^\alpha_{\Delta,t_0}(v_2 u_2)(t) I^\alpha_{\Delta,t_0} v_1(t) + I^\alpha_{\Delta,t_0} v_2(t) I^\alpha_{\Delta,t_0}(v_1 u_1)(t)}{\left(1 - I^\alpha_{\Delta,t_0} v_1(t)\right)\left(1 - I^\alpha_{\Delta,t_0} v_2(t)\right)},$$

$t \in [t_0, a),$

or

$$I^\alpha_{\Delta,t_0}(y_2 v_2)(t)$$

$$\leq \frac{I^\alpha_{\Delta,t_0}(v_2 u_2)(t) - I^\alpha_{\Delta,t_0}(v_2 u_2)(t) I^\alpha_{\Delta,t_0} v_1(t) + I^\alpha_{\Delta,t_0} v_2(t) I^\alpha_{\Delta,t_0}(v_1 u_1)(t)}{1 - I^\alpha_{\Delta,t_0} v_1(t) - I^\alpha_{\Delta,t_0} v_2(t)}$$

$$= g_2(t), \quad t \in [t_0, a).$$

Consequently,

$$I^\alpha_{\Delta,t_0}(y_1 v_1)(t) \leq g_1(t),$$
$$I^\alpha_{\Delta,t_0}(y_2 v_2)(t) \leq g_2(t), \quad t \in [t_0, a). \tag{3.17}$$

By the last system and the system (3.15), we find

$$y_1(t) \leq u_1(t) + g_1(t) + g_2(t)$$
$$y_2(t) \leq u_2(t) + g_1(t) + g_2(t), \quad t \in [t_0, a).$$

Now, we apply the system (3.17) to the system (3.15) and obtain

$$y_1(t) \leq u_1(t) + g_2(t) + I^\alpha_{\Delta,t_0}(v_1 y_1)(t),$$
$$y_2(t) \leq u_2(t) + g_1(t) + I^\alpha_{\Delta,t_0}(v_2 y_2)(t), \quad t \in [t_0, a).$$

Now, we apply Theorem 3.2 and arrive at

$$y_1(t) \leq u_1(t) + g_2(t) + \sum_{k=1}^\infty \left(I^\alpha_{\Delta,t_0} v_1(t)\right)^{k-1} I^\alpha_{\Delta,t_0}((v_1 + g_2)v_1)(t),$$

$$y_2(t) \leq u_2(t) + g_1(t) + \sum_{k=1}^\infty \left(I^\alpha_{\Delta,t_0} v_2(t)\right)^{k-1} I^\alpha_{\Delta,t_0}(v_2(v_2 + g_1))(t), \quad t \in [t_0, a).$$

This completes the proof. $\qquad\square$

3.4   Pachpatte Fractional Inequalities

In this section, we deduct some Pachpatte Riemann–Liouville fractional dynamic inequalities on time scales. We start with the following result.

Theorem 3.10. *Let $\alpha > 0$, $h, y : [t_0, a) \to [0, \infty)$ be Δ-integrable, $f, g : [t_0, a) \to [0, \infty)$ be Δ-integrable and non-decreasing with*

$$f(t) + g(t) I^\alpha_{\Delta, t_0} f(t) \leq B, \quad t \in [t_0, a),$$

for some positive constant B, and $(y, B, [t_0, a))$ is an α-fractional admissible triple. If

$$y(t) \leq h(t) + I^\alpha_{\Delta, t_0}(fy)(t) + I^\alpha_{\Delta, t_0}\left(f I^\alpha_{\Delta, t_0}(gy)\right)(t), \quad t \in [t_0, a),$$

then

$$y(t) \leq h(t) + \sum_{k=1}^{\infty} \left(f(t) + g(t) I^\alpha_{\Delta, t_0} f(t)\right)^k I^{k\alpha}_{\Delta, t_0} h(t), \quad t \in [t_0, a).$$

Proof.   Since f and g are non-decreasing on $[t_0, a)$, we have

$$
\begin{aligned}
y(t) &\leq h(t) + f(t) I^\alpha_{\Delta, t_0} y(t) + I^\alpha_{\Delta, t_0} f(t) I^\alpha_{\Delta, t_0}(gy)(t) \\
&\leq h(t) + f(t) I^\alpha_{\Delta, t_0} y(t) + g(t) I^\alpha_{\Delta, t_0} f(t) I^\alpha_{\Delta, t_0} y(t) \\
&= h(t) + \left(f(t) + g(t) I^\alpha_{\Delta, t_0} f(t)\right) I^\alpha_{\Delta, t_0} y(t), \quad t \in [t_0, a).
\end{aligned}
$$

Now, we apply Theorem 3.1 and arrive at

$$y(t) \leq h(t) + \sum_{k=1}^{\infty} \left(f(t) + g(t) I^\alpha_{\Delta, t_0} f(t)\right)^k I^{k\alpha}_{\Delta, t_0} h(t), \quad t \in [t_0, a).$$

This completes the proof.   □

Theorem 3.11. *Let $g, y : [t_0, a) \to [0, \infty)$ be Δ-integrable, $f_1, f_2, f_3 : [t_0, a) \to [0, \infty)$ be Δ-integrable and non-decreasing with*

$$f_1(t) + f_2(t) I^\alpha_{\Delta, t_0} f_1(t) + f_3(t) I^\alpha_{\Delta, t_0} f_2(t) I^\alpha_{\Delta, t_0} f_1(t) \leq B, \quad t \in [t_0, a),$$

for some non-negative constant B. Let also $[y, B, [t_0, a))$ be an α-fractional admissible triple. If

$$y(t) \leq g(t) + I^\alpha_{\Delta,t_0}(fy)(t) + I^\alpha_{\Delta,t_0}\left(f_1 I^\alpha_{\Delta,t_0}(f_2 y)\right)(t)$$
$$+ I^\alpha_{\Delta,t_0}\left(f_1 I^\alpha_{\Delta,t_0}\left(f_2 I^\alpha_{\Delta,t_0}(f_3 y)\right)\right)(t), \quad t \in [t_0, a),$$

then

$$y(t) \leq g(t) + \sum_{k=1}^\infty \left(f_1(t) + f_2(t)I^\alpha_{\Delta,t_0}f_1(t)\right.$$
$$\left. + f_3(t)I^\alpha_{\Delta,t_0}f_1(t)I^\alpha_{\Delta,t_0}f_2(t)\right)^k I^\alpha_{\Delta,t_0}g(t), \quad t \in [t_0, a).$$

Proof. Since f_1, f_2 and f_3 are non-decreasing functions on $[t_0, a)$, we get

$$y(t) \leq g(t) + I^\alpha_{\Delta,t_0}(fy)(t) + I^\alpha_{\Delta,t_0}\left(f_1 I^\alpha_{\Delta,t_0}(f_2 y)\right)(t)$$
$$+ I^\alpha_{\Delta,t_0}\left(f_1 I^\alpha_{\Delta,t_0}\left(f_2 I^\alpha_{\Delta,t_0}(f_3 y)\right)\right)(t)$$
$$\leq g(t) + f_1(t)I^\alpha_{\Delta,t_0}y(t) + I^\alpha_{\Delta,t_0}f_1(t)I^\alpha_{\Delta,t_0}(f_2 y)(t)$$
$$+ I^\alpha_{\Delta,t_0}\left(f_1 I^\alpha_{\Delta,t_0}f_2 I^\alpha_{\Delta,t_0}(f_3 y)\right)$$
$$\leq g(t) + f_1(t)I^\alpha_{\Delta,t_0}y(t) + f_2(t)I^\alpha_{\Delta,t_0}f_1(t)I^\alpha_{\Delta,t_0}y(t)$$
$$+ I^\alpha_{\Delta,t_0}f_1(t)I^\alpha_{\Delta,t_0}I^\alpha_{\Delta,t_0}f_2(t)I^\alpha_{\Delta,t_0}(f_3 y)(t)$$
$$\leq g(t) + \left(f_1(t) + f_2(t)I^\alpha_{\Delta,t_0}f_1(t)\right)I^\alpha_{\Delta,t_0}y(t)$$
$$+ f_3(t)I^\alpha_{\Delta,t_0}f_1(t)I^\alpha_{\Delta,t_0}f_2(t)I^\alpha_{\Delta,t_0}y(t)$$
$$= g(t) + \left(f_1(t) + f_2(t)I^\alpha_{\Delta,t_0}f_1(t)\right.$$
$$\left. + f_3(t)I^\alpha_{\Delta,t_0}f_1(t)I^\alpha_{\Delta,t_0}f_2(t)\right)I^\alpha_{\Delta,t_0}y(t), \quad t \in [t_0, a),$$

i.e.,

$$y(t) \leq g(t) + \left(f_1(t) + f_2(t)I^\alpha_{\Delta,t_0}f_1(t)\right.$$
$$\left. + f_3(t)I^\alpha_{\Delta,t_0}f_1(t)I^\alpha_{\Delta,t_0}f_2(t)\right)I^\alpha_{\Delta,t_0}y(t), \quad t \in [t_0, a).$$

Now, we apply Theorem 3.1 and find

$$y(t) \leq g(t) + \sum_{k=1}^{\infty} \left(f_1(t) + f_2(t) I_{\Delta,t_0}^{\alpha} f_1(t) \right.$$

$$+ f_3(t) I_{\Delta,t_0}^{\alpha} f_1(t) I_{\Delta,t_0}^{\alpha} f_2(t) \Big)^k I_{\Delta,t_0}^{\alpha} g(t), \quad t \in [t_0, a).$$

This completes the proof. $\qquad\qquad\qquad\qquad\qquad\qquad\qquad\square$

Theorem 3.12. *Let* $g_1, y : [t_0, a) \to [0, \infty)$ *be* Δ-*integrable,* $g_2 :$ $[t_0, a) \to (0, \infty)$ *be* Δ-*integrable and non-decreasing,* $f_1, f_2, f_3 :$ $[t_0, a) \to [0, \infty)$ *be* Δ-*integrable and non-decreasing with*

$$f_1(t) + f_2(t) I_{\Delta,t_0}^{\alpha} f_1(t) + f_3(t) I_{\Delta,t_0}^{\alpha} f_2(t) I_{\Delta,t_0}^{\alpha} f_1(t) \leq B, \quad t \in [t_0, a),$$

for some non-negative constant B. *Let also* $\left(\frac{y}{g_2}, B, [t_0, a) \right)$ *be an* α-*fractional admissible triple. If*

$$y(t) \leq g_1(t) g_2(t) + I_{\Delta,t_0}^{\alpha}(fy)(t) + I_{\Delta,t_0}^{\alpha} \left(f_1 I_{\Delta,t_0}^{\alpha}(f_2 y) \right)(t)$$

$$+ I_{\Delta,t_0}^{\alpha} \left(f_1 I_{\Delta,t_0}^{\alpha} \left(f_2 I_{\Delta,t_0}^{\alpha}(f_3 y) \right) \right)(t), \quad t \in [t_0, a),$$

then

$$y(t) \leq g_2(t) \left(g_1(t) + \sum_{k=1}^{\infty} \left(f_1(t) + f_2(t) I_{\Delta,t_0}^{\alpha} f_1(t) + f_3(t) I_{\Delta,t_0}^{\alpha} \right. \right.$$

$$\times \; f_1(t) I_{\Delta,t_0}^{\alpha} f_2(t) \Big)^k I_{\Delta,t_0}^{\alpha} g_1(t) \Bigg), \quad t \in [t_0, a).$$

Proof. We divide both sides of the given inequality by $g_2(t)$, $t \in$ $[t_0, a)$, and using g_2 as a non-decreasing function on $[t_0, a)$, we get

$$\frac{y(t)}{g_2(t)} \leq g_1(t) + \frac{1}{g_2(t)} I_{\Delta,t_0}^{\alpha}(fy)(t) + \frac{1}{g_2(t)} I_{\Delta,t_0}^{\alpha} \left(f_1 I_{\Delta,t_0}^{\alpha}(f_2 y) \right)(t)$$

$$+ \frac{1}{g_2(t)} I_{\Delta,t_0}^{\alpha} \left(f_1 I_{\Delta,t_0}^{\alpha} \left(f_2 I_{\Delta,t_0}^{\alpha}(f_3 y) \right) \right)(t)$$

$$\leq g_1(t) + I_{\Delta,t_0}^{\alpha} \left(f \frac{y}{g_2} \right)(t) + I_{\Delta,t_0}^{\alpha} \left(f_1 I_{\Delta,t_0}^{\alpha} \left(f_2 \frac{y}{g_2} \right) \right)(t)$$

$$+ I_{\Delta,t_0}^{\alpha} \left(f_1 I_{\Delta,t_0}^{\alpha} \left(f_2 I_{\Delta,t_0}^{\alpha} \left(f_3 \frac{y}{g_2} \right) \right) \right)(t), \quad t \in [t_0, a).$$

We set

$$z(t) = \frac{y(t)}{g_2(t)}, \quad t \in [t_0, a).$$

Then, by the last inequality, we arrive at the inequality

$$w(t) \leq g_1(t) + I^\alpha_{\Delta,t_0}\,(fz)\,(t) + I^\alpha_{\Delta,t_0}\left(f_1 I^\alpha_{\Delta,t_0}\,(f_2 z)\right)(t)$$
$$+ I^\alpha_{\Delta,t_0}\left(f_1 I^\alpha_{\Delta,t_0}\left(f_2 I^\alpha_{\Delta,t_0}\,(f_3 z)\right)\right)(t), \quad t \in [t_0, a).$$

Now, we apply Theorem 3.1 and find

$$z(t) \leq g_1(t) + \sum_{k=1}^\infty \left(f_1(t) + f_2(t) I^\alpha_{\Delta,t_0} f_1(t)\right.$$
$$+ f_3(t) I^\alpha_{\Delta,t_0} f_1(t) I^\alpha_{\Delta,t_0} f_2(t)\Big)^k I^\alpha_{\Delta,t_0} g_1(t), \quad t \in [t_0, a),$$

or

$$\frac{y(t)}{g_2(t)} \leq g_1(t) + \sum_{k=1}^\infty \left(f_1(t) + f_2(t) I^\alpha_{\Delta,t_0} f_1(t)\right.$$
$$+ f_3(t) I^\alpha_{\Delta,t_0} f_1(t) I^\alpha_{\Delta,t_0} f_2(t)\Big)^k I^\alpha_{\Delta,t_0} g_1(t), \quad [t_0, a),$$

whereupon

$$y(t) \leq g_2(t) \left(g_1(t) + \sum_{k=1}^\infty \left(f_1(t) + f_2(t) I^\alpha_{\Delta,t_0} f_1(t)\right.\right.$$
$$+ f_3(t) I^\alpha_{\Delta,t_0} f_1(t) I^\alpha_{\Delta,t_0} f_2(t)\Big)^k I^\alpha_{\Delta,t_0} g_1(t)\bigg), \quad t \in [t_0, a).$$

This completes the proof. □

3.5 Existence and Uniqueness of the Solutions

Let $\alpha > 0$ and $m = -\overline{[-\alpha]}$. Consider the Cauchy problem:

$$D^\alpha_{\Delta,t_0} y(t) = f(t, y(t)), \quad t > t_0, \tag{3.18}$$

$$D^{\alpha-k}_{\Delta,t_0} y(t_0) = b_k, \quad k \in \{1, \ldots, m\}, \tag{3.19}$$

where $f : \mathbb{T} \times \mathbb{R} \to \mathbb{R}$ is a given function, b_k, $k \in \{1, \ldots, m\}$, are given constants. Let $a \in \mathbb{T}$, $t_0 < a$. With $L_\Delta[t_0, a)$, we will denote the space of Δ-Lebesgue summable functions in $[t_0, a)$. We define the space

$$L_\Delta^\alpha[t_0, a) = \left\{ y \in L_\Delta[t_0, a) : D_{\Delta, t_0}^\alpha \in L_\Delta[t_0, a) \right\}.$$

Theorem 3.13. *Let G be an open set in $\mathbb{R}$ and $f : [t_0, a] \times G \to \mathbb{R}$ be a given function such that $f(\cdot, y) \in L_\Delta[t_0, a)$ for any $y \in G$. If $y \in L_\Delta[t_0, a)$, then the problem (3.18), (3.19) is equivalent to the equation*

$$y(t) = \sum_{k=1}^m b_k h_{\alpha-k}(t, t_0) + I_{\Delta, t_0}^\alpha f(t, y(t)), \quad t \in [t_0, a). \qquad (3.20)$$

Proof.

(1) Let $y \in L_\Delta[a, b)$ be a solution of the problem (3.18), (3.19). Then we apply I_{Δ, t_0}^α of both sides of (3.18) and using (3.19), we get

$$I_{\Delta, t_0}^\alpha D_{\Delta, t_0}^\alpha y(t) = y(t) - \sum_{k=1}^m h_{\alpha-k}(t, t_0) D_{\Delta, t_0}^{\alpha-k} y(t_0)$$

$$= y(t) - \sum_{k=1}^m b_k h_{\alpha-k}(t, t_0)$$

$$= I_{\Delta, t_0}^\alpha f(t, y(t)), \quad t \in [t_0, a),$$

i.e., y satisfies (3.20).

(2) Let $y \in L_\Delta[a, b)$ satisfies equation (3.20). Then we apply D_{Δ, t_0}^α of both sides of (3.20) and get

$$D_{\Delta, t_0}^\alpha y(t) = D_{\Delta, t_0}^\alpha \left(\sum_{k=1}^m b_k h_{\alpha-k}(t, t_0) + I_{\Delta, t_0}^\alpha f(t, y(t)) \right)$$

$$= D_{\Delta, t_0}^\alpha \left(\sum_{k=1}^m h_{\alpha-k}(t, t_0) b_k \right) + D_{\Delta, t_0}^\alpha I_{\Delta, t_0}^\alpha f(t, y(t))$$

$$= \sum_{k=1}^m D_{\Delta, t_0}^\alpha (h_{\alpha-k}(t, t_0)) b_k + f(t, y(t))$$

$$= f(t, y(t)), \quad t \in [t_0, a),$$

i.e., y satisfies (3.18). Also, for $j \in \{1, \ldots, m\}$, we have

$$
D_{\Delta,t_0}^{\alpha-j} y(t) = D_{\Delta,t_0}^{\alpha-j} \left(\sum_{k=1}^{m} h_{\alpha-k}(t,t_0) b_k + I_{\Delta,t_0}^{\alpha} f(t, y(t)) \right)
$$

$$
= D_{\Delta,t_0}^{\alpha-j} \left(\sum_{k=1}^{m} h_{\alpha-k}(t,t_0) b_k \right) + D_{\Delta,t_0}^{\alpha-j} I_{\Delta,t_0}^{\alpha} f(t, y(t))
$$

$$
= \sum_{k=1}^{m} D_{\Delta,t_0}^{\alpha-j} \left(h_{\alpha-k}(t,t_0) \right) b_k + I_{\Delta,t_0}^{j} f(t, y(t)), \quad t \in [t_0, a).
$$

Hence,

$$
D_{\Delta,t_0}^{\alpha-j} y(t_0) = \left(\sum_{k=1}^{m} D_{\Delta,t_0}^{\alpha-j} \left(h_{\alpha-k}(t,t_0) \right) b_k + I_{\Delta,t_0}^{j} f(t, y(t)) \right) \Big|_{t=t_0}
$$

$$
= b_j, \quad j \in \{1, \ldots, m\}.
$$

Therefore, y satisfies (3.19). This completes the proof. $\square$

Suppose

$$
|f(t, y_1(t)) - f(t, y_2(t))| \leq A |y_1(t) - y_2(t)|, \quad t \in [t_0, a], \quad (3.21)
$$

where $A > 0$ is a constant which does not depend on $t \in [t_0, a]$.

Theorem 3.14. *Let f be as in Theorem 3.13, which satisfies the Lipschitzian-type condition (3.21) and $|f(t, y)| \leq M$ on $[t_0, a] \times G$ for some positive constant M. Let also $_\Delta F_{\alpha,1}(A, t, t_0)$ be defined on $[t_0, a]$. Then the Cauchy problem (3.18), (3.19) has a unique solution, $y \in L_\Delta^\alpha[t_0, a)$.*

Proof. We define the sequence $\{y_l(t)\}_{l \in \mathbb{N}_0}$, $t \in [t_0, a)$, as follows:

$$
y_0(t) = \sum_{k=1}^{m} h_{\alpha-k}(t,t_0) b_k,
$$

$$
y_l(t) = y_0(t) + I_{\Delta,t_0}^{\alpha} f(t, y_{l-1}(t)), \quad t \in [t_0, a), \quad l \in \mathbb{N}.
$$

We have

$$|y_1(t) - y_0(t)| = \left| I^\alpha_{\Delta, t_0} f\left(t, y_0(t)\right) \right|$$
$$\le I^\alpha_{\Delta, t_0} |f\left(t, y_0(t)\right)|$$
$$\le M I^\alpha_{\Delta, t_0} h_0(t, t_0)$$
$$= M h_\alpha(t, t_0), \quad t \in [t_0, a).$$

Assume that

$$|y_{l-1}(t) - y_{l-2}(t)| \le M A^{l-2} h_{(l-1)\alpha}(t, t_0), \quad t \in [t_0, a), \qquad (3.22)$$

for some $l \in \mathbb{N}$, $l \ge 2$. We will prove that

$$|y_l(t) - y_{l-1}(t)| \le M A^{l-1} h_{l\alpha}(t, t_0), \quad t \in [t_0, a).$$

Really, we have

$$|y_l(t) - y_{l-1}(t)| = \left| y_0 + I^\alpha_{\Delta, t_0} f\left(t, y_{l-1}(t)\right) - y_0 - I^\alpha_{\Delta, t_0} f\left(t, y_{l-2}(t)\right) \right|$$
$$= \left| I^\alpha_{\Delta, t_0} \left(f\left(t, y_{l-1}(t)\right) - f\left(t, y_{l-2}(t)\right) \right) \right|$$
$$\le I^\alpha_{\Delta, t_0} |f\left(t, y_{l-1}(t)\right) - f\left(t, y_{l-2}(t)\right)|$$
$$\le A I^\alpha_{\Delta, t_0} |y_{l-1}(t) - y_{l-2}(t)|$$
$$\le M A^{l-1} I^\alpha_{\Delta, t_0} h_{(l-1)\alpha}(t, t_0)$$
$$= M A^{l-1} h_{l\alpha}(t, t_0), \quad t \in [t_0, a).$$

Therefore, (3.22) holds for any $l \in \mathbb{N}$, $l \ge 2$. Note that

$$\left| \lim_{l \to \infty} \left(y_l(t) - y_0(t) \right) \right| = \left| \sum_{l=1}^{\infty} \left(y_l(t) - y_{l-1}(t) \right) \right|$$
$$\le \sum_{l=1}^{\infty} |y_l(t) - y_{l-1}(t)|$$
$$\le M \sum_{l=1}^{\infty} A^{l-1} h_{l\alpha}(t, t_0)$$

$$= \frac{M}{A} \sum_{l=1}^{\infty} A^l h_{l\alpha}(t, t_0)$$

$$\leq \frac{M}{A} \sum_{l=1}^{\infty} A^l h_{l\alpha}(a, t_0)$$

$$= \frac{M}{A} \left({}_\Delta F_{\alpha,1}(A, a, t_0) - 1 \right)$$

$$< \infty, \quad t \in [t_0, a).$$

Consequently, the series

$$\sum_{l=1}^{\infty} (y_l(t) - y_{l-1}(t))$$

is uniformly convergent on $[t_0, a)$. Hence, there exists

$$\lim_{l \to \infty} (y_l(t) - y_0(t)) = \sum_{l=1}^{\infty} (y_l(t) - y_{l-1}(t)), \quad t \in [t_0, a),$$

and

$$y(t) = \lim_{l \to \infty} y_l(t), \quad t \in [t_0, a),$$

and $y(t)$, $t \in [t_0, a)$, satisfies (3.20). Hence, from Theorem 3.13, it follows that $y(t)$, $t \in [t_0, a)$, is a solution of the Cauchy problem (3.18), (3.19). Assume that the Cauchy problem (3.18), (3.19) has another solution, $z(t)$, $t \in [t_0, a)$. For this, we have

$$z(t) = y_0(t) + I^\alpha_{\Delta, t_0} f(t, z(t)), \quad t \in [t_0, a).$$

Note that

$$|y_0(t) - z(t)| = \left| I^\alpha_{\Delta, t_0} f(t, z(t)) \right|$$

$$\leq I^\alpha_{\Delta, t_0} |f(t, z(t))|$$

$$\leq M I^\alpha_{\Delta, t_0} h_0(t, t_0)$$

$$= M h_\alpha(t, t_0), \quad t \in [t_0, a).$$

Assume that

$$|y_{l-1}(t) - z(t)| \leq M A^{l-1} h_{l\alpha}(t, t_0), \quad t \in [t_0, a), \tag{3.23}$$

for some $l \in \mathbb{N}$. We will prove that

$$|y_l(t) - z(t)| \leq M A^l h_{(l+1)\alpha}(t, t_0), \quad t \in [t_0, a).$$

In fact, we have

$$
\begin{aligned}
|y_l(t) - z(t)| &= \left| I^\alpha_{\Delta, t_0} f\left(t, y_{l-1}(t)\right) - I^\alpha_{\Delta, t_0} f(t, z(t)) \right| \\
&= \left| I^\alpha_{\Delta, t_0} \left(f\left(t, y_{l-1}(t)\right) - f(t, z(t)) \right) \right| \\
&\leq I^\alpha_{\Delta, t_0} \left| f\left(t, y_{l-1}(t)\right) - f(t, z(t)) \right| \\
&\leq A I^\alpha_{\Delta, t_0} |y_{l-1}(t) - z(t)| \\
&\leq M A^l I^\alpha_{\Delta, t_0} h_{l\alpha}(t, t_0) \\
&= M A^l h_{(l+1)\alpha}(t, t_0), \quad t \in [t_0, a).
\end{aligned}
$$

Therefore, (3.23) holds for any $l \in \mathbb{N}$. Since

$$\lim_{l \to \infty} M A^l h_{(l+1)\alpha}(t, t_0) = 0, \quad t \in [t_0, a),$$

we conclude that

$$
\begin{aligned}
y(t) - z(t) &= \lim_{l \to \infty} \left(y_l(t) - z(t) \right) \\
&= 0, \quad t \in [t_0, a).
\end{aligned}
$$

This completes the proof. $\qquad\square$

3.6 The Dependency of the Solution upon the Initial Data

Let $\alpha \in (0, 1)$. In this section, we consider the Cauchy problems:

$$
\begin{aligned}
D^\alpha_{\Delta, t_0} y(t) &= f\left(t, y(t)\right), \quad t > t_0, \\
D^\alpha_{\Delta, t_0} y(t_0) &= \xi,
\end{aligned} \tag{3.24}
$$

and

$$D^\alpha_{\Delta,t_0} y(t) = f(t, y(t)), \quad t > t_0,$$
$$D^\alpha_{\Delta,t_0} y(t_0) = \eta, \tag{3.25}$$

where $f : \mathbb{T} \times \mathbb{R} \to \mathbb{R}$ is a given function, ξ and η are given constants. Let a and G be as in the previous section.

Theorem 3.15. *Suppose that f satisfies the Lipschitzian condition (3.21) and y and z are the solutions of the Cauchy problems (3.24) and (3.25), respectively. Let also $_\Delta F_{\alpha,\alpha}(A, \cdot, t_0)$ be defined on $[t_0, a)$. Then*

$$|y(t) - z(t)| \le |\xi - \eta|_\Delta F_{\alpha,\alpha}(A, b, t_0), \quad t \in [t_0, a).$$

Proof. Let

$$y_0(t) = \xi h_{\alpha-1}(t, t_0),$$
$$y_m(t) = y_0(t) + I^\alpha_{\Delta,t_0} f(t, y(t)),$$
$$z_0(t) = \eta h_{\alpha-1}(t, t_0),$$
$$z_m(t) = z_0(t) + I^\alpha_{\Delta,t_0} f(t, z(t)), \quad t \in [t_0, a), \quad m \in \mathbb{N}.$$

By the proof of Theorem 3.14, we have

$$y(t) = \lim_{m \to \infty} y_m(t),$$
$$z(t) = \lim_{m \to \infty} z_m(t), \quad t \in [t_0, a).$$

We have that

$$\begin{aligned}
|z_1(t) - y_1(t)| &= \left| z_0(t) + I^\alpha_{\Delta,t_0} f(t, z_0(t)) - z_0(t) - I^\alpha_{\Delta,t_0} f(t, y_0(t)) \right| \\
&= \left| z_0(t) - y_0(t) + I^\alpha_{\Delta,t_0} (f(t, z_0(t)) - f(t, y_0(t))) \right| \\
&\le |z_0(t) - y_0(t)| + \left| I^\alpha_{\Delta,t_0} (f(t, z_0(t)) - f(t, y_0(t))) \right| \\
&\le |z_0(t) - y_0(t)| + I^\alpha_{\Delta,t_0} |f(t, z_0(t)) - f(t, y_0(t))| \\
&\le |\xi - \eta| h_{\alpha-1}(t, t_0) + A I^\alpha_{\Delta,t_0} |z_0(t) - y_0(t)| \\
&\le |\xi - \eta| h_{\alpha-1}(t, t_0) + A |\xi - \eta| I^\alpha_{\Delta,t_0} h_{\alpha-1}(t, t_0) \\
&= |\xi - \eta| (h_{\alpha-1}(t, t_0) + A h_{\alpha+\alpha-1}(t, t_0)), \quad t \in [t_0, a).
\end{aligned}$$

Assume that

$$|z_m(t) - y_m(t)| \le |\xi - \eta| \sum_{j=0}^{m} A^j h_{j\alpha+\alpha-1}(t, t_0), \quad t \in [t_0, a), \quad (3.26)$$

for some $m \in \mathbb{N}$. We will prove that

$$|z_{m+1}(t) - y_{m+1}(t)| \le |\xi - \eta| \sum_{j=0}^{m+1} A^j h_{j\alpha+\alpha-1}(t, t_0), \quad t \in [t_0, a).$$

In fact, we have

$$|z_{m+1}(t) - y_{m+1}(t)|$$

$$= \left| z_0(t) + I^\alpha_{\Delta,t_0} f(t, z_m(t)) - y_0(t) - I^\alpha_{\Delta,t_0} f(t, y_m(t)) \right|$$

$$= \left| z_0(t) - y_0(t) + I^\alpha_{\Delta,t_0} (f(t, z_m(t)) - f(t, y_m(t))) \right|$$

$$\le |z_0(t) - y_0(t)| + \left| I^\alpha_{\Delta,t_0} (f(t, z_m(t)) - f(t, y_m(t))) \right|$$

$$\le |z_0(t) - y_0(t)| + I^\alpha_{\Delta,t_0} |f(t, z_m(t)) - f(t, y_m(t))|$$

$$\le |\xi - \eta| h_{\alpha-1}(t, t_0) + A I^\alpha_{\Delta,t_0} |z_m(t) - y_m(t)|$$

$$\le |\xi - \eta| h_{\alpha-1}(t, t_0) + A|\xi - \eta| I^\alpha_{\Delta,t_0} \sum_{j=0}^{m} A^j h_{j\alpha+\alpha-1}(t, t_0)$$

$$= |\xi - \eta| (h_{\alpha-1}(t, t_0) + \sum_{j=0}^{m} A^{j+1} h_{(j+1)\alpha+\alpha-1}(t, t_0))$$

$$= |\xi - \eta| \sum_{j=0}^{m+1} A^j h_{(j+1)\alpha+\alpha-1}(t, t_0), \quad t \in [t_0, a).$$

Therefore, (3.26) holds for any $m \in \mathbb{N}$. By (3.26), we get

$$|z(t) - y(t)| = \lim_{m \to \infty} |z_m(t) - y_m(t)|$$

$$\le |\xi - \eta| \sum_{j=0}^{\infty} A^j h_{j\alpha+\alpha-1}(t, t_0)$$

$$= |\xi - \eta|_\Delta F_{\alpha,\alpha}(A, t, t_0)$$

$$\le |\xi - \eta|_\Delta F_{\alpha,\alpha}(A, a, t_0), \quad t \in [t_0, a).$$

This completes the proof. $\square$

Chapter 4

Fractional Young and Hölder Inequalities

The main aim of this chapter is to investigate the fractional analogues of some Hölder and Young inequalities and their reverse inequalities.

Let $\mathbb{T}$ be a time scale with forward jump operator and delta differentiation operator σ and Δ, respectively. Let also $t_0, a \in \mathbb{T}$, $t_0 < a$. For $\alpha > 0$ and $p > 0$, define the set

$$L^{\alpha,p}([t_0, a)) = \{f : [t_0, a) \to \mathbb{R} : I^\alpha_{\Delta,t_0}(f^p)(t) < \infty, \quad t \in [t_0, a)\}.$$

4.1 Fractional Young Inequalities

In this section, we deduct some fractional analogues of the classical Young inequalities. We start with the following result.

Theorem 4.1. *Let* $\alpha > 0, p_1, p_2 > 0$ *be such that* $\frac{1}{p_1} + \frac{1}{p_2} = 1$. *Let also* $f_1, f_2 : [t_0, a) \to [0, \infty)$, $f_1 \in L^{\alpha,p_1}([t_0, a))$, $f_2 \in L^{\alpha,p_2}([t_0, a))$, $f_1 f_2 \in L^{\alpha,1}([t_0, a))$. *Then*

$$I^\alpha_{\Delta,t_0}(f_1 f_2)(t) \leq \frac{1}{p_1} I^\alpha_{\Delta,t_0}(f_1^{p_1})(t) + \frac{1}{p_2} I^\alpha_{\Delta,t_0}(f_2^{p_2})(t), \quad t \in [t_0, a).$$

Proof. Applying the classical Young inequality (see Theorem A.1 in Appendix A), we get

$$I_{\Delta,t_0}^{\alpha}(f_1 f_2)(t)$$

$$= \int_{t_0}^{t} h_{\alpha-1}(t,\sigma(s))f_1(s)f_2(s)\Delta s$$

$$= \int_{t_0}^{t} \left((h_{\alpha-1}(t,\sigma(s)))^{\frac{1}{p_1}} f_1(s)\right)\left((h_{\alpha-1}(t,\sigma(s)))^{\frac{1}{p_2}} f_2(s)\right)\Delta s$$

$$\leq \frac{1}{p_1}\int_{t_0}^{t} h_{\alpha-1}(t,\sigma(s))(f_1(s))^{p_1}\Delta s$$

$$+ \frac{1}{p_2}\int_{t_0}^{t} h_{\alpha-1}(t,\sigma(s))(f_2(s))^{p_2}\Delta s$$

$$= \frac{1}{p_1}I_{\Delta,t_0}^{\alpha}\left(f_1^{p_1}\right)(t) + \frac{1}{p_2}I_{\Delta,t_0}^{\alpha}\left(f_2^{p_2}\right)(t), \quad t \in [t_0,a).$$

This completes the proof. □

Theorem 4.2. *Let $\epsilon, \alpha > 0, p_1, p_2 > 0$ be such that $\frac{1}{p_1} + \frac{1}{p_2} = 1$. Let also $f_1, f_2 : [t_0,a) \to [0,\infty)$, $f_1 \in L^{\alpha,p_1}([t_0,a))$, $f_2 \in L^{\alpha,p_2}([t_0,a))$, $f_1 f_2 \in L^{\alpha,1}([t_0,a))$. Then*

$$I_{\Delta,t_0}^{\alpha}(f_1 f_2)(t) \leq \epsilon I_{\Delta,t_0}^{\alpha}\left(f_1^{p_1}\right)(t) + C(\epsilon)I_{\Delta,t_0}^{\alpha}\left(f_2^{p_2}\right)(t), \quad t \in [t_0,a),$$

where

$$C(\epsilon) = \frac{1}{p_2(\epsilon p_1)^{\frac{p_2}{p_1}}}.$$

Proof. Applying the classical Young inequality with ϵ (see Theorem A.1 in Appendix A), we arrive at the inequality

$$I_{\Delta,t_0}^{\alpha}(f_1 f_2)(t)$$

$$= \int_{t_0}^{t} h_{\alpha-1}(t,\sigma(s))f_1(s)f_2(s)\Delta s$$

$$= \int_{t_0}^{t} \left((h_{\alpha-1}(t,\sigma(s)))^{\frac{1}{p_1}} f_1(s)\right)\left((h_{\alpha-1}(t,\sigma(s)))^{\frac{1}{p_2}} f_2(s)\right)\Delta s$$

$$\leq \epsilon \int_{t_0}^{t} h_{\alpha-1}(t, \sigma(s))(f_1(s))^{p_1} \Delta s$$

$$+ C(\epsilon) \int_{t_0}^{t} h_{\alpha-1}(t, \sigma(s))(f_2(s))^{p_2} \Delta s$$

$$= \epsilon I_{\Delta,t_0}^{\alpha} \left(f_1^{p_1} \right)(t) + C(\epsilon) I_{\Delta,t_0}^{\alpha} \left(f_2^{p_2} \right)(t), \quad t \in [t_0, a).$$

This completes the proof. $\qquad\square$

Theorem 4.3. *Let* $\alpha, p_1, p_2 > 0$ *be such that* $\frac{1}{p_1} + \frac{1}{p_2} = 1$. *Let also* $f_1, f_2 : [t_0, a) \to [0, \infty)$, $f_1 \in L^{\alpha, p_1}([t_0, a))$, $f_2 \in L^{:\alpha, p_2}([t_0, a))$, $f_1 f_2 \in L^{\alpha, 1}([t_0, a))$, $f_1^{\frac{p_1}{2}} f_2^{\frac{p_2}{2}} \in L^{\alpha, 1}([t_0, a))$. *Then*

$$I_{\Delta,t_0}^{\alpha}(f_1 f_2)(t) \leq \left(\frac{1}{p_1} - r_0 \right) I_{\Delta,t_0}^{\alpha} \left(f_1^{p_1} \right)(t) + \left(\frac{1}{p_2} - r_0 \right) I_{\Delta,t_0}^{\alpha} \left(f_2^{p_2} \right)(t)$$

$$+ 2 r_0 I_{\Delta,t_0}^{\alpha} \left(f_1^{\frac{p_1}{2}} f_2^{\frac{p_2}{2}} \right)(t), \quad t \in [t_0, a),$$

where

$$r_0 = \min \left\{ \frac{1}{p_1}, \frac{1}{p_2} \right\}.$$

Proof. We apply the refined Young inequality (see Theorem A.7 in Appendix A) and get

$$I_{\Delta,t_0}^{\alpha}(f_1 f_2)(t)$$

$$= \int_{t_0}^{t} h_{\alpha-1}(t, \sigma(s)) f_1(s) f_2(s) \Delta s$$

$$= \int_{t_0}^{t} \left((h_{\alpha-1}(t, \sigma(s)))^{\frac{1}{p_1}} f_1(s) \right) \left((h_{\alpha-1}(t, \sigma(s)))^{\frac{1}{p_2}} f_2(s) \right) \Delta s$$

$$\leq \left(\frac{1}{p_1} - r_0 \right) \int_{t_0}^{t} h_{\alpha-1}(t, \sigma(s))(f_1(s))^{p_1} \Delta s$$

$$+ \left(\frac{1}{p_2} - r_0 \right) \int_{t_0}^{t} h_{\alpha-1}(t, \sigma(s))(f_2(s))^{p_2} \Delta s$$

$$+ 2 r_0 \int_{t_0}^{t} (h_{\alpha-1}(t, \sigma(s)))^{\frac{1}{2}} (f_1(s))^{\frac{p_1}{2}}$$

$$\times (h_{\alpha-1}(t, \sigma(s)))^{\frac{1}{2}} (f_2(s))^{\frac{p_2}{2}} \Delta s$$

$$= \left(\frac{1}{p_1} - r_0 \right) I^\alpha_{\Delta,t_0} \left(f_1^{p_1} \right)(t) + \left(\frac{1}{p_2} - r_0 \right) I^\alpha_{\Delta,t_0} \left(f_2^{p_2} \right)(t)$$

$$+ 2r_0 \int_{t_0}^t h_{\alpha-1}(t,\sigma(s)) \left(f_1(s) \right)^{\frac{p_1}{2}} \left(f_2(s) \right)^{\frac{p_2}{2}} \Delta s$$

$$\leq \left(\frac{1}{p_1} - r_0 \right) I^\alpha_{\Delta,t_0} \left(f_1^{p_1} \right)(t) + \left(\frac{1}{p_2} - r_0 \right) I^\alpha_{\Delta,t_0} \left(f_2^{p_2} \right)(t)$$

$$+ 2r_0 I^\alpha_{\Delta,t_0} \left(f_1^{\frac{p_1}{2}} f_2^{\frac{p_2}{2}} \right)(t), \quad t \in [t_0, a).$$

This completes the proof.   □

Theorem 4.4. *Let* $\alpha, p_1, p_2 > 0$, $\frac{1}{p_1} + \frac{1}{p_2} = 1$. *Let also* $f_1 : [t_0, a) \to (0, \infty)$, $f_2 : [t_0, a) \to [0, \infty)$ *and*

$$f_1^{p_1}, \quad f_2^{p_2}, \quad f_1 f_2 \left(S \left(\left(\frac{f_2^{p_2}}{f_1^{p_1}} \right)^r \right) \right)^2 \in L^{\alpha,1}([t_0, a)).$$

Then

$$I^\alpha_{\Delta,t_0} \left(f_1 f_2 \left(S \left(\left(\frac{f_2^{p_2}}{f_1^{p_1}} \right)^r \right) \right)^2 \right)(t)$$

$$\leq \frac{1}{p_1} I^\alpha_{\Delta,t_0} \left(f_1^{p_1} \right)(t) + \frac{1}{p_2} I^\alpha_{\Delta,t_0} \left(f_2^{p_2} \right)(t), \quad t \in [t_0, a),$$

where S *is the Specht ratio.*

Proof.   We apply the refined Young inequality (see Theorem A.7 in Appendix A) and get

$$I^\alpha_{\Delta,t_0} \left(f_1 f_2 \left(S \left(\left(\frac{f_2^{p_2}}{f_1^{p_1}} \right)^r \right) \right)^2 \right)(t)$$

$$= \int_{t_0}^t h_{\alpha-1}(t,\sigma(s)) f_1(s) f_2(s) \left(S \left(\left(\frac{f_2^{p_2}}{f_1^{p_1}} \right)^r \right) \right)^2 (s) \Delta s$$

$$= \int_{t_0}^t \left((h_{\alpha-1}(t,\sigma(s)))^{\frac{1}{p_1}} f_1(s) S \left(\left(\frac{f_2^{p_2}}{f_1^{p_1}} \right)^r \right)(s) \right)$$

$$\times \left((h_{\alpha-1}(t,\sigma(s)))^{\frac{1}{p_2}} f_2(s) S \left(\left(\frac{f_2^{p_2}}{f_1^{p_1}} \right)^r \right)(s) \right) \Delta s$$

$$\leq \frac{1}{p_1} \int_{t_0}^{t} h_{\alpha-1}(t, \sigma(s)) \, (f_1(s))^{p_1} \, \Delta s$$

$$+ \frac{1}{p_2} \int_{t_0}^{t} h_{\alpha-1}(t, \sigma(s)) \, (f_2(s))^{p_2} \, \Delta s$$

$$\leq \frac{1}{p_1} I_{\Delta,t_0}^{\alpha} \left(f_1^{p_1} \right)(t) + \frac{1}{p_2} I_{\Delta,t_0}^{\alpha} \left(f_2^{p_2} \right)(t), \quad t \in [t_0, a).$$

This completes the proof. □

4.2 Reverse Fractional Young Inequalities

In this section, we deduct some fractional analogues of some reverse Young inequalities.

Theorem 4.5. *Let* $\alpha, p_1, p_2 > 0$ *be such that* $\frac{1}{p_1} + \frac{1}{p_2} = 1$. *Let also* $f_1 : [t_0, a) \to [0, \infty)$, $f_2 : [t_0, a) \to (0, \infty)$ *and*

$$S\left(\frac{f_1^{p_1}}{f_2^{p_2}} \right) f_1 f_2, \quad f_1^{p_1}, f_2^{p_2} \in L^{\alpha,1}([t_0, a)).$$

Then

$$I_{\Delta,t_0} \left(S\left(\frac{f_1^{p_1}}{f_2^{p_2}} \right) f_1 f_2 \right)(t)$$

$$\geq \frac{1}{p_1} I_{\Delta,t_0}^{\alpha} \left(f_1^{p_1} \right)(t) + \frac{1}{p_2} I_{\Delta,t_0}^{\alpha} \left(f_2^{p_2} \right)(t), \quad t \in [t_0, a),$$

where S *denotes the Specht ratio.*

Proof. We apply the reverse Young inequality (see Theorem A.8 in Appendix A) and get

$$I_{\Delta,t_0}^{\alpha} \left(S\left(\frac{f_1^{p_1}}{f_2^{p_2}} \right) f_1 f_2 \right)(t)$$

$$= \int_{t_0}^{t} h_{\alpha-1}(t, \sigma(s)) S\left(\frac{f_1^{p_1}}{f_2^{p_2}} \right)(s) f_1(s) f_2(s) \Delta s$$

$$= \int_{t_0}^t S\left(\frac{f_1^{p_1}}{f_2^{p_2}}\right)(s)((h_{\alpha-1}(t,\sigma(s)))^{\frac{1}{p_1}} f_1(s))$$

$$\times ((h_{\alpha-1}(t,\sigma(s)))^{\frac{1}{p_2}} f_2(s))\Delta s$$

$$\geq \frac{1}{p_1} \int_{t_0}^t h_{\alpha-1}(t,\sigma(s))(f_1(s))^{p_1}\Delta s$$

$$+ \frac{1}{p_2} \int_{t_0}^t h_{\alpha-1}(t,\sigma(s))(f_2(s))^{p_2}\Delta s$$

$$= \frac{1}{p_1} I_{\Delta,t_0}^\alpha \left(f_1^{p_1}\right)(t) + \frac{1}{p_2} I_{\Delta,t_0}^\alpha \left(f_2^{p_2}\right)(t), \quad t \in [t_0, a).$$

This completes the proof. □

Theorem 4.6. *Let* $\alpha, p_1, p_2 > 0$ *be such that* $\frac{1}{p_1} + \frac{1}{p_2} = 1$. *Let also* $f_1 : [t_0, a) \to [0, \infty)$, $f_2 : [t_0, a) \to (0, \infty)$ *and*

$$S\left(\frac{f_1^{p_1}}{f_2^{p_2}}\right) f_1 f_2, \quad f_1^{p_1}, \quad f_2^{p_2}, \quad f_1^{p_1} f_2^{p_2} \in L^{\alpha,1}([t_0, a)).$$

Then

$$I_{\Delta,t_0}^\alpha \left(S\left(\frac{f_1^{p_1}}{f_2^{p_2}}\right) f_1 f_2\right)(t)$$

$$\geq \left(\frac{1}{p_1} - r\right) I_{\Delta,t_0}^\alpha \left(f_1^{p_1}\right)(t) + \left(\frac{1}{p_2} - r\right) I_{\Delta,t_0}^\alpha \left(f_2^{p_2}\right)(t)$$

$$+ 2r I_{\Delta,t_0}^\alpha \left(f_1^{\frac{p_1}{2}} f_2^{\frac{p_2}{2}}\right)(t), \quad t \in [t_0, a),$$

where

$$r = \min\left\{\frac{1}{p_1}, \frac{1}{p_2}\right\}$$

and S *denotes the Specht ratio.*

Proof. We apply the reverse Young inequality (see Theorem A.9 in Appendix A) and get

$$
I^\alpha_{\Delta,t_0} \left(S \left(\frac{f_1^{p_1}}{f_2^{p_2}} \right) f_1 f_2 \right) (t)
$$

$$
= \int_{t_0}^t h_{\alpha-1}(t, \sigma(s)) S \left(\frac{f_1^{p_1}}{f_2^{p_2}} \right) (s) f_1(s) f_2(s) \Delta s
$$

$$
= \int_{t_0}^t S \left(\frac{f_1^{p_1}}{f_2^{p_2}} \right) (s) \left((h_{\alpha-1}(t, \sigma(s)))^{\frac{1}{p_1}} f_1(s) \right)
$$

$$
\times \left((h_{\alpha-1}(t, \sigma(s)))^{\frac{1}{p_2}} f_2(s) \right) \Delta s
$$

$$
\geq \left(\frac{1}{p_1} - r \right) \int_{t_0}^t h_{\alpha-1}(t, \sigma(s)) (f_1(s))^{p_1} \Delta s
$$

$$
+ \left(\frac{1}{p_2} - r \right) \int_{t_0}^t h_{\alpha-1}(t, \sigma(s)) (f_2(s))^{p_2} \Delta s
$$

$$
+ 2r \int_{t_0}^t (h_{\alpha-1}(t, \sigma(s)))^1 2(f_1(s))^{\frac{p_1}{2}} (h_{\alpha-1}(t, \sigma(s)))^{\frac{1}{2}} (f_2(s))^{\frac{p_2}{2}} \Delta s
$$

$$
= \left(\frac{1}{p_1} - r \right) I^\alpha_{\Delta,t_0} (f_1^{p_1}) (t) + \left(\frac{1}{p_2} - r \right) I^\alpha_{\Delta,t_0} (f_2^{p_2}) (t)
$$

$$
+ 2r \int_{t_0}^t h_{\alpha-1}(t, \sigma(s)) (f_1(s))^{\frac{p_1}{2}} (f_2(s))^{\frac{p_2}{2}} \Delta s
$$

$$
= \left(\frac{1}{p_1} - r \right) I^\alpha_{\Delta,t_0} (f_1^{p_1}) (t) + \left(\frac{1}{p_2} - r \right) I^\alpha_{\Delta,t_0} (f_2^{p_2}) (t)
$$

$$
+ 2r I^\alpha_{\Delta,t_0} \left(f_1^{\frac{p_1}{2}} f_2^{\frac{p_2}{2}} \right) (t), \quad t \in [t_0, a).
$$

This completes the proof. □

4.3 Fractional Hölder Inequalities

In this section, we deduct a fractional analogue of the fractional Hölder inequality.

Theorem 4.7. *Let $\alpha, p_1, p_2 > 0$ be such that $\frac{1}{p_1} + \frac{1}{p_2} = 1$. Let also $f_1, f_2 : [t_0, a) \to [0, \infty)$ and $f_1 f_2, f_1^{p_1}, f_2^{p_2} \in L^{\alpha,1}([t_0, a))$. Then*

$$I^{\alpha}_{\Delta, t_0}(f_1 f_2)(t) \le \left(I^{\alpha}_{\Delta, t_0}\left(f_1^{p_1}\right)(t)\right)^{\frac{1}{p_1}} \left(I^{\alpha}_{\Delta, t_0}\left(f_2^{p_2}\right)(t)\right)^{\frac{1}{p_2}}, \quad t \in [t_0, a).$$

Proof. We apply the classical Hölder inequality and find

$$I^{\alpha}_{\Delta, t_0}(f_1 f_2)(t)$$

$$= \int_{t_0}^{t} h_{\alpha-1}(t, \sigma(s)) f_1(s) f_2(s) \Delta s$$

$$= \int_{t_0}^{t} \left((h_{\alpha-1}(t, \sigma(s)))^{\frac{1}{p_1}} f_1(s)\right) \left((h_{\alpha-1}(t, \sigma(s)))^{\frac{1}{p_2}} f_2(s)\right) \Delta s$$

$$\le \left(\int_{t_0}^{t} h_{\alpha-1}(t, \sigma(s))(f_1(s))^{p_1} \Delta s\right)^{\frac{1}{p_1}}$$

$$\times \left(\int_{t_0}^{t} h_{\alpha-1}(t, \sigma(s))(f_2(s))^{p_2} \Delta s\right)^{\frac{1}{p_2}}$$

$$= \left(I^{\alpha}_{\Delta, t_0}\left(f_1^{p_1}\right)(t)\right)^{\frac{1}{p_1}} \left(I^{\alpha}_{\Delta, t_0}\left(f_2^{p_2}\right)(t)\right)^{\frac{1}{p_2}}, \quad t \in [t_0, a).$$

This completes the proof. □

Theorem 4.8. *Let $\alpha, p_1, p_2, r > 0$ be such that*

$$\frac{1}{p_1} + \frac{1}{p_2} = \frac{1}{r}.$$

Let also $f_1, f_2 : [t_0, a) \to [0, \infty)$ and

$$f_1^r f_2^r, \quad f_1^{p_1}, \quad f_2^{p_2} \in L^{\alpha,1}([t_0, a)).$$

Then

$$\left(I^{\alpha}_{\Delta, t_0}\left(f_1^r f_2^r\right)(t)\right)^{\frac{1}{r}}$$

$$\le \left(I^{\alpha}_{\Delta, t_0}\left(f_1^{p_1}\right)(t)\right)^{\frac{1}{p_1}} \left(I^{\alpha}_{\Delta, t_0}\left(f_2^{p_2}\right)(t)\right)^{\frac{1}{p_2}}, \quad t \in [t_0, a).$$

Proof. Note that

$$\frac{r}{p_1} + \frac{r}{p_2} = 1.$$

Then

$$I^\alpha_{\Delta,t_0}\left(f_1^r f_2^r\right)(t)$$

$$= \int_{t_0}^t h_{\alpha-1}(t, \sigma(s))\, (f_1(s))^r\, (f_2(s))^r\, \Delta s$$

$$= \int_{t_0}^t \left((h_{\alpha-1}(t, \sigma(s)))^{\frac{r}{p_1}}\, (f_1(s))^r\right) \left((h_{\alpha-1}(t, \sigma(s)))^{\frac{r}{p_2}}\, (f_2(s))^r\right) \Delta s$$

$$\leq \left(\int_{t_0}^t h_{\alpha-1}(t, \sigma(s))\, (f_1(s))^{p_1}\, \Delta s\right)^{\frac{r}{p_1}}$$

$$\times \left(\int_{t_0}^t h_{\alpha-1}(t, \sigma(s))\, (f_2(s))^{p_2}\, \Delta s\right)^{\frac{r}{p_2}}$$

$$= \left(I^\alpha_{\Delta,t_0}\left(f_1^{p_1}\right)(t)\right)^{\frac{1}{p_1}} \left(I^\alpha_{\Delta,t_0}\left(f_2^{p_2}\right)(t)\right)^{\frac{1}{p_2}}, \quad t \in [t_0, a).$$

This completes the proof. $\qquad\square$

4.4 Reverse Fractional Hölder Inequalities

In this section, we deduct the fractional analogues of some reverse Hölder inequalities. We start with the following result.

Theorem 4.9. *Let* $p_1, p_2 > 1$ *be such that* $\frac{1}{p_1} + \frac{1}{p_2} = 1$, $f_1, f_2 :$ $[t_0, a) \to (0, \infty)$ *be such that*

$$f_1^{p_1}, \quad f_2^{p_2}, \quad S\left(\frac{f_1^{p_1} \|f_2\|_{p_2}^{p_2}}{f_2^{p_2} \|f_1\|_{p_1}^{p_1}}\right) f_1 f_2 \in L^{\alpha,1}([t_0, a)).$$

Then

$$I^\alpha_{\Delta,t_0}\left(S\left(\frac{f_1^{p_1} \|f_2\|_{p_2}^{p_2}}{f_2^{p_2} \|f_1\|_{p_1}^{p_1}}\right) f_1 f_2\right)(t)$$

$$\geq \left(I_{\Delta,t_0}\left(f_1^{p_1}\right)(t)\right)^{\frac{1}{p_1}} \left(I^\alpha_{\Delta,t_0}\left(f_2^{p_2}\right)(t)\right)^{\frac{1}{p_2}}, \quad t \in [t_0, a).$$

Proof.   We apply the reverse Hölder inequality (see Theorem A.13 in Appendix A) and get

$$
I_{\Delta,t_0}^{\alpha}\left(S\left(\frac{f_1^{p_1}\|f_2\|_{p_2}^{p_2}}{f_2^{p_2}\|f_1\|_{p_1}^{p_1}}\right)f_1f_2\right)(t)
$$

$$
=\int_{t_0}^{t} f_{\alpha-1}(t,\sigma(s))S\left(\frac{f_1^{p_1}\|f_2\|_{p_2}^{p_2}}{f_2^{p_2}\|f_1\|_{p_1}^{p_1}}\right)(s)f_1(s)f_2(s)\Delta s
$$

$$
=\int_{t_0}^{t} S\left(\frac{f_1^{p_1}\|f_2\|_{p_2}^{p_2}}{f_2^{p_2}\|f_1\|_{p_1}^{p_1}}\right)(s)\left((f_{\alpha-1}(t,\sigma(s)))^{\frac{1}{p_1}}f_1(s)\right)
$$

$$
\times\left((f_{\alpha-1}(t,\sigma(s)))^{\frac{1}{p_2}}f_2(s)\right)\Delta s
$$

$$
\geq\left(\int_{t_0}^{t} f_{\alpha-1}(t,\sigma(s))\,(f_1(s))^{p_1}\,\Delta s\right)^{\frac{1}{p_1}}
$$

$$
\times\left(\int_{t_0}^{t} f_{\alpha-1}(t,\sigma(s))(f_2(s))^{p_2}\Delta s\right)^{\frac{1}{p_2}}
$$

$$
=\left(I_{\Delta,t_0}^{\alpha}\left(f_1^{p_1}\right)(t)\right)^{\frac{1}{p_1}}\left(I_{\Delta,t_0}^{\alpha}\left(f_2^{p_2}\right)(t)\right)^{\frac{1}{p_2}},\quad t\in[t_0,a).
$$

This completes the proof.   □

Theorem 4.10. *Let* $p_1,p_2>1$ *be such that* $\frac{1}{p_1}+\frac{1}{p_2}=1$, $f_1,f_2:$ $[t_0,a)\to(0,\infty)$ *be such that*

$$
f_1f_2,\quad f_1^{p_1},f_2^{p_2}\in L^{\alpha,1}([t_0,a)).
$$

Let also there exist constants $M>m>0$ *so that*

$$
0<m\leq\frac{(f_1(t))^{p_1}}{(f_2(t))^{p_2}}\leq M<\infty,\quad t\in[t_0,a).
$$

Then

$$
\left(I_{\Delta,t_0}^{\alpha}\left(f_1^{p_1}\right)(t)\right)^{\frac{1}{p_1}}\left(I_{\Delta,t_0}^{\alpha}\left(f_2^{p_2}\right)(t)\right)^{\frac{1}{p_2}}\leq\left(\frac{m}{M}\right)^{-\frac{1}{p_1p_2}}I_{\Delta,t_0}^{\alpha}(f_1f_2)(t),
$$

$$
t\in[t_0,a).
$$

Proof. Note that

$$0 < m \leq \frac{h_{\alpha-1}(t,\sigma(s))(f_1(t))^{p_1}}{h_{\alpha-1}(t,\sigma(s))(f_2(t))^{p_2}} \leq M < \infty, \quad t,s \in [t_0,a), \quad s \leq t.$$

We apply the reverse Hölder inequality (see Theorem A.14 in Appendix A) and find

$$\left(\frac{m}{M}\right)^{-\frac{1}{p_1 p_2}} I_{\Delta,t_0}^{\alpha}(f_1 f_2)(t)$$

$$= \left(\frac{m}{M}\right)^{-\frac{1}{p_1 p_2}} \int_{t_0}^{t} h_{\alpha-1}(t,\sigma(s)) f_1(s) f_2(s) \Delta s$$

$$= \left(\frac{m}{M}\right)^{-\frac{1}{p_1 p_2}} \int_{t_0}^{t} \left((h_{\alpha-1}(t,\sigma(s)))^{\frac{1}{p_1}} f_1(s) \right)$$

$$\times \left((h_{\alpha-1}(t,\sigma(s)))^{\frac{1}{p_1}} f_2(s) \right) \Delta s$$

$$\geq \left(\int_{t_0}^{t} h_{\alpha-1}(t,\sigma(s)) (f_1(s))^{p_1} \Delta s \right)^{\frac{1}{p_1}}$$

$$\times \left(\int_{t_0}^{t} h_{\alpha-1}(t,\sigma(s)) (f_2(s))^{p_2} \Delta s \right)^{\frac{1}{p_2}}$$

$$\geq \left(I_{\Delta,t_0}^{\alpha} (f_1^{p_1})(t) \right)^{\frac{1}{p_1}} \left(I_{\Delta,t_0}^{\alpha} (f_2^{p_2})(t) \right)^{\frac{1}{p_2}}, \quad t \in [t_0,a).$$

This completes the proof. $\square$

Theorem 4.11. *Let $p_1, p_2, p_3 > 0$ be such that*

$$\frac{1}{p_1} + \frac{1}{p_2} + \frac{1}{p_3} = 1$$

and $f_1, f_2, f_3 : [t_0,a) \to (0,\infty)$ be such that

$$f_1^{p_1}, \quad f_2^{p_2}, \quad f_3^{p_3}, \quad f_1 f_2 f_3 \in L^{\alpha,1}([t_0,a)),$$

and

$$0 < m \leq \frac{(f_1(t))^{\frac{p_1}{r}}}{(f_2(t))^{\frac{p_2}{r}}} \leq M < \infty,$$

$$0 < m \leq \frac{(f_1(t)f_2(t))^r}{(f_3(t))^{p_3}} \leq M < \infty, \quad t \in [t_0,a),$$

for some positive constants $M > m$ and a positive constant r so that $\frac{1}{p_1} + \frac{1}{p_2} = \frac{1}{r}$. Then

$$\left(I^\alpha_{\Delta,t_0}\left(f_1^{p_1}\right)(t)\right)^{\frac{1}{p_1}} \left(I^\alpha_{\Delta,t_0}\left(f_2^{p_2}\right)(t)\right)^{\frac{1}{p_2}} \left(I^\alpha_{\Delta,t_0}\left(f_3^{p_3}\right)(t)\right)^{\frac{1}{p_3}}$$

$$\le \left(\frac{m}{M}\right)^{-\left(\frac{1}{rp_3} + \frac{r^2}{p_1 p_2}\right)} I^\alpha_{\Delta,t_0}\left(f_1 f_2 f_3\right)(t), \quad t \in [t_0, a).$$

Proof. We apply the reverse Hölder inequality (see Theorem A.15 in Appendix A) and get

$$\left(\frac{m}{M}\right)^{-\left(\frac{1}{rp_3} + \frac{r^2}{p_1 p_2}\right)} I^\alpha_{\Delta,t_0}(f_1 f_2 f_3)(t)$$

$$= \left(\frac{m}{M}\right)^{-\left(\frac{1}{rp_3} + \frac{r^2}{p_1 p_2}\right)} \int_{t_0}^t h_{\alpha-1}(t, \sigma(s)) f_1(s) f_2(s) f_3(s) \Delta s$$

$$= \left(\frac{m}{M}\right)^{-\left(\frac{1}{rp_3} + \frac{r^2}{p_1 p_2}\right)} \int_{t_0}^t \left((h_{\alpha-1}(t, \sigma(s)))^{\frac{1}{p_1}} f_1(s)\right)$$

$$\times \left((h_{\alpha-1}(t, \sigma(s)))^{\frac{1}{p_2}} f_2(s)\right) \left((h_{\alpha-1}(t, \sigma(s)))^{\frac{1}{p_3}} f_3(s)\right) \Delta s$$

$$\ge \left(\int_{t_0}^t h_{\alpha-1}(t, \sigma(s))(f_1(s))^{p_1} \Delta s\right)^{\frac{1}{p_1}}$$

$$\times \left(\int_{t_0}^t h_{\alpha-1}(t, \sigma(s))(f_2(s))^{p_2} \Delta s\right)^{\frac{1}{p_2}}$$

$$\times \left(\int_{t_0}^t h_{\alpha-1}(t, \sigma(s))(f_3(s))^{p_3} \Delta s\right)^{\frac{1}{p_3}}$$

$$= \left(I^\alpha_{\Delta,t_0}\left(f_1^{p_1}\right)(t)\right)^{\frac{1}{p_1}} \left(I^\alpha_{\Delta,t_0}\left(f_2^{p_2}\right)(t)\right)^{\frac{1}{p_2}} \left(I^\alpha_{\Delta,t_0}\left(f_3^{p_3}\right)(t)\right)^{\frac{1}{p_3}},$$

$$t \in [t_0, a).$$

This completes the proof. $\qquad\qquad\qquad\qquad\qquad\qquad\qquad\quad \square$

Chapter 5

Fractional Inequalities for Convex Functions

In this chapter, we introduce the notion of exponentially convex functions on time scales and then establish Hermite–Hadamard-type inequalities for this class of functions. As a special case, we derive this double inequality in the context of classical notion of exponentially convex functions and convex functions. Moreover, we prove some new integral inequalities for n-times continuously differentiable functions with exponentially convex first Δ-derivative. Some Ostrowski-type inequalities on arbitrary time scales are deducted. Then strongly r-convex functions are defined and some upper bounds for the delta-Riemann–Liouville fractional integral are deducted. Next, we attempt to prove some upper bounds for the delta-Riemann–Liouville fractional integral of functions which are n-times rd-continuously Δ-differentiable with exponentially s-convexity property in the second sense on an interval in some time scales.

Let $\mathbb{T}$ be a time scale with forward jump operator and delta differentiation operator σ and Δ, respectively. Let also $a, b \in \mathbb{T}$, $a < b$, and $x_0 \in \mathbb{T}$.

5.1 Inequalities for Exponentially Convex Functions

First, we will give a definition for time scales exponentially convex functions.

Definition 5.1. We say that a function $f : [a,b] \to \mathbb{R}$ is an exponentially convex function if

$$f(t) \leq \frac{b-t}{b-a} \cdot \frac{f(a)}{e_\beta(a,x_0)} + \frac{t-a}{b-a} \cdot \frac{f(b)}{e_\beta(b,x_0)} \tag{5.1}$$

for any $t \in [a,b]$ and for some $\beta \in \mathcal{R}_+$. If (5.11) holds in the reversed sense, then we say that f is exponentially concave function.

Example 5.1. Let $f : \mathbb{T} \to [0,\infty)$ be a convex function, $x_0 = a$, $\beta < 0$ is such that $1 + \mu(t)\beta > 0$, $t \in \mathbb{T}$. Then $0 < e_\beta(b,a) < 1$ and

$$f(t) \leq \frac{(b-t)f(a)}{b-a} + \frac{(t-a)f(b)}{b-a}$$

$$= \frac{b-t}{b-a} \cdot \frac{f(a)}{e_\beta(a,a)} + \frac{(t-b)f(b)}{b-a}$$

$$\leq \frac{b-t}{b-a} \cdot \frac{f(a)}{e_\beta(a,a)} + \frac{t-a}{b-a} \cdot \frac{f(b)}{e_\beta(b,a)}, \quad t \in \mathbb{T}.$$

Example 5.2. Let $\mathbb{T} = \mathbb{Z}$, $a = -1$, $b = 3$ and $x_0 = \frac{a+b}{2} = 1$. Then $f(t) = -t^3$ is exponentially convex on $[-1,3]$ with $\beta = 2$.

Remark 5.1. If $\mathbb{T} = \mathbb{R}$, since $e_\alpha(t,t_0) = e^{\alpha(t-t_0)}$, $t \in \mathbb{R}$, by (5.11), we obtain

$$f(t) \leq \frac{b-t}{b-a} \cdot \frac{f(a)}{e^{\beta(a-x_0)}} + \frac{t-a}{b-a} \cdot \frac{f(b)}{e^{\beta(b-x_0)}}, \quad t \in \mathbb{R}.$$

If $x_0 = 0$ in the above inequality, we get the classical definition for exponentially convex functions, which is introduced by Awan *et al.* [3]. Also, if we consider $\beta = 0$, then this new class of convex functions reduces to the class of classical convex functions. Hence, we conclude that every convex function is an exponentially convex function, but it does not hold vice versa.

We will use the following notations:

$$x_\alpha = \frac{1}{b-a} \int_a^b t \diamond_\alpha t,$$

$$x_{\alpha,\alpha} = \frac{1}{b-a} \int_a^b t^2 \diamond_\alpha t,$$

$$h_2(a,b) = \int_a^b (b-t)\Delta t,$$

$$h_2(b,a) = \int_a^b (t-a)\Delta t.$$

We start here with our first main result.

Theorem 5.1. *Let $\beta \in \mathcal{R}_+$, $f : [a,b] \to \mathbb{R}$, $f \in \mathcal{C}([a,b])$ be positive and exponentially convex function. Then*

$$\frac{1}{b-a} \int_a^b f(t) \diamond_\alpha t \le \frac{b-x_\alpha}{b-a} \cdot \frac{f(a)}{e_\beta(a,x_0)} + \frac{x_\alpha-a}{b-a} \cdot \frac{f(b)}{e_\beta(b,x_0)}. \qquad (5.2)$$

If, in addition, f is convex, then

$$f(x_\alpha) \le \frac{1}{b-a} \int_a^b f(t) \diamond_\alpha t. \qquad (5.3)$$

Proof. By taking the diamond-α integral side by side in (5.11), we obtain

$$\int_a^b f(t) \diamond_\alpha t \le \frac{f(a)}{e_\beta(a,x_0)(b-a)} \int_a^b (b-s) \diamond_\alpha s$$

$$+ \frac{f(b)}{(b-a)e_\beta(b,x_0)} \int_a^b (t-a) \diamond_\alpha t$$

$$= \frac{f(a)}{e_\beta(a,x_0)(b-a)} \left(b(b-a) - \int_a^b s \diamond_\alpha s \right)$$

$$+ \frac{f(b)}{(b-a)e_\beta(b,x_0)} \left(\int_a^b t \diamond_\alpha t - a(b-a) \right)$$

$$= \frac{f(a)}{e_\beta(a,x_0)}(b-x_\alpha) + \frac{f(b)}{e_\beta(b,x_0)}(x_\alpha-a),$$

whereupon

$$\frac{1}{b-a}\int_a^b f(t) \diamond_\alpha t \leq \frac{b-x_\alpha}{b-a} \cdot \frac{f(a)}{e_\beta(a,x_0)} + \frac{x_\alpha - a}{b-a} \cdot \frac{f(b)}{e_\beta(b,x_0)}.$$

Now, suppose that f is convex. Then by the diamond-α Jensen inequality (see Theorem B.1 in Appendix B), we find

$$f(x_\alpha) = f\left(\frac{\int_a^b s \diamond_\alpha s}{b-a}\right) \leq \frac{\int_a^b f(s) \diamond_\alpha s}{b-a}.$$

This completes the proof.   □

Let now $\mathbb{T} = \mathbb{R}$ in the above theorem. Then, we have $x_\alpha = \frac{b+a}{2}$. Also, if $x_0 = 0$, according to (5.2) and by some simple computations, we obtain

$$\frac{1}{b-a}\int_a^b f(t)dt \leq \frac{1}{2}\left(\frac{f(a)}{e^{\beta a}} + \frac{f(b)}{e^{\beta b}}\right), \tag{5.4}$$

which is the second inequality of Hermite–Hadamard like inequality for the classical exponentially convex functions.

Let $\beta = 0$ in (5.4). Then

$$\frac{1}{b-a}\int_a^b f(t)dt \leq \frac{f(a)+f(b)}{2}. \tag{5.5}$$

On the other hand, by (5.3), we derive

$$f\left(\frac{a+b}{2}\right) \leq \frac{1}{b-a}\int_a^b f(t)dt. \tag{5.6}$$

So, by (5.5) and (5.6), we recover the classical Hermite–Hadamard inequality.

We now present the following identities which are needed in our next result:

$$\frac{1}{b-a}\int_a^b (b-t)^2 \diamond_\alpha t$$

$$= \frac{1}{b-a}\int_a^b (b^2 - 2bt + t^2) \diamond_\alpha t$$

$$= \frac{b^2}{b-a} \int_a^b \diamond_\alpha t - \frac{2b}{b-a} \int_a^b t \diamond_\alpha t + \frac{1}{b-a} \int_a^b t^2 \diamond_\alpha t$$

$$= b^2 - 2bx_\alpha + x_{\alpha,\alpha},$$

$$\frac{1}{b-a} \int_a^b (t-a)^2 \diamond_\alpha t$$

$$= \frac{1}{b-a} \int_a^b (t^2 - 2at + a^2) \diamond_\alpha t$$

$$= \frac{1}{b-a} \int_a^b t^2 \diamond_\alpha t - \frac{2a}{b-a} \int_a^b t \diamond_\alpha t + \frac{a^2}{b-a} \int_a^b \diamond_\alpha t$$

$$= x_{\alpha,\alpha} - 2ax_\alpha + a^2,$$

and

$$\frac{1}{b-a} \int_a^b (b-t)(t-a) \diamond_\alpha t = \frac{1}{b-a} \int_a^b (bt - ab - t^2 + at) \diamond_\alpha t$$

$$= \frac{1}{b-a} \int_a^b (-t^2 + (a+b)t - ab) \diamond_\alpha t$$

$$= -x_{\alpha,\alpha} + (a+b)x_\alpha - ab.$$

Our next result reads as follows.

Theorem 5.2. *Let* $f, g : [a, b] \to [0, \infty)$, $f, g \in C([a, b])$ *be exponentially convex functions. Then*

$$\int_a^b f(t)g(t) \diamond_\alpha t$$

$$\leq \frac{f(a)g(a)}{(b-a)e_{\beta \oplus \beta}(a, x_0)} (b^2 - 2bx_\alpha + x_{\alpha,\alpha})$$

$$+ \frac{f(a)g(b) + f(b)g(a)}{e_\beta(a, x_0)e_\beta(b, x_0)(b-a)} (-x_{\alpha,\alpha} + (a+b)x_\alpha - ab)$$

$$+ \frac{f(b)g(b)}{(b-a)e_{\beta \oplus \beta}(b, x_0)} (a^2 - 2ax_\alpha + x_{\alpha,\alpha}).$$

Proof.   Since f and g are exponentially convex functions, we have

$$f(t) \leq \frac{b-t}{b-a} \cdot \frac{f(a)}{e_\beta(a,x_0)} + \frac{t-a}{b-a} \cdot \frac{f(b)}{e_\beta(b,x_0)},$$

$$g(t) \leq \frac{b-t}{b-a} \cdot \frac{g(a)}{e_\beta(a,x_0)} + \frac{t-a}{b-a} \cdot \frac{g(b)}{e_\beta(b,x_0)}, \quad t \in [a,b].$$

Then

$$f(t)g(t)$$
$$\leq \left(\frac{b-t}{b-a} \cdot \frac{f(a)}{e_\beta(a,x_0)} + \frac{t-a}{b-a} \cdot \frac{f(b)}{e_\beta(b,x_0)} \right)$$
$$\times \left(\frac{b-t}{b-a} \cdot \frac{g(a)}{e_\beta(a,x_0)} + \frac{t-a}{b-a} \cdot \frac{g(b)}{e_\beta(b,x_0)} \right)$$
$$= \frac{(b-t)^2}{(b-a)^2} \cdot \frac{f(a)g(a)}{e_{\beta\oplus\beta}(a,x_0)} + \frac{(b-t)(t-a)}{(b-a)^2} \cdot \frac{f(a)g(b)}{e_\beta(a,x_0)e_\beta(b,x_0)}$$
$$+ \frac{(b-t)(t-a)}{(b-a)^2} \cdot \frac{f(b)g(a)}{e_\beta(a,x_0)e_\beta(b,x_0)} + \frac{(t-a)^2}{(b-a)^2} \cdot \frac{f(b)g(b)}{e_{\beta\oplus\beta}(b,x_0)}$$
$$= \frac{(b-t)^2}{(b-a)^2} \cdot \frac{f(a)g(a)}{e_{\beta\oplus\beta}(a,x_0)} + \frac{(b-t)(t-a)}{(b-a)^2} \cdot \frac{f(a)g(b)+f(b)g(a)}{e_\beta(a,x_0)e_\beta(b,x_0)}$$
$$+ \frac{(t-a)^2}{(b-a)^2} \cdot \frac{f(b)g(b)}{e_{\beta\oplus\beta}(b,x_0)}, \quad t \in [a,b].$$

By taking the diamond-α integral in the last inequality, we find

$$\int_a^b f(t)g(t) \diamond_\alpha t$$
$$\leq \frac{f(a)g(a)}{(b-a)^2 e_{\beta\oplus\beta}(a,x_0)} \int_a^b (b-t)^2 \diamond_\alpha t$$
$$+ \frac{f(a)g(b)+f(b)g(a)}{e_\beta(a,x_0)e_\beta(b,x_0)(b-a)^2} \int_a^b (b-t)(t-a) \diamond_\alpha t$$
$$+ \frac{f(b)g(b)}{(b-a)^2 e_{\beta\oplus\beta}(b,x_0)} \int_a^b (t-a)^2 \diamond_\alpha t$$

$$= \frac{f(a)g(a)}{(b-a)e_{\beta\oplus\beta}(a,x_0)}(b^2 - 2bx_\alpha + x_{\alpha,\alpha})$$

$$+ \frac{f(a)g(b) + f(b)g(a)}{e_\beta(a,x_0)e_\beta(b,x_0)(b-a)}(-x_{\alpha,\alpha} + (a+b)x_\alpha - ab)$$

$$+ \frac{f(b)g(b)}{(b-a)e_{\beta\oplus\beta}(b,x_0)}(a^2 - 2ax_\alpha + x_{\alpha,\alpha}).$$

This completes the proof. □

Theorem 5.3. *Let $a > 0$, $f \in C^1_{rd}([a,b])$ and $|f^\Delta|$ be an exponentially convex function. Then*

$$\left| \frac{bf(b) - af(a)}{b-a} - \frac{1}{b-a} \int_a^b f(t)\Delta t \right| \leq \frac{\sigma(b)|f^\Delta(a)|}{(b-a)^2 e_\beta(a,x_0)} h_2(a,b)$$

$$+ \frac{\sigma(b)|f^\Delta(b)|}{(b-a)^2 e_\beta(b,x_0)} h_2(b,a).$$

Proof. Observe that

$$\int_a^b \sigma(t) f^\Delta(t)\Delta t = bf(b) - af(a) - \int_a^b f(t)\Delta t. \tag{5.7}$$

Hence, using $|f^\Delta|$ as an exponentially convex function, we obtain

$$\left| \frac{bf(b) - af(a)}{b-a} - \frac{1}{b-a} \int_a^b f(t)\Delta t \right|$$

$$= \frac{1}{b-a} \left| \int_a^b \sigma(t) f^\Delta(t)\Delta t \right|$$

$$\leq \frac{1}{b-a} \int_a^b \sigma(t) |f^\Delta(t)|\Delta t$$

$$\leq \frac{\sigma(b)}{b-a} \int_a^b |f^\Delta(t)|\Delta t$$

$$\leq \frac{\sigma(b)|f^\Delta(a)|}{(b-a)^2 e_\beta(a,x_0)} \int_a^b (b-t)\Delta t$$

$$+ \frac{\sigma(b)|f^{\Delta}(b)|}{(b-a)^2 e_{\beta}(b, x_0)} \int_a^b (t-a)\Delta t$$

$$= \frac{\sigma(b)|f^{\Delta}(a)|}{(b-a)^2 e_{\beta}(a, x_0)} h_2(a, b)$$

$$+ \frac{\sigma(b)|f^{\Delta}(b)|}{(b-a)^2 e_{\beta}(b, x_0)} h_2(b, a).$$

This completes the proof. □

Remark 5.2. Let $[a, b] \subset \mathbb{T} = \mathbb{R}$ and $x_0 = 0$. Then $f^{\Delta} = f'$ and $\sigma(b) = b$. Hence, according to the above theorem, we derive

$$\left| \frac{bf(b) - af(a)}{b-a} - \frac{1}{b-a} \int_a^b f(t)dt \right| \leq \frac{b}{2} \left(\frac{|f'(a)|}{e^{\beta a}} + \frac{|f'(b)|}{e^{\beta b}} \right),$$

for $f \in \mathcal{C}^1([a, b])$, whose $|f'|$ is an exponentially convex function in the classical sense.

Theorem 5.4. *Let $a > 0$, $f \in \mathcal{C}^1_{rd}([a, b])$, $|f^{\Delta}|^q$ be an exponentially convex function, where $q \geq 1$. Then*

$$\left| \frac{bf(b) - af(a)}{b-a} - \frac{1}{b-a} \int_a^b f(t)\Delta t \right|$$

$$\leq \frac{2^{\frac{1}{q}} \sigma(b)}{(b-a)^{\frac{1}{q}}} \left(\frac{|f^{\Delta}(a)|}{(e_{\beta}(a, x_0))^{\frac{1}{q}}} (h_2(a, b))^{\frac{1}{q}} \right.$$

$$\left. + \frac{|f^{\Delta}(b)|}{(e_{\beta}(b, x_0))^{\frac{1}{q}}} (h_2(b, a))^{\frac{1}{q}} \right).$$

Proof. Take $p \geq 1$ so that $\frac{1}{p} + \frac{1}{q} = 1$. By (5.7) and applying the Hölder inequality, we get

$$\left| \frac{bf(b) - af(a)}{b-a} - \frac{1}{b-a} \int_a^b f(t)\Delta t \right|$$

$$= \frac{1}{b-a} \left| \int_a^b \sigma(t) f^{\Delta}(t) \Delta t \right|$$

$$\leq \frac{1}{b-a} \int_a^b \sigma(t)|f^\Delta(t)|\Delta t$$

$$\leq \frac{1}{b-a} \left(\int_a^b (\sigma(t))^p \Delta t \right)^{\frac{1}{p}} \left(\int_a^b |f^\Delta(t)|^q \Delta t \right)^{\frac{1}{q}}$$

$$\leq \frac{(b-a)^{\frac{1}{p}} \sigma(b)}{b-a} \left(\frac{|f^\Delta(a)|^q}{(b-a)e_\beta(a,x_0)} \int_a^b (b-t)\Delta t \right.$$

$$\left. + \frac{|f^\Delta(b)|^q}{(b-a)e_\beta(b,x_0)} \int_a^b (t-a)\Delta t \right)^{\frac{1}{q}}$$

$$= \frac{(b-a)^{\frac{1}{p}} \sigma(b)}{b-a} \left(\frac{|f^\Delta(a)|^q}{(b-a)e_\beta(a,x_0)} h_2(a,b) \right.$$

$$\left. + \frac{|f^\Delta(b)|^q}{(b-a)e_\beta(b,x_0)} h_2(b,a) \right)^{\frac{1}{q}}$$

$$\leq \frac{2^{\frac{1}{q}} \sigma(b)}{(b-a)^{\frac{1}{q}}} \left(\frac{|f^\Delta(a)|}{(b-a)^{\frac{1}{q}} (e_\beta(a,x_0))^{\frac{1}{q}}} (h_2(a,b))^{\frac{1}{q}} \right.$$

$$\left. + \frac{|f^\Delta(b)|}{(b-a)^{\frac{1}{q}} (e_\beta(b,x_0))^{\frac{1}{q}}} (h_2(b,a))^{\frac{1}{q}} \right)$$

$$= \frac{2^{\frac{1}{q}} \sigma(b)}{(b-a)^{\frac{2}{q}}} \left(\frac{|f^\Delta(a)|}{(e_\beta(a,x_0))^{\frac{1}{q}}} (h_2(a,b))^{\frac{1}{q}} + \frac{|f^\Delta(b)|}{(e_\beta(b,x_0))^{\frac{1}{q}}} (h_2(b,a))^{\frac{1}{q}} \right).$$

This completes the proof. $\qquad\square$

Now, we provide another result which is the special case of Theorem 5.4 for $|f^\Delta| \geq 1$, but by using another approach.

Theorem 5.5. *Let $a > 0$, $f \in C_{rd}^1([a,b])$, $|f^\Delta|^q$ be an exponentially convex function, where $q \geq 1$, and $|f^\Delta| \geq 1$ on $[a,b]$. Then*

$$\left| \frac{bf(b) - af(a)}{b-a} - \frac{1}{b-a} \int_a^b f(t)\Delta t \right|$$

$$\leq \frac{\sigma(b)}{(b-a)^2} \left(\frac{|f^\Delta(a)|^q}{e_\beta(a,x_0)} h_2(a,b) + \frac{|f^\Delta(b)|^q}{e_\beta(b,x_0)} h_2(b,a) \right).$$

Proof.   By (5.7), we find

$$\left| \frac{bf(b) - af(a)}{b - a} - \frac{1}{b - a} \int_a^b f(t)\Delta t \right|$$

$$= \frac{1}{b - a} \left| \int_a^b \sigma(t) f^\Delta(t)\Delta t \right|$$

$$\leq \frac{1}{b - a} \int_a^b \sigma(t) |f^\Delta(t)|\Delta t$$

$$\leq \frac{\sigma(b)}{b - a} \int_a^b |f^\Delta(t)|\Delta t$$

$$\leq \frac{\sigma(b)}{b - a} \int_a^b |f^\Delta(t)|^q \Delta t$$

$$\leq \frac{\sigma(b)}{b - a} \left(\frac{|f^\Delta(a)|^q}{(b - a)e_\beta(a, x_0)} \int_a^b (b - t)\Delta t \right.$$

$$\left. + \frac{|f^\Delta(b)|^q}{(b - a)e_\beta(b, x_0)} \int_a^b (t - a)\Delta t \right)$$

$$= \frac{\sigma(b)}{b - a} \left(\frac{|f^\Delta(a)|^q}{(b - a)e_\beta(a, x_0)} h_2(a, b) + \frac{|f^\Delta(b)|^q}{(b - a)e_\beta(b, x_0)} h_2(b, a) \right).$$

This completes the proof.   □

For a function $f \in C_{rd}^{n-1}([a, b])$, define

$$I(a, b, n, f) = \int_a^b f(s)\Delta s - \sum_{k=1}^n h_k(b, a) f^{\Delta^{k-1}}(a),$$

where $h_k(b, a)$, $k \in \{1, \ldots, n\}$, are the coefficients of the Taylor expansion.

We prove some upper bounds for this identity integral in the following theorem.

Theorem 5.6. *Let $a > 0$, $f \in C^n_{rd}([a,b])$, $|f^{\Delta^n}|^q$ be an exponentially convex function, where $q \geq 1$, and $|f^{\Delta^n}| \geq 1$ on $[a,b]$. Then*

$$\frac{1}{b-a}|I(a,b,n,f)|$$

$$\leq \frac{(b-a)^{n-2}}{n!}\left(\frac{|f^{\Delta^n}(a)|^q}{e_\beta(a,x_0)}h_2(a,b) + \frac{|f^{\Delta^n}(b)|^q}{e_\beta(b,x_0)}h_2(b,a)\right).$$

Proof. By the Taylor formula, we have

$$\frac{1}{b-a}|I(a,b,n,f)| = \frac{1}{b-a}\left|\int_a^{\rho^n(b)} h_n(b,\sigma(\tau))f^{\Delta^n}(\tau)\Delta\tau\right|$$

$$\leq \frac{1}{b-a}\int_a^{\rho^n(b)} h_n(b,\sigma(\tau))|f^{\Delta^n}(\tau)|\Delta\tau$$

$$\leq \frac{1}{b-a}\int_a^b \frac{(b-\sigma(\tau))^n}{n!}|f^{\Delta^n}(\tau)|\Delta\tau$$

$$\leq \frac{(b-a)^{n-1}}{n!}\int_a^b |f^{\Delta^n}(\tau)|\Delta\tau. \tag{5.8}$$

Hence, using $|f^{\Delta^n}| \geq 1$ on $[a,b]$ and $|f^{\Delta^n}|^q$ as an exponentially convex function, we get

$$\frac{1}{b-a}|I(a,b,n,f)|$$

$$\leq \frac{(b-a)^{n-1}}{n!}\int_a^b |f^{\Delta^n}(\tau)|^q\Delta\tau$$

$$\leq \frac{(b-a)^{n-1}}{n!}\left(\frac{|f^{\Delta^n}(a)|^q}{(b-a)e_\beta(a,x_0)}\int_a^b (b-t)\Delta t\right.$$

$$\left. + \frac{|f^{\Delta^n}(b)|^q}{(b-a)e_\beta(b,x_0)}\int_a^b (t-a)\Delta t\right)$$

$$= \frac{(b-a)^{n-2}}{n!}\left(\frac{|f^{\Delta^n}(a)|^q}{e_\beta(a,x_0)}h_2(a,b) + \frac{|f^{\Delta^n}(b)|^q}{e_\beta(b,x_0)}h_2(b,a)\right).$$

This completes the proof. $\square$

Theorem 5.7. *Let $a > 0$, $f \in C_{rd}^n([a, b])$, $|f^{\Delta^n}|^q$ is an exponentially convex function, where $q \geq 1$. Then*

$$\frac{1}{b-a}|I(a, b, n, f)| \leq \frac{2^{\frac{1}{q}}(b-a)^{n-\frac{2}{q}}}{n!}\left(\frac{|f^{\Delta^n}(a)|}{(e_\beta(a, x_0))^{\frac{1}{q}}}(h_2(a, b))^{\frac{1}{q}}\right.$$

$$\left. + \frac{|f^{\Delta^n}(b)|}{(e_\beta(b, x_0))^{\frac{1}{q}}}(h_2(b, a))^{\frac{1}{q}}\right).$$

Proof.   Let $p \geq 1$ be chosen so that $\frac{1}{p} + \frac{1}{q} = 1$. Applying (5.8) and using $|f^{\Delta^n}|^q$ as an exponentially convex function, we get

$$\frac{1}{b-a}|I(a, b, n, f) \leq \frac{(b-a)^{n-1}}{n!}\int_a^b |f^{\Delta^n}(t)|\Delta t$$

$$\leq \frac{(b-a)^{n-1}}{(n-1)!}\left(\int_a^b \Delta t\right)^{\frac{1}{p}}\left(\int_a^b |f^{\Delta^n}(t)|^q \Delta t\right)^{\frac{1}{q}}$$

$$= \frac{(b-a)^{n-1+\frac{1}{p}}}{n!}\left(\frac{|f^{\Delta^n}(a)|^q}{(b-a)e_\beta(a, x_0)}\int_a^b(b-t)\Delta t\right.$$

$$\left. + \frac{|f^{\Delta^n}(b)|^q}{(b-a)e_\beta(b, x_0)}\int_a^b(t-a)\Delta t\right)^{\frac{1}{q}}$$

$$= \frac{(b-a)^{n-\frac{1}{q}}}{n!}\left(\frac{|f^{\Delta^n}(a)|^q}{(b-a)e_\beta(a, x_0)}h_2(a, b)\right.$$

$$\left. + \frac{|f^{\Delta^n}(b)|^q}{(b-a)e_\beta(b, x_0)}h_2(b, a)\right)^{\frac{1}{q}}$$

$$\leq \frac{2^{\frac{1}{q}}(b-a)^{n-\frac{1}{q}}}{n!}\left(\frac{|f^{\Delta^n}(a)|}{(b-a)^{\frac{1}{q}}(e_\beta(a, x_0))^{\frac{1}{q}}}(h_2(a, b))^{\frac{1}{q}}\right.$$

$$\left. + \frac{|f^{\Delta^n}(b)|}{(b-a)^{\frac{1}{q}}(e_\beta(b, x_0))^{\frac{1}{q}}}(h_2(b, a))^{\frac{1}{q}}\right)$$

$$= \frac{2^{\frac{1}{q}}(b-a)^{n-\frac{2}{q}}}{n!} \left(\frac{|f^{\Delta^n}(a)|}{(e_\beta(a,x_0))^{\frac{1}{q}}} (h_2(a,b))^{\frac{1}{q}} \right.$$

$$\left. + \frac{|f^{\Delta^n}(b)|}{(e_\beta(b,x_0))^{\frac{1}{q}}} (h_2(b,a))^{\frac{1}{q}} \right).$$

This completes the proof. □

5.2 Examples

In this section, we illustrate our main results in this chapter on the time scales $\mathbb{T} = h\mathbb{Z}$, $h > 0$, and $\mathbb{T} = r^{\mathbb{N}_0}$, $r > 1$, $x_0 \leq a < b$, $x_0, a, b \in \mathbb{T}$.

5.2.1 *The time scale* $\mathbb{T} = h\mathbb{Z}$, $h > 0$

Let $\mathbb{T} = h\mathbb{Z}$, $h > 0$. In this case, we have

$$\sigma(t) = t + h,$$

$$\mu(t) = h,$$

$$\rho(t) = t - h,$$

$$\nu(t) = t - \rho(t)$$

$$= h, \quad t \in \mathbb{T},$$

$$x_\alpha = \frac{1}{2}(b + a + h(1 - 2\alpha)),$$

$$x_{\alpha,\alpha} = \frac{1}{6} \left(2b^2 + 2ab + 2a^2 + h^2 + 3h(b+a)(1 - 2\alpha) \right),$$

$$h_2(a,b) = \frac{b-a}{2}(b - a + h),$$

$$h_2(b,a) = \frac{b-a}{2}(b - a - h),$$

$$\beta \oplus \beta = \beta(2 + h\beta),$$

$$e_\beta(a, x_0) = (1 + \beta h)^{\frac{a-x_0}{h}},$$

$$e_\beta(b, x_0) = (1 + \beta h)^{\frac{b-x_0}{h}},$$

$$e_{\beta \oplus \beta}(a, x_0) = (1 + \beta h(2 + h\beta))^{\frac{a-x_0}{h}},$$

$$e_{\beta \oplus \beta}(b, x_0) = (1 + \beta h(2 + h\beta))^{\frac{b-x_0}{h}}.$$

Then, by Theorems 5.1–5.7, we get the following corollaries, respectively.

Corollary 5.1. *Let $\beta \in \mathcal{R}_+$, $f : [a, b] \to \mathbb{R}$, $f \in C([a, b])$ be positive and exponentially convex function. Then*

$$\frac{1}{b-a} \int_a^b f(t) \diamond_\alpha t \le \frac{b-a-h(1-2\alpha)}{2(b-a)} \cdot \frac{f(a)}{(1+\beta h)^{\frac{a-x_0}{h}}}$$

$$+ \frac{b-a+h(1-2\alpha)}{2(b-a)} \cdot \frac{f(b)}{(1+\beta h)^{\frac{b-x_0}{h}}}.$$

If, in addition, f is convex, then

$$f\left(\frac{1}{2}(b+a+h(1-2\alpha))\right) \le \frac{1}{b-a} \int_a^b f(t) \diamond_\alpha t. \qquad (5.9)$$

Corollary 5.2. *Let $f, g : [a, b] \to [0, \infty)$, $f, g \in C([a, b])$ be exponentially convex functions. Then*

$$\int_a^b f(t)g(t) \diamond_\alpha t \le \frac{2b - 2a + \frac{h^2}{b-a} - 3h(1-2\alpha)}{6\left(1 + \beta h(2 + h\beta)\right)^{\frac{a-x_0}{h}}} f(a)g(a)$$

$$+ \frac{2b - 2a + \frac{h^2}{b-a} + 3h(1-2\alpha)}{6\left(1 + \beta h(2 + h\beta)\right)^{\frac{b-x_0}{h}}} f(b)g(b)$$

$$+ \frac{f(a)g(b) + f(b)g(a)}{6(b-a)\left(1 + \beta h\right)^{\frac{a+b-2x_0}{h}}} \left((b-a)^2 - h^2\right).$$

Corollary 5.3. *Let* $a > 0$, $f \in \mathcal{C}^1_{rd}([a,b])$ *and* $|f^\Delta|$ *be exponentially convex function. Then*

$$\left| \frac{bf(b) - af(a)}{b - a} - \frac{1}{b - a} \int_a^b f(t)\Delta t \right|$$

$$\leq \frac{b + h}{2(b - a)(1 + \beta h)^{\frac{a - x_0}{h}}} \left(|f^\Delta(a)|(b - a + h) \right.$$

$$\left. + \frac{|f^\Delta(b)|}{(1 + \beta h)^{\frac{b - a}{h}}}(b - a - h) \right).$$

Corollary 5.4. *Let* $a > 0$, $f \in \mathcal{C}^1_{rd}([a,b])$, $|f^\Delta|^q$ *be an exponentially convex function, where* $q \geq 1$. *Then*

$$\left| \frac{bf(b) - af(a)}{b - a} - \frac{1}{b - a} \int_a^b f(t)\Delta t \right|$$

$$\leq \frac{b + h}{(1 + \beta h)^{\frac{a - x_0}{qh}}} \left(|f^\Delta(a)|(b - a + h)^{\frac{1}{q}} \right.$$

$$\left. + \frac{|f^\Delta(b)|}{(1 + \beta h)^{\frac{b - a}{qh}}}(b - a - h)^{\frac{1}{q}} \right).$$

Corollary 5.5. *Let* $a > 0$, $f \in \mathcal{C}^1_{rd}([a,b])$, $|f^\Delta|^q$ *be an exponentially convex function, where* $q \geq 1$, *and* $|f^\Delta| \geq 1$ *on* $[a,b]$. *Then*

$$\left| \frac{bf(b) - af(a)}{b - a} - \frac{1}{b - a} \int_a^b f(t)\Delta t \right|$$

$$\leq \frac{b + h}{2(b - a)(1 + \beta h)^{\frac{a - x_0}{h}}} \left(|f^\Delta(a)|^q(b - a + h) \right.$$

$$\left. + \frac{|f^\Delta(b)|^q}{(1 + \beta h)^{\frac{b - a}{h}}}(b - a - h) \right).$$

Corollary 5.6. *Let $a > 0$, $f \in C_{rd}^n([a,b])$, $|f^{\Delta^n}|^q$ be an exponentially convex function, where $q \geq 1$, and $|f^{\Delta^n}| \geq 1$ on $[a,b]$. Then*

$$\frac{1}{b-a} \; |I(a,b,n,f)|$$

$$\leq \frac{(b-a)^{n-1}}{2n!(1+\beta h)^{\frac{a-x_0}{h}}} \left(|f^{\Delta^n}(a)|^q (b-a+h) \right.$$

$$\left. + \frac{|f^{\Delta^n}(b)|^q}{(1+\beta h)^{\frac{b-a}{h}}} (b-a-h) \right).$$

Corollary 5.7. *Let $a > 0$, $f \in C_{rd}^n([a,b])$, $|f^{\Delta^n}|^q$ be an exponentially convex function, where $q \geq 1$. Then*

$$\frac{1}{b-a} \; |I(a,b,n,f)|$$

$$\leq \frac{(b-a)^{n-\frac{1}{q}}}{n!(1+\beta h)^{\frac{a-x_0}{qh}}} \left(|f^{\Delta^n}(a)|(b-a+h)^{\frac{1}{q}} \right.$$

$$\left. + \frac{|f^{\Delta^n}(b)|}{(1+\beta h)^{\frac{b-a}{qh}}} (b-a-h)^{\frac{1}{q}} \right).$$

5.2.2   The time scale $\mathbb{T} = r^{\mathbb{N}_0}$, $r > 1$

Let $\mathbb{T} = r^{\mathbb{N}_0}$, $r > 1$, $1 < x_0 \leq a < b$, $x_0, a, b \in \mathbb{T}$. In this case, we have

$$\sigma(t) = rt,$$

$$\mu(t) = (r-1)t, \quad t \in \mathbb{T},$$

$$\rho(t) = \frac{1}{r}t,$$

$$\nu(t) = t - \rho(t)$$

$$= \frac{r-1}{r}t, \quad t \in \mathbb{T}, \quad t > 1,$$

$$x_\alpha = \frac{b+a}{r+1}(r + \alpha(1-r)),$$

$$x_{\alpha,\alpha} = \frac{a^2 + ab + b^2}{r^2 + r + 1}(r^2 + \alpha(1 - r^2)),$$

$$h_2(a, b) = \frac{(b - a)(br - a)}{r + 1},$$

$$h_2(b, a) = \frac{b - a}{r + 1}(b - ar),$$

$$\beta \oplus \beta = \beta(2 + (r - 1)t\beta),$$

$$e_\beta(a, x_0) = \prod_{s \in [x_0, a)} (1 + (r - 1)\beta s),$$

$$e_\beta(b, x_0) = \prod_{s \in [x_0, b)} (1 + (r - 1)\beta s),$$

$$e_{\beta \oplus \beta}(a, x_0) = \prod_{s \in [x_0, a)} (1 + (r - 1)\beta(2 + (r - 1)s\beta)s),$$

$$e_{\beta \oplus \beta}(b, x_0) = \prod_{s \in [x_0, b)} (1 + (r - 1)\beta(2 + (r - 1)s\beta)s).$$

Then, by Theorems 5.1–5.7, we get the following corollaries, respectively.

Corollary 5.8. *Let* $\beta \in \mathcal{R}_+$, $f : [a, b] \to \mathbb{R}$, $f \in \mathcal{C}([a, b])$ *be positive and exponentially convex function. Then*

$$\frac{1}{b - a} \int_a^b f(t) \diamond_\alpha t \le \frac{1}{(b - a)(1 + r) \prod_{s \in [x_0, a)} (1 + (r - 1)\beta s)}$$

$$\times \left((b - ar - (b + a)\alpha(1 - r))f(a) + (br - a \right.$$

$$\left. + \alpha(b + a)(1 - r)) \frac{f(b)}{\prod_{s \in [a, b)} (1 + (r - 1)\beta s)} \right).$$

If, in addition, f *is convex, then*

$$f\left(\frac{b + a}{r + 1}(r + \alpha(1 - r)) \right) \le \frac{1}{b - a} \int_a^b f(t) \diamond_\alpha t. \tag{5.10}$$

Corollary 5.9. *Let $f, g : [a, b] \to [0, \infty)$, $f, g \in \mathcal{C}([a, b])$ be exponentially convex functions. Then*

$$\int_a^b f(t)g(t) \diamond_\alpha t$$

$$\leq \frac{f(a)g(a)}{(b-a)\prod_{s \in [x_0, a)}(1 + (r-1)\beta(2 + (r-1)s\beta)s)}$$

$$\times \left(b^2 - \frac{2b(b+a)}{r+1}(r + \alpha(1-r)) \right.$$

$$\left. + \frac{a^2 + ab + b^2}{r^2 + r + 1}(r^2 + \alpha(1 - r^2)) \right)$$

$$+ \frac{f(a)g(b) + f(b)g(a)}{(b-a)\prod_{s \in [x_0, a)}(1 + (r-1)\beta s)\prod_{s \in [x_0, b)}(1 + (r-1)\beta s)}$$

$$\times \left(-\frac{a^2 + ab + b^2}{r^2 + r + 1}(r^2 + \alpha(1 - r^2)) \right.$$

$$\left. + \frac{(a+b)^2}{r+1}(r + \alpha(1-r)) - ab \right)$$

$$+ \frac{f(b)g(b)}{(b-a)\prod_{s \in [x_0, b)}(1 + (r-1)\beta(2 + (r-1)s\beta)s)}$$

$$\times \left(a^2 - \frac{2a(b+a)}{r+1}(r + \alpha(1-r)) \right.$$

$$\left. + \frac{a^2 + ab + b^2}{r^2 + r + 1}(r^2 + \alpha(1 - r^2)) \right).$$

Corollary 5.10. *Let $a > 0$, $f \in \mathcal{C}^1_{rd}([a, b])$ and $|f^\Delta|$ be an exponentially convex function. Then*

$$\left| \frac{bf(b) - af(a)}{b - a} - \frac{1}{b - a} \int_a^b f(t)\Delta t \right|$$

$$\leq \frac{rb}{(r+1)(b-a)\prod_{s \in [x_0, a)}(1 + (r-1)\beta s)}$$

$$\times \left(|f^\Delta(a)|(br - a) + \frac{|f^\Delta(b)|}{\prod_{s \in [a, b)}(1 + (r-1)\beta s)}(b - ar) \right).$$

Corollary 5.11. *Let $a > 0$, $f \in C_{rd}^1([a, b])$, $|f^\Delta|^q$ be an exponentially convex function, where $q \geq 1$. Then*

$$\left| \frac{bf(b) - af(a)}{b - a} - \frac{1}{b - a} \int_a^b f(t)\Delta t \right|$$

$$\leq \frac{2^{\frac{1}{q}} rb}{(r+1)^{\frac{1}{q}} \left(\prod_{s \in [x_0, a)} (1 + (r-1)\beta s) \right)^{\frac{1}{q}}}$$

$$\times \left(|f^\Delta(a)|(br - a)^{\frac{1}{q}} + \frac{|f^\Delta(b)|}{\left(\prod_{s \in [a, b)} (1 + (r-1)\beta s) \right)^{\frac{1}{q}}} (b - ar)^{\frac{1}{q}} \right).$$

Corollary 5.12. *Let $a > 0$, $f \in C_{rd}^1([a, b])$, $|f^\Delta|^q$ be an exponentially convex function, where $q \geq 1$, and $|f^\Delta| \geq 1$ on $[a, b]$. Then*

$$\left| \frac{bf(b) - af(a)}{b - a} - \frac{1}{b - a} \int_a^b f(t)\Delta t \right|$$

$$\leq \frac{rb}{(r+1)(b-a) \prod_{s \in [x_0, a)} (1 + (r-1)\beta s)}$$

$$\times \left(|f^\Delta(a)|^q (br - a) + \frac{|f^\Delta(b)|^q}{\prod_{s \in [a, b)} (1 + (r-1)\beta s)} (b - ar) \right).$$

Corollary 5.13. *Let $a > 0$, $f \in C_{rd}^n([a, b])$, $|f^{\Delta^n}|^q$ be an exponentially convex function, where $q \geq 1$, and $|f^{\Delta^n}| \geq 1$ on $[a, b]$. Then*

$$\frac{1}{b - a} |I(a, b, n, f)|$$

$$\leq \frac{(b-a)^{n-1}}{(r+1)n! \prod_{s \in [x_0, a)} (1 + (r-1)\beta s)}$$

$$\times \left(|f^{\Delta^n}(a)|^q (br - a) + \frac{|f^{\Delta^n}(b)|^q}{\prod_{s \in [a, b)} (1 + (r-1)\beta s)} (b - ar) \right).$$

Corollary 5.14. *Let* $a > 0$, $f \in C_{rd}^n([a,b])$, $|f^{\Delta^n}|^q$ *be an exponentially convex function, where* $q \geq 1$. *Then*

$$\frac{1}{b-a}|I(a,b,n,f)|$$

$$\leq \frac{2^{\frac{1}{q}}(b-a)^{n-\frac{1}{q}}}{n!(r+1)^{\frac{1}{q}}\left(\prod_{s\in[x_0,a)}(1+(r-1)\beta s)\right)^{\frac{1}{q}}}$$

$$\times \left(|f^{\Delta^n}(a)|(br-a)^{\frac{1}{q}} + \frac{|f^{\Delta^n}(b)}{\left(\prod_{s\in[a,b)}(1+(r-1)\beta s)\right)^{\frac{1}{q}}}(b-ar)^{\frac{1}{q}}\right).$$

5.3   Ostrowski-Type Inequalities

Definition 5.2. Let $s \in (0,1]$. A function $f : [a,b] \to \mathbb{R}$ is called exponentially s-convex in the second sense if

$$f(t) \leq \left(\frac{b-t}{b-a}\right)^s \frac{f(a)}{e_\beta(a,x_0)} + \left(\frac{t-a}{b-a}\right)^s \frac{f(b)}{e_\beta(b,x_0)} \qquad (5.11)$$

for any $t \in [a,b]$ and for some $\beta \in \mathcal{R}_+$. If (5.11) holds in the reverse sense, then we say that f is exponentially s-concave in the second sense.

Throughout this section, without loss of generality, suppose that $s \in (0,1)$. We start with the following technical lemma.

Lemma 5.1. *We have*

$$\int_a^b (b-t)^s \Delta t \leq (b-a)^{1-s}(h_2(a,b))^s,$$

$$\int_a^b (t-a)^s \Delta t \leq (b-a)^{1-s}(h_2(b,a))^s.$$

Proof. We have

$$\int_a^b (b-t)^s \Delta t \leq \left(\int_a^b (b-t) \Delta t \right)^s \left(\int_a^b \Delta t \right)^{1-s}$$

$$= (b-a)^{1-s} \left(\int_b^a (t-b) \Delta t \right)^s$$

$$= (b-a)^{1-s} (h_2(a,b))^s$$

and

$$\int_a^b (t-a)^s \Delta t \leq \left(\int_a^b (t-a) \Delta t \right)^s \left(\int_a^b \Delta t \right)^{1-s}$$

$$= (b-a)^{1-s} (h_2(b,a))^s.$$

This completes the proof. □

Theorem 5.8. *Let $f : [a,b] \to \mathbb{R}$ be an exponentially s-convex function. Then*

$$\frac{1}{b-a} \int_a^b f(t) \Delta t \leq \frac{f(a)}{e_\beta(a,x_0)(b-a)^{2s}} (h_2(a,b))^s$$

$$+ \frac{f(b)}{e_\beta(b,x_0)(b-a)^{2s}} (h_2(b,a))^s.$$

Proof. Applying the definition for exponentially s-convex function in the second sense and Lemma 5.1, we find

$$\int_a^b f(t) \Delta t \leq \frac{f(a)}{e_\beta(a,x_0)(b-a)^s} \int_a^b (b-t)^s \Delta t$$

$$+ \frac{f(b)}{e_\beta(b,x_0)(b-a)^s} \int_a^b (t-a)^s \Delta t$$

$$\leq \frac{f(a)(b-a)}{e_\beta(a,x_0)(b-a)^{2s}} (h_2(a,b))^s$$

$$+ \frac{f(b)(b-a)}{e_\beta(b,x_0)(b-a)^{2s}} (h_2(b,a))^s.$$

Hence,

$$\frac{1}{b-a}\int_a^b f(t)\Delta t \leq \frac{f(a)}{e_\beta(a,x_0)(b-a)^{2s}}(h_2(a,b))^s$$

$$+\frac{f(b)}{e_\beta(b,x_0)(b-a)^{2s}}(h_2(b,a))^s.$$

This completes the proof.   □

Theorem 5.9. *Let* $f:[a,b]\to\mathbb{R}$, $|f|\geq 1$ *on* $[a,b]$ *and* $|f|^q$ *be an exponentially s-convex function for some* $q\geq 1$. *Then*

$$\frac{1}{b-a}\int_a^b f(t)\Delta t \leq \frac{|f(a)|^q}{e_\beta(a,x_0)(b-a)^{2s}}(h_2(a,b))^s$$

$$+\frac{|f(b)|^q}{e_\beta(b,x_0)(b-a)^{2s}}(h_2(b,a))^s.$$

Proof.   Using Lemma 5.1, we arrive at

$$\int_a^b f(t)\Delta t \leq \int_a^b |f(t)|\Delta t$$

$$\leq \int_a^b |f(t)|^q \Delta t$$

$$\leq \frac{|f(a)|^q}{e_\beta(a,x_0)(b-a)^s}\int_a^b (b-t)^s\Delta t$$

$$+\frac{|f(b)|^q}{e_\beta(b,x_0)(b-a)^s}\int_a^b (t-a)^s\Delta t$$

$$\leq \frac{|f(a)|^q(b-a)}{e_\beta(a,x_0)(b-a)^{2s}}(h_2(a,b))^s$$

$$+\frac{|f(b)|^q(b-a)}{e_\beta(b,x_0)(b-a)^{2s}}(h_2(b,a))^s.$$

Hence,

$$\frac{1}{b-a}\int_a^b f(t)\Delta t \leq \frac{|f(a)|^q}{e_\beta(a,x_0)(b-a)^{2s}}(h_2(a,b))^s$$

$$+ \frac{|f(b)|^q}{e_\beta(b,x_0)(b-a)^{2s}}(h_2(b,a))^s.$$

This completes the proof. □

Theorem 5.10. *Let* $f : [a,b] \to \mathbb{R}$, $|f|^q$ *be an exponentially s-convex function for some* $q \geq 1$. *Then*

$$\frac{1}{b-a}\int_a^b f(t)\Delta t \leq \frac{2^{\frac{1}{q}}}{(b-a)^{\frac{2s}{q}}}\left(\frac{|f(a)|}{(e_\beta(a,x_0))^{\frac{1}{q}}}(h_2(a,b))^{\frac{s}{q}}\right.$$

$$\left.+ \frac{|f(b)|}{(e_\beta(b,x_0))^{\frac{1}{q}}}(h_2(b,a))^{\frac{s}{q}}\right).$$

Proof. We have

$$\int_a^b f(t)\Delta t \leq \int_a^b |f(t)|\Delta t$$

$$\leq (b-a)^{\frac{1}{p}}\left(\int_a^b |f(t)|^q\Delta t\right)^{\frac{1}{q}}$$

$$\leq (b-a)^{\frac{1}{p}}\left(\frac{|f(a)|^q}{(b-a)^s e_\beta(a,x_0)}\int_a^b (b-t)^s\Delta t\right.$$

$$\left.+ \frac{|f(b)|^q}{(b-a)^s e_\beta(b,x_0)}\int_a^b (a-t)^s\Delta t\right)^{\frac{1}{q}}$$

$$\leq (b-a)^{\frac{1}{p}}\left(\frac{(b-a)|f(a)|^q}{(b-a)^{2s}e_\beta(a,x_0)}(h_2(a,b))^s\right.$$

$$\left.+ \frac{(b-a)|f(b)|^q}{(b-a)^{2s}e_\beta(b,x_0)}(h_2(b,a))^s\right)^{\frac{1}{q}}$$

$$\leq 2^{\frac{1}{q}}(b-a)^{\frac{1}{p}}\left((b-a)^{\frac{1}{q}}\frac{|f(a)|}{(b-a)^{\frac{2s}{q}}(e_\beta(a,x_0))^{\frac{1}{q}}}(h_2(a,b))^{\frac{s}{q}}\right.$$

$$\left.+\frac{(b-a)^{\frac{1}{q}}|f(b)|}{(b-a)^{\frac{2s}{q}}(e_\beta(b,x_0))^{\frac{1}{q}}}(h_2(b,a))^{\frac{s}{q}}\right),$$

whereupon we get the desired result. This completes the proof. □

Lemma 5.2. *We have*

$$\int_a^b \sigma(t)(b-t)^s \Delta t \leq (b-a)^s(g_2(b,a)+a(b-a)),$$

$$\int_a^b \sigma(t)(t-a)^s \Delta t \leq (b-a)^s(g_2(b,a)+a(b-a)).$$

Proof.   We have

$$\int_a^b \sigma(t)(b-t)^s \Delta t \leq (b-a)^s \int_a^b \sigma(t)\Delta t$$

$$= (b-a)^s \int_a^b (\sigma(t)-a)\Delta t + a(b-a)^{1+s}$$

$$= (b-a)^s g_2(b,a) + a(b-a)^{1+s}$$

$$= (b-a)^s(g_2(b,a)+a(b-a))$$

and

$$\int_a^b \sigma(t)(t-a)^s \Delta t \leq (b-a)^s \int_a^b \sigma(t)\Delta t$$

$$\leq (b-a)^s(g_2(b,a)+a(b-a)).$$

This completes the proof.   □

Let

$$G(a,b) = (b-a)^s(g_2(b,a) + a(b-a)).$$

Then, by Lemma 5.2, we have

$$\int_a^b \sigma(t)(b-t)^s \Delta t \le G(a,b), \quad \int_a^b \sigma(t)(t-a)^s \Delta t \le G(a,b).$$

Theorem 5.11. *Let* $f \in C_{rd}^1([a,b])$. *If* $|f^\Delta|$ *is exponentially s-convex function on* $[a,b]$, *then*

$$\left| \frac{bf(b) - af(a)}{b-a} - \frac{1}{b-a} \int_a^b f(t)\Delta t \right|$$

$$\le \frac{G(a,b)}{(b-a)^{1+s}} \left(\frac{|f^\Delta(a)|}{e_\beta(a,x_0)} + \frac{|f^\Delta(b)|}{e_\beta(b,x_0)} \right).$$

Proof. Using $|f^\Delta|$ as exponentially s-convex function and Lemma 5.2, we get

$$\left| \frac{bf(b) - af(a)}{b-a} - \frac{1}{b-a} \int_a^b f(t)\Delta t \right|$$

$$= \frac{1}{b-a} \left| \int_a^b \sigma(t)f^\Delta(t)\Delta t \right|$$

$$\le \frac{1}{b-a} \int_a^b \sigma(t)|f^\Delta(t)|\Delta t$$

$$\le \frac{|f^\Delta(a)|}{(b-a)^{1+s}e_\beta(a,x_0)} \int_a^b \sigma(t)(b-t)^s \Delta t$$

$$+ \frac{|f^\Delta(b)|}{(b-a)^{1+s}e_\beta(b,x_0)} \int_a^b \sigma(t)(t-a)^s \Delta t$$

$$\le \frac{G(a,b)}{(b-a)^{1+s}} \left(\frac{|f^\Delta(a)|}{e_\beta(a,x_0)} + \frac{|f^\Delta(b)|}{e_\beta(b,x_0)} \right).$$

This completes the proof. $\qquad\square$

Theorem 5.12. *Let $f \in C^1_{rd}([a,b])$. If $|f^\Delta| \geq 1$ on $[a,b]$ and $|f^\Delta|^q$ is an exponentially s-convex function on $[a,b]$ for some $q \geq 1$, then*

$$\left| \frac{bf(b) - af(a)}{b - a} - \frac{1}{b - a} \int_a^b f(t)\Delta t \right|$$

$$\leq \frac{G(a,b)}{(b-a)^{1+s}} \left(\frac{|f^\Delta(a)|^q}{e_\beta(a,x_0)} + \frac{|f^\Delta(b)|^q}{e_\beta(b,x_0)} \right).$$

Proof. We have

$$\left| \frac{bf(b) - af(a)}{b - a} - \frac{1}{b - a} \int_a^b f(t)\Delta t \right|$$

$$= \frac{1}{b - a} \left| \int_a^b \sigma(t) f^\Delta(t)\Delta t \right|$$

$$\leq \frac{1}{b - a} \int_a^b \sigma(t) |f^\Delta(t)|\Delta t$$

$$\leq \frac{1}{b - a} \int_a^b \sigma(t) |f^\Delta(t)|^q \Delta t$$

$$\leq \frac{|f^\Delta(a)|^q}{(b-a)^{1+s} e_\beta(a,x_0)} \int_a^b \sigma(t)(b-t)^s \Delta t$$

$$+ \frac{|f^\Delta(b)|^q}{(b-a)^{1+s} e_\beta(b,x_0)} \int_a^b \sigma(t)(t-a)^s \Delta t$$

$$\leq \frac{G(a,b)}{(b-a)^{1+s}} \left(\frac{|f^\Delta(a)|^q}{e_\beta(a,x_0)} + \frac{|f^\Delta(b)|^q}{e_\beta(b,x_0)} \right).$$

This completes the proof. □

Theorem 5.13. *Let $f \in C^1_{rd}([a,b])$ and $|f^\Delta|^q$ be an exponentially s-convex function for some $q \geq 1$. Then*

$$\left| \frac{bf(b) - af(a)}{b - a} - \frac{1}{b - a} \int_a^b f(t)\Delta t \right|$$

$$\leq \frac{2^{\frac{1}{q}} \sigma(b)}{(b-a)^{\frac{2s}{q}}} \left(\frac{|f^\Delta(a)|}{(e_\beta(a,x_0))^{\frac{1}{q}}} (h_2(a,b))^{\frac{s}{q}} + \frac{|f^\Delta(b)|}{(e_\beta(b,x_0))^{\frac{1}{q}}} (h_2(b,a))^{\frac{s}{q}} \right).$$

Proof. Let $\frac{1}{p} + \frac{1}{q} = 1$. Then

$$
\left| \frac{bf(b) - af(a)}{b-a} - \frac{1}{b-a} \int_a^b f(t)\Delta t \right|
$$

$$
\leq \frac{1}{b-a} \int_a^b \sigma(t) |f^\Delta(t)| \Delta t
$$

$$
\leq \frac{\sigma(b)}{b-a} \int_a^b |f^\Delta(t)| \Delta t
$$

$$
\leq \frac{\sigma(b)}{b-a} (b-a)^{\frac{1}{p}} \left(\int_a^b |f^\Delta(t)|^q \Delta t \right)^{\frac{1}{q}}
$$

$$
\leq \frac{\sigma(b)}{(b-a)^{\frac{1}{q}}} \left(\frac{|f^\Delta(a)|^q}{(b-a)^s e_\beta(a, x_0)} \int_a^b (b-t)^s \Delta t \right.
$$

$$
\left. + \frac{|f^\Delta(b)|^q}{(b-a)^s e_\beta(b, x_0)} \int_a^b (t-a)^s \Delta t \right)^{\frac{1}{q}}
$$

$$
\leq \frac{\sigma(b)}{(b-a)^{\frac{s+1}{q}}} \left(\frac{|f^\Delta(a)|^q}{e_\beta(a, x_0)} (b-a)^{1-s} (h_2(a,b))^s \right.
$$

$$
\left. + \frac{|f^\Delta(b)|^q}{e_\beta(b, x_0)} (b-a)^{1-s} (h_2(b,a))^s \right)^{\frac{1}{q}}
$$

$$
= \frac{\sigma(b)}{(b-a)^{\frac{2s}{q}}} \left(\frac{|f^\Delta(a)|^q}{e_\beta(a, x_0)} (h_2(a,b))^s + \frac{|f^\Delta(b)|^q}{e_\beta(b, x_0)} (h_2(b,a))^s \right)^{\frac{1}{q}}
$$

$$
\leq \frac{2^{\frac{1}{q}} \sigma(b)}{(b-a)^{\frac{2s}{q}}} \left(\frac{|f^\Delta(a)|}{(e_\beta(a, x_0))^{\frac{1}{q}}} (h_2(a,b))^{\frac{s}{q}} + \frac{|f^\Delta(b)|}{(e_\beta(b, x_0))^{\frac{1}{q}}} (h_2(b,a))^{\frac{s}{q}} \right).
$$

This completes the proof. □

For a function $f \in C_{rd}^{n-1}([a,b])$, define

$$
I(a, b, n, f) = \int_a^b f(s)\Delta s - \sum_{k=1}^n h_k(b,a) f^{\Delta^{k-1}}(a).
$$

Theorem 5.14. *Let $a > 0$, $f \in C^n_{rd}([a,b])$, $|f^{\Delta^n}|^q$ be an exponentially s-convex function, where $q \geq 1$, and $|f^{\Delta^n}| \geq 1$ on $[a,b]$. Then*

$$\frac{1}{b-a}|I(a,b,n,f)|$$

$$\leq \frac{(b-a)^{n-2s}}{n!}\left(\frac{|f^{\Delta^n}(a)|^q}{e_\beta(a,x_0)}(h_2(a,b))^s + \frac{|f^{\Delta^n}(b)|^q}{e_\beta(b,x_0)}(h_2(b,a))^s\right).$$

Proof.   By the Taylor formula, we have

$$\frac{1}{b-a}|I(a,b,n,f)|$$

$$= \frac{1}{b-a}\left|\int_a^{\rho^n(b)} h_n(b,\sigma(\tau))f^{\Delta^n}(\tau)\Delta\tau\right|$$

$$\leq \frac{1}{b-a}\int_a^{\rho^n(b)} h_n(b,\sigma(\tau))|f^{\Delta^n}(\tau)|\Delta\tau$$

$$\leq \frac{1}{b-a}\int_a^b \frac{(b-\sigma(\tau))^n}{n!}|f^{\Delta^n}(\tau)|\Delta\tau$$

$$\leq \frac{(b-a)^{n-1}}{n!}\int_a^b |f^{\Delta^n}(\tau)|^q\Delta\tau$$

$$\leq \frac{(b-a)^{n-1}}{n!}\left(\frac{|f^{\Delta^n}(a)|^q}{(b-a)^s e_\beta(a,x_0)}\int_a^b (b-t)^s\Delta t\right.$$

$$\left. + \frac{|f^{\Delta^n}(b)|^q}{(b-a)^s e_\beta(b,x_0)}\int_a^b (t-a)^s\Delta t\right)$$

$$= \frac{(b-a)^{n-2s}}{n!}\left(\frac{|f^{\Delta^n}(a)|^q}{e_\beta(a,x_0)}(h_2(a,b))^s + \frac{|f^{\Delta^n}(b)|^q}{e_\beta(b,x_0)}(h_2(b,a))^s\right).$$

This completes the proof.   □

Theorem 5.15. *Let $a > 0$, $f \in C^n_{rd}([a, b])$, $|f^{\Delta^n}|^q$ be an exponentially s-convex function, where $q \geq 1$. Then*

$$\frac{1}{b-a}|I(a,b,n,f)| \leq \frac{2^{\frac{1}{q}}(b-a)^{n-\frac{2s}{q}}}{n!}\left(\frac{|f^{\Delta^n}(a)|}{(e_\beta(a,x_0))^{\frac{1}{q}}}(h_2(a,b))^{\frac{s}{q}}\right.$$

$$\left. + \frac{|f^{\Delta^n}(b)|}{(e_\beta(b,x_0))^{\frac{1}{q}}}(h_2(b,a))^{\frac{s}{q}}\right).$$

Proof. Let $p \geq 1$ be chosen so that $\frac{1}{p} + \frac{1}{q} = 1$. Then

$$\frac{1}{b-a}|I(a,b,n,f)|$$

$$\leq \frac{(b-a)^{n-1}}{n!}\int_a^b |f^{\Delta^n}(t)|\Delta t$$

$$\leq \frac{(b-a)^{n-1}}{(n-1)!}\left(\int_a^b \Delta t\right)^{\frac{1}{p}}\left(\int_a^b |f^{\Delta^n}(t)|^q\Delta t\right)^{\frac{1}{q}}$$

$$= \frac{(b-a)^{n-1+\frac{1}{p}}}{n!}\left(\frac{|f^{\Delta^n}(a)|^q}{(b-a)^s e_\beta(a,x_0)}\int_a^b (b-t)^s\Delta t\right.$$

$$\left. + \frac{|f^{\Delta^n}(b)|^q}{(b-a)^s e_\beta(b,x_0)}\int_a^b (t-a)^s\Delta t\right)^{\frac{1}{q}}$$

$$= \frac{(b-a)^{n-\frac{1}{q}}}{n!}\left(\frac{|f^{\Delta^n}(a)|^q}{(b-a)^{2s-1}e_\beta(a,x_0)}(h_2(a,b))^s\right.$$

$$\left. + \frac{|f^{\Delta^n}(b)|^q}{(b-a)^{2s-1}e_\beta(b,x_0)}(h_2(b,a))^s\right)^{\frac{1}{q}}$$

$$\leq \frac{2^{\frac{1}{q}}(b-a)^{n-\frac{1}{q}}}{n!}\left(\frac{|f^{\Delta^n}(a)|}{(b-a)^{\frac{2s-1}{q}}(e_\beta(a,x_0))^{\frac{1}{q}}}(h_2(a,b))^{\frac{s}{q}}\right.$$

$$\left. + \frac{|f^{\Delta^n}(b)|}{(b-a)^{\frac{2s-1}{q}}(e_\beta(b,x_0))^{\frac{1}{q}}}(h_2(b,a))^{\frac{s}{q}}\right)$$

$$= \frac{2^{\frac{1}{q}}(b-a)^{n-\frac{2s}{q}}}{n!} \left(\frac{|f^{\Delta^n}(a)|}{(e_\beta(a,x_0))^{\frac{1}{q}}} (h_2(a,b))^{\frac{s}{q}} \right.$$

$$\left. + \frac{|f^{\Delta^n}(b)|}{(e_\beta(b,x_0))^{\frac{1}{q}}} (h_2(b,a))^{\frac{s}{q}} \right).$$

This completes the proof. $\qquad\qquad\qquad\qquad\qquad\qquad\qquad\qquad\square$

5.4   Inequalities for Strongly r-Convex Functions

We start the investigations in this section with the following useful lemma.

Lemma 5.3. *Let* $f \in C_{rd}^{m-1}([a,b])$, $\alpha > 2$, $m-1 < \alpha < m$, $\nu = m-\alpha$. *Then*

$$\int_a^t f(s)\Delta s = \sum_{k=0}^{m-1} h_k(t,a) f^{\Delta^k}(a)$$

$$- \int_a^t f^{\Delta^{m-1}}(u)\mu(u)h_{\alpha-2}(t,\sigma(u))h_\nu(u,\sigma(u))\Delta u$$

$$+ \int_a^t h_{\alpha-2}(t,\sigma(\tau)) \left(\int_a^\tau h_\nu(\tau,\sigma(u)) f^{\Delta^{m-1}}(u)\Delta u \right) \Delta \tau,$$

$$t \in [a,b].$$

Proof.   Let

$$g(t) = \int_0^t f(s)\Delta s, \quad t \in [a,b].$$

Then

$$g^{\Delta^k}(t) = f^{\Delta^{k-1}}(t), \quad k \in \{1,\ldots,m\},$$

$$\Delta_a^{\alpha-1} g(t) = \int_a^t h_\nu(t,\sigma(u)) f^{\Delta^{m-1}}(u)\Delta u, \quad t \in [a,b].$$

We apply the fractional Taylor formula for the function g and get the desired result. This completes the proof. $\qquad\qquad\qquad\qquad\square$

For $\alpha \geq 2$, $\nu \geq 0$, $p \geq 1$, denote

$$H(\alpha, \nu, t, a) = \int_a^t h_{\alpha-2}(t, \sigma(\tau)) \int_a^\tau h_\nu(\tau, \sigma(u))(b-u)(u-a)\Delta u \Delta \tau,$$

$$G(\alpha, \nu, p, t, a) = \left(\int_a^t (h_\nu(t, \sigma(u)))^p \Delta u \right)^{\frac{1}{p}},$$

$$\Psi(\nu, t, a) = \int_a^t h_\nu(t, \sigma(u))(b-u)(u-a)\Delta u, \quad t \in [a, b].$$

Definition 5.3. Let $\alpha, \beta > 1$, $f \in C_{rd}([a, b])$. The integral

$$E(f, \alpha, \beta, t) = \int_a^t f(u)\mu(u)h_{\alpha-1}(t, \sigma(u))h_{\beta-1}(u, \sigma(u))\Delta u, \quad t \in [a, b],$$

is called the forward graininess deviation functional of f.

Definition 5.4. Let $\alpha > 2$ and $m - 1 < \alpha < m$, $m \in \mathbb{N}$, $\nu = m - \alpha$. For a function $f \in C_{rd}^m([a, b])$, define

$$\Delta_a^{\alpha-1} f(t) = D_a^{\nu+1} f^{\Delta^m}(t)$$

$$= \int_a^t h_\nu(t, \sigma(u))f^{\Delta^m}(u)\Delta u, \quad t \in [a, b].$$

Definition 5.5. A positive function $f : I \to \mathbb{R}$ is called strongly r-convex function with modulus c on $[a, b]$ if for each $x, y \in [a, b]$ and $\lambda \in [0, 1]$, we have

$$f(\lambda x + (1 - \lambda)y) \leq (\lambda(f(x))^r + (1 - \lambda)(f(y))^r)^{\frac{1}{r}}$$
$$- c\lambda(1 - \lambda)(x - y)^2, \quad r \neq 0.$$

If we take $c = 0$, we get the definition of r-convexity of the function f.

Note that, if $f : I \to \mathbb{R}$ is positive strongly r-convex function with modulus c, we have

$$f(t) \leq \left(\frac{b-t}{b-a}(f(a))^r + \frac{t-a}{b-a}(f(b))^r \right)^{\frac{1}{r}} - c(b-t)(t-a), \quad t \in [a, b].$$

Theorem 5.16. *Let $r > 0$, $\alpha > 2$, $m - 1 < \alpha < m$, $\nu = m - \alpha$, $q \geq 1$, $f \in C^m_{rd}([a, b])$, $|f^{\Delta^m}| \geq 1$ on $[a, b]$ and $|f^{\Delta^m}|^q$ be a strongly r-convex function with modulus c on $[a, b]$, $f^{\Delta^k}(a) = 0$, $k \in \{0, 1, \ldots, m - 1\}$. Then*

$$|B(t)| \leq 2^{\frac{1}{r}} \left(|f^{\Delta^m}(a)|^q + |f^{\Delta^m}(b)|^q \right) h_{\alpha + \nu}(t, a)$$
$$- cH(\alpha, \nu, t, a), \quad t \in [a, b].$$

Proof. Since $|f^{\Delta^m}|^q$ is a strongly r-convex function on $[a, b]$, we have

$$|f^{\Delta^m}(t)|^q \leq \left(\frac{b - t}{b - a} |f^{\Delta^m}(a)|^{qr} + \frac{t - a}{b - a} |f^{\Delta^m}(b)|^{qr} \right)^{\frac{1}{r}}$$
$$- c(b - t)(t - a), \quad t \in [a, b].$$

From here, we get

$$|B(t)| = \left| \int_a^t h_{\alpha - 2}(t, \sigma(\tau)) \left(\int_a^\tau h_\nu(\tau, \sigma(u)) f^{\Delta^m}(u) \Delta u \right) \Delta \tau \right|$$

$$\leq \int_a^t h_{\alpha - 2}(t, \sigma(\tau)) \int_a^\tau h_\nu(\tau, \sigma(u)) |f^{\Delta^m}(u)| \Delta u \Delta \tau$$

$$\leq \int_a^t h_{\alpha - 2}(t, \sigma(\tau)) \int_a^\tau h_\nu(\tau, \sigma(u)) |f^{\Delta^m}(u)|^q \Delta u \Delta \tau$$

$$\leq \int_a^t h_{\alpha - 2}(t, \sigma(\tau)) \int_a^\tau h_\nu(\tau, \sigma(u))$$

$$\times \left(\left(\frac{b - u}{b - a} |f^{\Delta^m}(a)|^{qr} + \frac{u - a}{b - a} |f^{\Delta^m}(b)|^{qr} \right)^{\frac{1}{r}} \right.$$
$$\left. - c(b - u)(u - a) \right) \Delta u \Delta \tau$$

$$\leq \int_a^t h_{\alpha - 2}(t, \sigma(\tau)) \int_a^\tau h_\nu(\tau, \sigma(u))$$

$$\times \left(|f^{\Delta^m}(a)|^{qr} + |f^{\Delta^m}(b)|^{qr} \right)^{\frac{1}{r}} \Delta u \Delta \tau$$

$$- c \int_a^t h_{\alpha - 2}(t, \sigma(\tau)) \int_a^\tau h_\nu(\tau, \sigma(u))(b - u)(u - a) \Delta u \Delta \tau$$

$$\leq 2^{\frac{1}{r}} \left(|f^{\Delta^m}(a)|^q + |f^{\Delta^m}(b)|^q \right) \int_a^t h_{\alpha-2}(t, \sigma(\tau))$$

$$\times \int_a^\tau h_\nu(\tau, \sigma(u)) \Delta u \Delta \tau - cH(\alpha, \nu, t, a)$$

$$= 2^{\frac{1}{r}} \left(|f^{\Delta^m}(a)|^q + |f^{\Delta^m}(b)|^q \right) \int_a^t h_{\alpha-2}(t, \sigma(\tau)) h_{\nu+1}(\tau, a) \Delta \tau$$

$$- cH(\alpha, \nu, t, a)$$

$$= 2^{\frac{1}{r}} \left(|f^{\Delta^m}(a)|^q + |f^{\Delta^m}(b)|^q \right) h_{\alpha+\nu}(t, a)$$

$$- cH(\alpha, \nu, t, a), \quad t \in [a, b].$$

This completes the proof. □

Theorem 5.17. *Let* $r > 0$, $\alpha > 2$, $m - 1 < \alpha < m$, $\nu = m - \alpha$, $p, q \geq 1$, $\frac{1}{p} + \frac{1}{q} = 1$, $f \in C_{rd}^m([a, b])$, $|f^{\Delta^m}|^q$ *be a strongly* r-*convex function with modulus* c *on* $[a, b]$, $f^{\Delta^k}(a) = 0$, $k \in \{0, 1, \ldots, m - 1\}$. *Then*

$$|B(t)| \leq 2^{\frac{1}{q}} (b - a)^{\frac{1}{q}} G(\alpha, \nu, p, t, a)$$

$$\times \left(2^{\frac{r+1}{rq}} \left(|f^{\Delta^m}(a)| + |f^{\Delta^m}(b)| \right) + c^{\frac{1}{q}} (h_2(b, a))^{\frac{1}{q}} \right)$$

$$\times h_{\alpha-1}(t, a), \quad t \in [a, b].$$

Proof. Since $|f^{\Delta^m}|^q$ is a strongly r-convex function on $[a, b]$, we have

$$|f^{\Delta^m}(t)|^q \leq \left(\frac{b-t}{b-a} |f^{\Delta^m}(a)|^{qr} + \frac{t-a}{b-a} |f^{\Delta^m}(b)|^{qr} \right)^{\frac{1}{r}}$$

$$- c(b - t)(t - a), \quad t \in [a, b].$$

Then

$$|B(t)| = \left| \int_a^t h_{\alpha-2}(t, \sigma(\tau)) \left(\int_a^\tau h_\nu(\tau, \sigma(u)) f^{\Delta^m}(u) \Delta u \right) \Delta \tau \right|$$

$$\leq \int_a^t h_{\alpha-2}(t, \sigma(\tau)) \left(\int_a^\tau h_\nu(\tau, \sigma(u)) |f^{\Delta^m}(u)| \Delta u \right) \Delta \tau$$

$$\leq \int_a^t h_{\alpha-2}(t,\sigma(\tau)) \left(\left(\int_a^\tau (h_\nu(\tau,\sigma(u)))^p \Delta u \right)^{\frac{1}{p}} \right.$$

$$\left. \times \left(\int_a^\tau |f^{\Delta^m}(u)|^q \Delta u \right)^{\frac{1}{q}} \right) \Delta\tau$$

$$\leq \int_a^t h_{\alpha-2}(t,\sigma(\tau)) \left(\left(\int_a^t (h_\nu(t,\sigma(u)))^p \Delta u \right)^{\frac{1}{p}} \right.$$

$$\left. \times \left(\int_a^\tau |f^{\Delta^m}(u)|^q \Delta u \right)^{\frac{1}{q}} \right) \Delta\tau$$

$$= G(\alpha,\nu,p,t,a) \int_a^t h_{\alpha-2}(t,\sigma(\tau)) \left(\int_a^\tau |f^{\Delta^m}(u)|^q \Delta u \right)^{\frac{1}{q}} \Delta\tau$$

$$\leq G(\alpha,\nu,p,t,a) \int_a^t h_{\alpha-2}(t,\sigma(\tau))$$

$$\times \left(\int_a^\tau \left(\left(\frac{b-u}{b-a} |f^{\Delta^m}(a)|^{qr} + \frac{u-a}{b-a} |f^{\Delta^m}(b)|^{qr} \right)^{\frac{1}{r}} \right. \right.$$

$$\left. \left. - c(b-u)(u-a) \right) \Delta u \right)^{\frac{1}{q}} \Delta\tau$$

$$\leq G(\alpha,\nu,p,t,a) \int_a^t h_{\alpha-2}(t,\sigma(\tau))$$

$$\times \left(\int_a^\tau \left(\left(|f^{\Delta^m}(a)|^{qr} + |f^{\Delta^m}(b)|^{qr} \right)^{\frac{1}{r}} \right. \right.$$

$$\left. \left. + c(b-a)(u-a) \right) \Delta u \right)^{\frac{1}{q}} \Delta\tau$$

$$\leq G(\alpha,\nu,p,t,a) \int_a^t h_{\alpha-2}(t,\sigma(\tau))$$

$$\times \left(2^{\frac{1}{r}} \left(|f^{\Delta^m}(a)|^q + |f^{\Delta^m}(b)|^q \right)(b-a) \right.$$

$$\left. + c(b-a) \int_a^\tau (u-a)\Delta u \right)^{\frac{1}{q}} \Delta\tau$$

$$= G(\alpha,\nu,p,t,a)(b-a)^{\frac{1}{q}} \int_a^t h_{\alpha-2}(t,\sigma(\tau))$$

$$\times \left(2^{\frac{1}{r}} \left(|f^{\Delta^m}(a)|^q + |f^{\Delta^m}(b)|^q \right) + ch_2(\tau,a) \right)^{\frac{1}{q}} \Delta\tau$$

$$\leq 2^{\frac{1}{q}}(b-a)^{\frac{1}{q}}G(\alpha,\nu,p,t,a)\int_a^t h_{\alpha-2}(t,\sigma(\tau))$$

$$\times\left(2^{\frac{1}{rq}}\left(|f^{\Delta^m}(a)|^q+|f^{\Delta^m}(b)|^q\right)^{\frac{1}{q}}+c^{\frac{1}{q}}(h_2(\tau,a))^{\frac{1}{q}}\right)\Delta\tau$$

$$\leq 2^{\frac{1}{q}}(b-a)^{\frac{1}{q}}G(\alpha,\nu,p,t,a)\int_a^t h_{\alpha-2}(t,\sigma(\tau))$$

$$\times\left(2^{\frac{r+1}{rq}}\left(|f^{\Delta^m}(a)|+|f^{\Delta^m}(b)|\right)+c^{\frac{1}{q}}(h_2(\tau,a))^{\frac{1}{q}}\right)\Delta\tau$$

$$\leq 2^{\frac{1}{q}}(b-a)^{\frac{1}{q}}G(\alpha,\nu,p,t,a)\left(2^{\frac{r+1}{rq}}\left(|f^{\Delta^m}(a)|+|f^{\Delta^m}(b)|\right)\right.$$

$$\left.+c^{\frac{1}{q}}(h_2(b,a))^{\frac{1}{q}}\right)h_{\alpha-1}(t,a),\quad t\in[a,b].$$

This completes the proof. $\qquad\square$

Theorem 5.18. *Let* $r>0$, $\alpha>2$, $m-1<\alpha<m$, $\nu=m-\alpha$, $q\geq 1$, $f\in C_{rd}^m([a,b])$, $|f^{\Delta^m}|\geq 1$ *and* $|f^{\Delta^m}|^q$ *be a strongly* r-*convex function with modulus* c *on* $[a,b]$. *Then*

$$|\Delta_a^{\alpha-1}f(t)|\leq 2^{\frac{1}{r}}\left(|f^{\Delta^m}(a)|^q+|f^{\Delta^m}(b)|^q\right)h_{\nu+1}(t,a)$$

$$-c\Psi(\nu,t,a),\quad t\in[a,b].$$

Proof. Since $|f^{\Delta^m}|^q$ is a strongly r-convex function on $[a,b]$, we have

$$|f^{\Delta^m}(t)|^q\leq\left(\frac{b-t}{b-a}|f^{\Delta^m}(a)|^{qr}+\frac{t-a}{b-a}|f^{\Delta^m}(b)|^{qr}\right)^{\frac{1}{r}}$$

$$-c(b-t)(t-a),\quad t\in[a,b].$$

Then, using $|f^{\Delta^m}|\geq 1$ on $[a,b]$, we have

$$|\Delta_a^{\alpha-1}f(t)|$$

$$=\left|\int_a^t h_\nu(t,\sigma(u))f^{\Delta^m}(u)\Delta u\right|$$

$$\leq\int_a^t h_\nu(t,\sigma(u))|f^{\Delta^m}(u)|\Delta u$$

$$\leq \int_a^t h_\nu(t, \sigma(u)) |f^{\Delta^m}(u)|^q \Delta u$$

$$\leq \int_a^t h_\nu(t, \sigma(u)) \left(\left(\frac{b-u}{b-a} |f^{\Delta^m}(a)|^{qr} + \frac{u-a}{b-a} |f^{\Delta^m}(b)|^{qr} \right)^{\frac{1}{r}} \right.$$

$$\left. - c(b-u)(u-a) \right) \Delta u$$

$$= \int_a^t h_\nu(t, \sigma(u)) \left(\frac{b-u}{b-a} |f^{\Delta^m}(a)|^{qr} + \frac{u-a}{b-a} |f^{\Delta^m}(b)|^{qr} \right)^{\frac{1}{r}} \Delta u$$

$$- c \int_a^t h_\nu(t, \sigma(u))(b-u)(u-a) \Delta u$$

$$= \int_a^t h_\nu(t, \sigma(u)) \left(\frac{b-u}{b-a} |f^{\Delta^m}(a)|^{qr} + \frac{u-a}{b-a} |f^{\Delta^m}(b)|^{qr} \right)^{\frac{1}{r}} \Delta u$$

$$- c\Psi(\nu, t, a)$$

$$\leq \int_a^t h_\nu(t, \sigma(u)) \left(|f^{\Delta^m}(a)|^{qr} + |f^{\Delta^m}(b)|^{qr} \right)^{\frac{1}{r}} \Delta u$$

$$- c\Psi(\nu, t, a)$$

$$\leq 2^{\frac{1}{r}} \left(|f^{\Delta^m}(a)|^q + |f^{\Delta^m}(b)|^q \right) \int_a^t h_\nu(t, \sigma(u)) \Delta u$$

$$- c\Psi(\nu, t, a)$$

$$= 2^{\frac{1}{r}} \left(|f^{\Delta^m}(a)|^q + |f^{\Delta^m}(b)|^q \right) h_{\nu+1}(t, a)$$

$$- c\Psi(\nu, t, a), \quad t \in [a, b].$$

This completes the proof. $\square$

Corollary 5.15. *Suppose that all conditions of Theorem 5.18 hold and $f^{\Delta^k}(a) = 0$, $k \in \{0, 1, \ldots, m-1\}$. Then*

$$|B(t)| \leq 2^{\frac{1}{r}} \left(|f^{\Delta^m}(a)|^q + |f^{\Delta^m}(b)|^q \right) h_{\alpha+\nu}(t, a)$$

$$+ c\Psi(\nu, b, a) h_{\alpha-1}(t, a), \quad t \in [a, b].$$

Proof. We have

$$|B(t)| = \left| \int_a^t h_{\alpha-2}(t, \sigma(\tau)) \Delta_a^{\alpha-1} f(\tau) \Delta \tau \right|$$

$$\leq \int_a^t h_{\alpha-2}(t, \sigma(\tau)) |\Delta_a^{\alpha-1} f(\tau)| \Delta \tau$$

$$\leq 2^{\frac{1}{r}} \left(|f^{\Delta^m}(a)|^q + |f^{\Delta^m}(b)|^q \right) \int_a^t h_{\alpha-2}(t, \sigma(\tau)) h_{\nu+1}(\tau, a) \Delta \tau$$

$$+ c\Psi(\nu, b, a) \int_a^t h_{\alpha-2}(t, \sigma(\tau)) \Delta \tau$$

$$= 2^{\frac{1}{r}} \left(|f^{\Delta^m}(a)|^q + |f^{\Delta^m}(b)|^q \right) h_{\alpha+\nu}(t, a)$$

$$+ c\Psi(\nu, b, a) h_{\alpha-1}(t, a), \quad t \in [a, b].$$

This completes the proof. $\square$

Theorem 5.19. *Let $r > 0$, $\alpha > 2$, $m-1 < \alpha < m$, $\nu = m - \alpha$, $q \geq 1$, $\frac{1}{p} + \frac{1}{q} = 1$, $f \in C_{rd}^m([a, b])$, $|f^{\Delta^m}|^q$ be a strongly r-convex function with modulus c on $[a, b]$. Then*

$$|\Delta_a^{\alpha-1} f(t)| \leq 2^{\frac{1}{q}} G(\alpha, \nu, p, t, a)(b - a)^{\frac{1}{q}}$$

$$\times \left(2^{\frac{r+1}{rq}} \left(|f^{\Delta^m}(a)| + |f^{\Delta^m}(b)| \right) + c^{\frac{1}{q}} (h_2(t, a))^{\frac{1}{q}} \right),$$

$$t \in [a, b].$$

Proof. Since $|f^{\Delta^m}|^q$ is a strongly r-convex function on $[a, b]$, we have

$$|f^{\Delta^m}(t)|^q \leq \left(\frac{b-t}{b-a} |f^{\Delta^m}(a)|^{qr} + \frac{t-a}{b-a} |f^{\Delta^m}(b)|^{qr} \right)^{\frac{1}{r}}$$

$$- c(b-t)(t-a), \quad t \in [a, b].$$

Then

$$|\Delta_a^{\alpha-1} f(t)|$$

$$= \left| \int_a^t h_\nu(t, \sigma(u)) f^{\Delta^m}(u) \Delta u \right|$$

$$\leq \int_a^t h_\nu(t, \sigma(u)) |f^{\Delta^m}(u)| \Delta u$$

$$\leq \left(\int_a^t (h_\nu(t,\sigma(u)))^p \Delta u \right)^{\frac{1}{p}} \left(\int_a^t |f^{\Delta^m}(u)|^q \Delta u \right)^{\frac{1}{q}}$$

$$= G(\alpha,\nu,p,t,a) \left(\int_a^t |f^{\Delta^m}(u)|^q \Delta u \right)^{\frac{1}{q}}$$

$$\leq G(\alpha,\nu,p,t,a) \left(\int_a^t \left(\frac{b-u}{b-a}|f^{\Delta^m}(a)|^{qr} + \frac{u-a}{b-a}|f^{\Delta^m}(b)|^{qr} \right)^{\frac{1}{r}} \right.$$

$$\left. - c(b-u)(u-a)\Delta u \right)^{\frac{1}{q}}$$

$$\leq G(\alpha,\nu,p,t,a) \left(\int_a^t \left(|f^{\Delta^m}(a)|^{qr} + |f^{\Delta^m}(b)|^{qr} \right)^{\frac{1}{r}} \Delta u \right.$$

$$\left. + c \int_a^t (b-u)(u-a)\Delta u \right)^{\frac{1}{q}}$$

$$\leq G(\alpha,\nu,p,t,a) \left(2^{\frac{1}{r}} \int_a^t \left(|f^{\Delta^m}(a)|^q + |f^{\Delta^m}(b)|^q \right) \Delta u \right.$$

$$\left. + c(b-a) \int_a^t (u-a)\Delta u \right)^{\frac{1}{q}}$$

$$= G(\alpha,\nu,p,t,a) \left(2^{\frac{1}{r}} \left(|f^{\Delta^m}(a)|^q + |f^{\Delta^m}(b)|^q \right)(b-a) \right.$$

$$\left. + c(b-a)h_2(t,a) \right)^{\frac{1}{q}}$$

$$= G(\alpha,\nu,p,t,a)(b-a)^{\frac{1}{q}} \left(2^{\frac{1}{r}} \left(|f^{\Delta^m}(a)|^q + |f^{\Delta^m}(b)|^q \right) \right.$$

$$\left. + ch_2(t,a) \right)^{\frac{1}{q}}$$

$$\leq 2^{\frac{1}{q}} G(\alpha,\nu,p,t,a)(b-a)^{\frac{1}{q}} \left(2^{\frac{1}{rq}} \left(|f^{\Delta^m}(a)|^q + |f^{\Delta^m}(b)|^q \right)^{\frac{1}{q}} \right.$$

$$\left. + c^{\frac{1}{q}}(h_2(t,a))^{\frac{1}{q}} \right)$$

$$\leq 2^{\frac{1}{q}} G(\alpha, \nu, p, t, a)(b-a)^{\frac{1}{q}} \left(2^{\frac{r+1}{rq}} \left(|f^{\Delta^m}(a)| + |f^{\Delta^m}(b)| \right) \right.$$

$$\left. + c^{\frac{1}{q}} (h_2(t,a))^{\frac{1}{q}} \right), \quad t \in [a,b].$$

This completes the proof. $\qquad \square$

Corollary 5.16. *Suppose that all conditions of Theorem 5.19 hold and* $f^{\Delta^k}(a) = 0$, $k \in \{0, 1, \dots, m-1\}$. *Then*

$$|B(t)| \leq 2^{\frac{2r+1}{rq}} (b-a)^{\frac{1+q}{q}} \left(|f^{\Delta^m}(a)| \right.$$

$$+ |f^{\Delta^m}(b)| \Big) G(\alpha, \nu, p, b, a) h_{\alpha-1}(t,a)$$

$$+ 2^{\frac{1}{q}} (b-a) c^{\frac{1}{q}} G(\alpha, \nu, p, b, a) G(\alpha, \alpha$$

$$- 2, p, b, a)(h_3(t,a))^{\frac{1}{q}}, \quad t \in [a,b].$$

Proof. We have

$$|B(t)| = \left| \int_a^t h_{\alpha-2}(t, \sigma(u)) \Delta_a^{\alpha-1} f(u) \Delta u \right|$$

$$\leq \int_a^t h_{\alpha-2}(t, \sigma(u)) |\Delta_a^{\alpha-1} f(u)| \Delta u$$

$$\leq 2^{\frac{2r+1}{rq}} (b-a)^{\frac{1}{q}} \left(|f^{\Delta^m}(a)| + |f^{\Delta^m}(b)| \right)$$

$$\times \int_a^t h_{\alpha-2}(t, \sigma(u)) G(\alpha, \nu, p, \tau, a) \Delta \tau$$

$$+ 2^{\frac{1}{q}} (b-a)^{\frac{1}{q}} c^{\frac{1}{q}} \int_a^t h_{\alpha-2}(t, \sigma(u)) G(\alpha, \nu, p, \tau, a)(h_2(\tau, a))^{\frac{1}{q}} \Delta \tau$$

$$\leq 2^{\frac{2r+1}{rq}} (b-a)^{\frac{1}{q}} \left(|f^{\Delta^m}(a)| + |f^{\Delta^m}(b)| \right) G(\alpha, \nu, p, b, a) h_{\alpha-1}(t,a)$$

$$+ 2^{\frac{1}{q}} (b-a)^{\frac{1}{q}} c^{\frac{1}{q}} G(\alpha, \nu, p, b, a) \left(\int_a^t (h_{\alpha-2}(t, \sigma(u)))^p \Delta \tau \right)^{\frac{1}{p}}$$

$$\times \left(\int_a^t h_2(\tau, a) \Delta \tau \right)^{\frac{1}{q}}$$

$$\leq 2^{\frac{2r+1}{rq}} (b-a)^{\frac{1}{q}} \left(|f^{\Delta^m}(a)| + |f^{\Delta^m}(b)| \right) G(\alpha, \nu, p, b, a) h_{\alpha-1}(t, a)$$

$$+ 2^{\frac{1}{q}} (b-a) c^{\frac{1}{q}} G(\alpha, \nu, p, b, a) G(\alpha, \alpha - 2, p, b, a)(h_3(t, a))^{\frac{1}{q}},$$

$$t \in [a, b].$$

This completes the proof.   □

5.5   Inequalities for Exponentially s-Convex Functions

Definition 5.6. Let $s \in (0, 1]$, $x_0 \in [a, b]$. A function $f : [a, b] \to \mathbb{R}$ is called exponentially s-convex in the second sense if

$$f(t) \leq \left(\frac{b-t}{b-a} \right)^s \frac{f(a)}{e_\beta(a, x_0)} + \left(\frac{t-a}{b-a} \right)^s \frac{f(b)}{e_\beta(b, x_0)} \tag{5.12}$$

for any $t \in [a, b]$ and for some $\beta \in \mathcal{R}_+$. If (5.11) holds in the reverse sense, then we say that f is exponentially s-concave in the second sense.

Throughout this section, without loss of generality, suppose that $s \in (0, 1)$.

Theorem 5.20. *Let* $\alpha > 2$, $m - 1 < \alpha < m$, $\nu = m - \alpha$, $f \in C_{rd}^m([a, b])$, $f^{\Delta^k}(a) = 0$, $k \in \{0, 1, \ldots, m-1\}$, $|f^{\Delta^m}|$ *be an exponentially s-convex function in the second sense on* $[a, b]$. *Then*

$$|B(t)| \leq \left(\frac{|f^{\Delta^m}(a)|}{e_\beta(a, x_0)} + \frac{|f^{\Delta^m}(b)|}{e_\beta(b, x_0)} \right) h_{\alpha+\nu}(t, a), \quad t \in [a, b], \tag{5.13}$$

and

$$|B(t)| \leq h_{\alpha+\nu-1}(t, a)(b-a)^{1-2s}$$

$$\times \left(\frac{|f^{\Delta^m}(a)|}{e_\beta(a, x_0)} (h_2(a, b))^s + \frac{|f^{\Delta^m}(b)|}{e_\beta(b, x_0)} (h_2(b, a))^s \right), \tag{5.14}$$

$$t \in [a, b].$$

Proof.   Since $|f^{\Delta^m}|$ is exponentially s-convex in the second sense on $[a, b]$, we have

$$|f^{\Delta^m}(t)| \leq \left(\frac{b-t}{b-a} \right)^s \frac{|f^{\Delta^m}(a)|}{e_\beta(a, x_0)} + \left(\frac{t-a}{b-a} \right)^s \frac{|f^{\Delta^m}(b)|}{e_\beta(b, x_0)}, \quad t \in [a, b]. \tag{5.15}$$

1. First, we will prove (5.13). We have

$$|B(t)| = \left| \int_a^t h_{\alpha-2}(t, \sigma(\tau)) \left(\int_a^\tau h_\nu(\tau, \sigma(u)) f^{\Delta^m}(u) \Delta u \right) \Delta \tau \right|$$

$$\leq \int_a^t h_{\alpha-2}(t, \sigma(\tau)) \int_a^\tau h_\nu(\tau, \sigma(u)) |f^{\Delta^m}(u)| \Delta u \Delta \tau$$

$$\leq \int_a^t h_{\alpha-2}(t, \sigma(\tau)) \int_a^\tau h_\nu(\tau, \sigma(u)) \left(\left(\frac{b-u}{b-a} \right)^s \frac{|f^{\Delta^m}(a)|}{e_\beta(a, x_0)} \right.$$

$$\left. + \left(\frac{u-a}{b-a} \right)^s \frac{|f^{\Delta^m}(b)|}{e_\beta(b, x_0)} \right) \Delta u \Delta \tau$$

$$\leq \left(\frac{|f^{\Delta^m}(a)|}{e_\beta(a, x_0)} + \frac{|f^{\Delta^m}(b)|}{e_\beta(b, x_0)} \right) \int_a^t h_{\alpha-2}(t, \sigma(\tau))$$

$$\times \int_a^\tau h_\nu(\tau, \sigma(u)) \Delta u \Delta \tau$$

$$= \left(\frac{|f^{\Delta^m}(a)|}{e_\beta(a, x_0)} + \frac{|f^{\Delta^m}(b)|}{e_\beta(b, x_0)} \right) \int_a^t h_{\alpha-2}(t, \sigma(\tau)) h_{\nu+1}(\tau, a) \Delta \tau$$

$$= \left(\frac{|f^{\Delta^m}(a)|}{e_\beta(a, x_0)} + \frac{|f^{\Delta^m}(b)|}{e_\beta(b, x_0)} \right) h_{\alpha+\nu}(t, a), \quad t \in [a, b].$$

2. Now, we will prove (5.14). We have

$$|B(t)| \leq \int_a^t h_{\alpha-2}(t, \sigma(\tau)) \int_a^\tau h_\nu(\tau, \sigma(u))$$

$$\times \left(\left(\frac{b-u}{b-a} \right)^s \frac{|f^{\Delta^m}(a)|}{e_\beta(a, x_0)} + \left(\frac{u-a}{b-a} \right)^s \frac{|f^{\Delta^m}(b)|}{e_\beta(b, x_0)} \right) \Delta u \Delta \tau$$

$$\leq \left(\int_a^t h_{\alpha-2}(t, \sigma(\tau)) h_\nu(\tau, a) \Delta \tau \right)$$

$$\times \left(\int_a^b \left(\left(\frac{b-u}{b-a} \right)^s \frac{|f^{\Delta^m}(a)|}{e_\beta(a, x_0)} + \left(\frac{u-a}{b-a} \right)^s \frac{|f^{\Delta^m}(b)|}{e_\beta(b, x_0)} \right) \Delta u \right)$$

$$\leq h_{\alpha+\nu-1}(t, a) \left(\frac{|f^{\Delta^m}(a)|}{e_\beta(a, x_0)} (b-a)^{1-2s} (h_2(a, b))^s \right.$$

$$\left. + \frac{|f^{\Delta^m}(b)|}{e_\beta(b, x_0)} (b-a)^{1-2s} (h_2(b, a))^s \right)$$

$$= h_{\alpha+\nu-1}(t,a)(b-a)^{1-2s}$$

$$\times \left(\frac{|f^{\Delta^m}(a)|}{e_\beta(a,x_0)}(h_2(a,b))^s + \frac{|f^{\Delta^m}(b)|}{e_\beta(b,x_0)}(h_2(b,a))^s \right), \quad t \in [a,b].$$

This completes the proof. $\qquad\square$

The next result can be stated as follows.

Theorem 5.21. *Let* $q \geq 1$, $\alpha \geq 2$, $m-1 < \alpha < m$, $\nu = m-\alpha$, $f \in C_{rd}^m([a,b])$, $|f^{\Delta^m}| \geq 1$ *on* $[a,b]$, $|f^{\Delta^m}|^q$ *be an exponentially s-convex function in the second sense on* $[a,b]$, $f^{\Delta^k}(a) = 0$, $k \in \{0,1,\dots,m-1\}$. *Then*

$$|B(t)| \leq \left(\frac{|f^{\Delta^m}(a)|^q}{e_\beta(a,x_0)} + \frac{|f^{\Delta^m}(b)|^q}{e_\beta(b,x_0)} \right) h_{\alpha+\nu}(t,a), \quad t \in [a,b], \quad (5.16)$$

and

$$|B(t)| \leq h_{\alpha+\nu-1}(t,a)(b-a)^{1-2s}$$

$$\times \left(\frac{|f^{\Delta^m}(a)|^q}{e_\beta(a,x_0)}(h_2(a,b))^s + \frac{|f^{\Delta^m}(b)|^q}{e_\beta(b,x_0)}(h_2(b,a))^s \right), \quad (5.17)$$

$$t \in [a,b].$$

Proof. Since $|f^{\Delta^m}|^q$ is exponentially s-convex function in the second sense on $[a,b]$, we have

$$|f^{\Delta^m}(t)|^q \leq \frac{(b-t)^s}{(b-a)^s} \frac{|f^{\Delta^m}(a)|^q}{e_\beta(a,x_0)} + \frac{(t-a)^s}{(b-a)^s} \frac{|f^{\Delta^m}(b)|^q}{e_\beta(b,x_0)}, \quad t \in [a,b].$$

$$(5.18)$$

1. First, we will prove (5.16). We have

$$|B(t)| = \left| \int_a^t h_{\alpha-2}(t,\sigma(\tau)) \left(\int_a^\tau h_\nu(\tau,\sigma(u)) f^{\Delta^m}(u)\Delta u \right) \Delta\tau \right|$$

$$\leq \int_a^t h_{\alpha-2}(t,\sigma(\tau)) \int_a^\tau h_\nu(\tau,\sigma(u)) |f^{\Delta^m}(u)|\Delta u\Delta\tau$$

$$\leq \int_a^t h_{\alpha-2}(t,\sigma(\tau)) \int_a^\tau h_\nu(\tau,\sigma(u)) |f^{\Delta^m}(u)|^q \Delta u\Delta\tau$$

$$\leq \int_a^t h_{\alpha-2}(t, \sigma(\tau)) \int_a^\tau h_\nu(\tau, \sigma(u)) \left(\frac{(b-u)^s}{(b-a)^s} \frac{|f^{\Delta^m}(a)|^q}{e_\beta(a, x_0)} \right.$$

$$\left. + \frac{(u-a)^s}{(b-a)^s} \frac{|f^{\Delta^m}(b)|^q}{e_\beta(b, x_0)} \right) \Delta u \Delta \tau$$

$$\leq \left(\frac{|f^{\Delta^m}(a)|^q}{e_\beta(a, x_0)} + \frac{|f^{\Delta^m}(b)|^q}{e_\beta(b, x_0)} \right) \int_a^t h_{\alpha-2}(t, \sigma(\tau))$$

$$\times \int_a^\tau h_\nu(\tau, \sigma(u)) \Delta u \Delta \tau$$

$$\leq \left(\frac{|f^{\Delta^m}(a)|^q}{e_\beta(a, x_0)} + \frac{|f^{\Delta^m}(b)|^q}{e_\beta(b, x_0)} \right) \int_a^t h_{\alpha-2}(t, \sigma(\tau)) h_{\nu+1}(\tau, a) \Delta \tau$$

$$= \left(\frac{|f^{\Delta^m}(a)|^q}{e_\beta(a, x_0)} + \frac{|f^{\Delta^m}(b)|^q}{e_\beta(b, x_0)} \right) h_{\alpha+\nu}(t, a), \quad t \in [a, b].$$

2. Now, we will prove (5.17). We find

$$|B(t)| \leq \int_a^t h_{\alpha-2}(t, \sigma(\tau)) \int_a^\tau h_\nu(\tau, \sigma(u)) \left(\frac{(b-u)^s}{(b-a)^s} \frac{|f^{\Delta^m}(a)|^q}{e_\beta(a, x_0)} \right.$$

$$\left. + \frac{(u-a)^s}{(b-a)^s} \frac{|f^{\Delta^m}(b)|^q}{e_\beta(b, x_0)} \right) \Delta u \Delta \tau$$

$$\leq \left(\int_a^t h_{\alpha-2}(t, \sigma(\tau)) h_\nu(\tau, a) \Delta \tau \right)$$

$$\times \left(\frac{|f^{\Delta^m}(a)|^q}{(b-a)^s e_\beta(a, x_0)} \int_a^b (b-u)^s \Delta u \right.$$

$$\left. + \frac{|f^{\Delta^m}(b)|^q}{(b-a)^s e_\beta(b, x_0)} \int_a^b (u-a)^s \Delta u \right)$$

$$\leq h_{\alpha+\nu-1}(t, a) \left(\frac{|f^{\Delta^m}(a)|^q}{e_\beta(a, x_0)} (b-a)^{1-2s} (h_2(a, b))^s \right.$$

$$\left. + \frac{|f^{\Delta^m}(b)|^q}{e_\beta(b, x_0)} (b-a)^{1-2s} (h_2(b, a))^s \right)$$

$$= h_{\alpha+\nu-1}(t,a)(b-a)^{1-2s}$$

$$\times \left(\frac{|f^{\Delta^m}(a)|^q}{e_\beta(a,x_0)}(h_2(a,b))^s + \frac{|f^{\Delta^m}(b)|^q}{e_\beta(b,x_0)}(h_2(b,a))^s \right),$$

$$t \in [a,b].$$

This completes the proof.

□

Theorem 5.22. *Let $\alpha \geq 2$, $m-1 < \alpha < m$, $\nu = m-\alpha$, $p,q \geq 1$, $\frac{1}{p}+\frac{1}{q} = 1$, $f \in C^m_{rd}([a,b])$, $|f^{\Delta^m}|$ be an exponentially s-convex function in the second sense on $[a,b]$, $f^{\Delta^k}(a) = 0$, $k \in \{0,1,\dots,m-1\}$. Then*

$$|B(t)| \leq (b-a)^{\frac{1-2s}{q}} 2^{\frac{1}{q}} G(\alpha,\nu,p,t,a) h_{\alpha-1}(t,a)$$

$$\times \left(\frac{|f^{\Delta^m}(a)|}{(e_\beta(a,x_0))^{\frac{1}{q}}}(h_2(a,b))^{\frac{s}{q}} + \frac{|f^{\Delta^m}(b)|}{(e_\beta(b,x_0))^{\frac{1}{q}}}(h_2(b,a))^{\frac{s}{q}} \right),$$

$$t \in [a,b],$$

where

$$G(\alpha,\nu,p,t,a) = \left(\int_a^t (h_\nu(t,\sigma(u)))^p \Delta u \right)^{\frac{1}{p}}, \quad t \in [a,b].$$

Proof. Since $|f^{\Delta^m}|^q$ is an exponentially s-convex function in the second sense on $[a,b]$, we have the inequality (5.15):

$$|B(t)| = \left| \int_a^t h_{\alpha-2}(t,\sigma(\tau)) \left(\int_a^\tau h_\nu(\tau,\sigma(u))f^{\Delta^m}(u)\Delta u \right) \Delta\tau \right|$$

$$\leq \int_a^t h_{\alpha-2}(t,\sigma(\tau)) \int_a^\tau h_\nu(\tau,\sigma(u))|f^{\Delta^m}(u)|\Delta u \Delta\tau$$

$$\leq \int_a^t h_{\alpha-2}(t,\sigma(\tau)) \left(\int_a^\tau (h_\nu(\tau,\sigma(u)))^p \Delta u \right)^{\frac{1}{p}}$$

$$\times \left(\int_a^\tau |f^{\Delta^m}(u)|^q \Delta u \right)^{\frac{1}{q}}$$

$$= \int_a^t h_{\alpha-2}(t, \sigma(\tau))G(\alpha, \nu, p, \tau, a)\left(\int_a^\tau |f^{\Delta^m}(u)|^q \Delta u\right)^{\frac{1}{q}} \Delta\tau$$

$$\leq G(\alpha, \nu, p, t, a)\int_a^t h_{\alpha-2}(t, \sigma(\tau))\left(\int_a^\tau |f^{\Delta^m}(u)|^q \Delta u\right)^{\frac{1}{q}} \Delta\tau$$

$$\leq G(\alpha, \nu, p, t, a)\int_a^t h_{\alpha-2}(t, \sigma(\tau))$$

$$\times \left(\int_a^\tau \left(\frac{(b-u)^s}{(b-a)^s}\frac{|f^{\Delta^m}(a)|^q}{e_\beta(a, x_0)} + \frac{(u-a)^s}{(b-a)^s}\frac{|f^{\Delta^m}(b)|^q}{e_\beta(b, x_0)}\right)\Delta u\right)^{\frac{1}{q}} \Delta\tau$$

$$\leq G(\alpha, \nu, p, t, a)\int_a^t h_{\alpha-2}(t, \sigma(\tau))\Delta\tau$$

$$\times \left(\frac{|f^{\Delta^m}(a)|^q}{e_\beta(a, x_0)(b-a)^s}\int_a^b (b-u)^s \Delta u\right.$$

$$\left. + \frac{|f^{\Delta^m}(b)|^q}{e_\beta(b, x_0)(b-a)^s}\int_a^b (u-a)^s \Delta u\right)^{\frac{1}{q}}$$

$$\leq G(\alpha, \nu, p, t, a)\left((b-a)^{1-2s}\frac{|f^{\Delta^m}(a)|^q}{e_\beta(a, x_0)}(h_2(a, b))^s\right.$$

$$\left. + (b-a)^{1-2s}\frac{|f^{\Delta^m}(b)|^q}{e_\beta(b, x_0)}(h_2(b, a))^s\right)^{\frac{1}{q}} h_{\alpha-1}(t, a)$$

$$\leq (b-a)^{\frac{1-2s}{q}}2^{\frac{1}{q}}G(\alpha, \nu, p, t, a)h_{\alpha-1}(t, a)$$

$$\times \left(\frac{|f^{\Delta^m}(a)|}{(e_\beta(a, x_0))^{\frac{1}{q}}}(h_2(a, b))^{\frac{s}{q}} + \frac{|f^{\Delta^m}(b)|}{(e_\beta(b, x_0))^{\frac{1}{q}}}(h_2(b, a))^{\frac{s}{q}}\right),$$

$$t \in [a, b],$$

where the last inequality results from $(x+y)^k \leq 2^k(x^k+y^k)$, $x, y \geq 0$, $k > 0$. This completes the proof. $\qquad\square$

Theorem 5.23. *Let* $\alpha > 2$, $m - 1 < \alpha < m$, $\nu = m - \alpha$, $f \in C_{rd}^m([a, b])$, $|f^{\Delta^m}|$ *be an exponential s-convex function in the second sense on* $[a, b]$. *Then*

$$|\Delta_a^{\alpha-1} f(t)| \leq \left(\frac{|f^{\Delta^m}(a)|}{e_\beta(a, x_0)} + \frac{|f^{\Delta^m}(b)|}{e_\beta(b, x_0)}\right)h_{\nu+1}(t, a), \quad t \in [a, b].$$

Proof.   Since $|f^{\Delta^m}|$ is exponentially s-convex in the second sense on $[a, b]$, we have the inequality (5.15). Then

$$
\begin{aligned}
&|\Delta_a^{\alpha-1} f(t)| \\
&= \left| \int_a^t h_\nu(t, \sigma(u)) f^{\Delta^m}(u) \Delta u \right| \\
&\leq \int_a^t h_\nu(t, \sigma(u)) |f^{\Delta^m}(u)| \Delta u \\
&\leq \int_a^t h_\nu(t, \sigma(u)) \left(\left(\frac{b-t}{b-a} \right)^s \frac{|f^{\Delta^m}(a)|}{e_\beta(a, x_0)} + \left(\frac{t-a}{b-a} \right)^s \frac{|f^{\Delta^m}(b)|}{e_\beta(b, x_0)} \right) \Delta u \\
&\leq \left(\frac{|f^{\Delta^m}(a)|}{e_\beta(a, x_0)} + \frac{|f^{\Delta^m}(b)|}{e_\beta(b, x_0)} \right) \int_a^t h_\nu(t, \sigma(u)) \Delta u \\
&= \left(\frac{|f^{\Delta^m}(a)|}{e_\beta(a, x_0)} + \frac{|f^{\Delta^m}(b)|}{e_\beta(b, x_0)} \right) h_{\nu+1}(t, a), \quad t \in [a, b].
\end{aligned}
$$

This completes the proof. $\square$

Theorem 5.24. *Let $q \geq 1$, $\alpha > 2$, $m - 1 < \alpha < m$, $\nu = m - \alpha$, $f \in C_{rd}^m([a, b])$, $|f^{\Delta^m}| \geq 1$ on $[a, b]$, $|f^{\Delta^m}|^q$ be an exponentially s-convex function in the second sense on $[a, b]$. Then*

$$
|\Delta_a^{\alpha-1} f(t)| \leq \left(\frac{|f^{\Delta^m}(a)|^q}{e_\beta(a, x_0)} + \frac{|f^{\Delta^m}(b)|^q}{e_\beta(b, x_0)} \right) h_{\nu+1}(t, a), \quad t \in [a, b].
$$

Proof.   Since $|f^{\Delta^m}|^q$ is exponentially s-convex in the second sense on $[a, b]$, we have the inequality (5.18). Then

$$
\begin{aligned}
&|\Delta_a^{\alpha-1} f(t)| \\
&= \left| \int_a^t h_\nu(t, \sigma(u)) f^{\Delta^m}(u) \Delta u \right| \\
&\leq \int_a^t h_\nu(t, \sigma(u)) |f^{\Delta^m}(u)| \Delta u
\end{aligned}
$$

$$\leq \int_a^t h_\nu(t, \sigma(u)) |f^{\Delta^m}(u)|^q \Delta u$$

$$\leq \int_a^t h_\nu(t, \sigma(u)) \left(\left(\frac{b-u}{b-a} \right)^s \frac{|f^{\Delta^m}(a)|^q}{e_\beta(a, x_0)} \right)$$

$$+ \left(\frac{u-a}{b-a} \right)^s \frac{|f^{\Delta^m}(b)|^q}{e_\beta(b, x_0)} \right) \Delta u$$

$$\leq \left(\frac{|f^{\Delta^m}(a)|^q}{e_\beta(a, x_0)} + \frac{|f^{\Delta^m}(b)|^q}{e_\beta(b, x_0)} \right) \int_a^t h_\nu(t, \sigma(u)) \Delta u$$

$$= \left(\frac{|f^{\Delta^m}(a)|^q}{e_\beta(a, x_0)} + \frac{|f^{\Delta^m}(b)|^q}{e_\beta(b, x_0)} \right) h_{\nu+1}(t, a), \quad t \in [a, b].$$

This completes the proof. □

Theorem 5.25. *Let* $p, q \geq 1$, $\frac{1}{p} + \frac{1}{q} = 1$, $\alpha > 2$, $m - 1 < \alpha < m$, $\nu = m - \alpha$, $f \in C_{rd}^m([a, b])$, $|f^{\Delta^m}|^q$ *be an exponentially s-convex function in the second sense on* $[a, b]$. *Then*

$$|\Delta_a^{\alpha-1} f(t)| \leq (b-a)^{\frac{1-2s}{q}} 2^{\frac{1}{q}} G(\alpha, \nu, p, t, a)$$

$$\times \left(\frac{|f^{\Delta^m}(a)|}{(e_\beta(a, x_0))^{\frac{1}{q}}} (h_2(a, b))^{\frac{s}{q}} + \frac{|f^{\Delta^m}(b)|}{(e_\beta(b, x_0))^{\frac{1}{q}}} (h_2(b, a))^{\frac{s}{q}} \right),$$

$$t \in [a, b],$$

where $G(\alpha, \nu, p, t, a)$, $t \in [a, b]$, *is defined as in Theorem* 5.22.

Proof. Since $|f^{\Delta^m}|^q$ is an exponentially s-convex function in the second sense on $[a, b]$, the inequality (5.18) holds. Then

$$|\Delta_a^{\alpha-1} f(t)| = \left| \int_a^t h_\nu(t, \sigma(u)) f^{\Delta^m}(u) \Delta u \right|$$

$$\leq \int_a^t h_\nu(t, \sigma(u)) |f^{\Delta^m}(u)| \Delta u$$

$$\leq \left(\int_a^t (h_\nu(t, \sigma(u)))^p \Delta u \right)^{\frac{1}{p}} \left(\int_a^t |f^{\Delta^m}(u)|^q \Delta u \right)^{\frac{1}{q}}$$

$$= G(\alpha, \nu, p, t, a) \left(\int_a^t |f^{\Delta^m}(u)|^q \Delta u \right)^{\frac{1}{q}}$$

$$\leq G(\alpha, \nu, p, t, a) \left(\int_a^t \left(\left(\frac{b-u}{b-a} \right)^s \frac{|f^{\Delta^m}(a)|^q}{e_\beta(a, x_0)} \right. \right.$$

$$\left. \left. + \left(\frac{u-a}{b-a} \right)^s \frac{|f^{\Delta^m}(b)|^q}{e_\beta(b, x_0)} \right) \Delta u \right)^{\frac{1}{q}}$$

$$\leq G(\alpha, \nu, p, t, a) \left((b-a)^{1-2s} \frac{|f^{\Delta^m}(a)|^q}{e_\beta(a, x_0)} (h_2(a, b))^s \right.$$

$$\left. + (b-a)^{1-2s} \frac{|f^{\Delta^m}(b)|^q}{e_\beta(b, x_0)} (h_2(b, a))^s \right)^{\frac{1}{q}}$$

$$\leq (b-a)^{\frac{1-2s}{q}} 2^{\frac{1}{q}} G(\alpha, \nu, p, t, a) \left(\frac{|f^{\Delta^m}(a)|}{(e_\beta(a, x_0))^{\frac{1}{q}}} (h_2(a, b))^{\frac{s}{q}} \right.$$

$$\left. + \frac{|f^{\Delta^m}(b)|}{(e_\beta(b, x_0))^{\frac{1}{q}}} (h_2(b, a))^{\frac{s}{q}} \right), \quad t \in [a, b].$$

This completes the proof.   □

Chapter 6

Opial-Type Inequalities

In this chapter, we introduce some opial-type inequalities on arbitrary time scales. Using the classical Young and Hölder inequalities and their refinements, we deduct the opial-type inequalities for Riemann–Liouville fractional derivatives, Poincaré inequalities and Ostrowski inequalities. Some of the results in this chapter can be found in Ref. [2].

Suppose that $\mathbb{T}$ is a time scale with forward jump operator σ and Δ, respectively. Let also $a, t_0 \in \mathbb{T}$, $t_0 < a$, $\alpha > 2$, $m - 1 < \alpha < m$, $m \in \mathbb{N}$, $\nu = m - \alpha$.

6.1 Opial-Type Inequalities for Riemann–Liouville Fractional Derivatives-I

In this section, we deduct some opial-type inequalities for Riemann–Liouville fractional derivatives.

Theorem 6.1. *Let* $f \in C_{rd}^m([t_0, a))$, $f^{\Delta^k}(t_0) = 0$, $k \in \{0, 1, \ldots, m - 1\}$, $p, q \geq 1$, $\frac{1}{p} + \frac{1}{q} = 1$. *Then*

$$\int_{t_0}^t |B(\tau)| \, |\Delta_{t_0}^{\alpha-1} f(\tau)|^q \, \Delta\tau$$

$$\leq \int_{t_0}^t |\Delta_{t_0}^{\alpha-1} f(\tau)|^q \, \Delta\tau \left(\frac{1}{p} \left(G(\alpha, \alpha - 2, p, a, t_0) \right)^p \right.$$

$$\left. + \frac{1}{q} \int_{t_0}^t |\Delta_{t_0}^{\alpha-1} f(\tau)|^q \, \Delta\tau \right), \quad t \in [t_0, a).$$

Proof.   By the definition of B and applying the Young inequality, we get

$$|B(t)| = \left| \int_{t_0}^{t} h_{\alpha-2}(t, \sigma(\tau)) \Delta_{t_0}^{\alpha-1} f(\tau) \Delta\tau \right|$$

$$\leq \int_{t_0}^{t} h_{\alpha-2}(t, \sigma(\tau)) \left| \Delta_{t_0}^{\alpha-1} f(\tau) \right| \Delta\tau$$

$$\leq \frac{1}{p} \int_{t_0}^{t} (h_{\alpha-2}(t, \sigma(\tau)))^p \, \Delta\tau + \frac{1}{q} \int_{t_0}^{t} \left| \Delta_{t_0}^{\alpha-1} f(\tau) \right|^q \Delta\tau$$

$$\leq \frac{1}{p} \int_{t_0}^{t} (h_{\alpha-2}(t, \sigma(\tau)))^p \, \Delta\tau$$

$$+ \frac{1}{q} \left(\int_{t_0}^{\sigma(t)} \left| \Delta_{t_0}^{\alpha-1} f(\tau) \right|^q \Delta\tau + \int_{t_0}^{t} \left| \Delta_{t_0}^{\alpha-1} f(\tau) \right|^q \Delta\tau \right)$$

$$= \frac{1}{p} (G(\alpha, \alpha - 2, p, t, t_0))^p$$

$$+ \frac{1}{q} \left(\int_{t_0}^{\sigma(t)} \left| \Delta_{t_0}^{\alpha-1} f(\tau) \right|^q \Delta\tau + \int_{t_0}^{t} \left| \Delta_{t_0}^{\alpha-1} f(\tau) \right|^q \Delta\tau \right)$$

$$\leq \frac{1}{p} (G(\alpha, \alpha - 2, p, a, t_0))^p$$

$$+ \frac{1}{q} \left(\int_{t_0}^{\sigma(t)} \left| \Delta_{t_0}^{\alpha-1} f(\tau) \right|^q \Delta\tau + \int_{t_0}^{t} \left| \Delta_{t_0}^{\alpha-1} f(\tau) \right|^q \Delta\tau \right),$$

$$t \in [t_0, a).$$

Let

$$z(t) = \int_{t_0}^{t} \left| \Delta_{t_0}^{\alpha-1} f(\tau) \right|^q \Delta\tau, \quad t \in [t_0, a).$$

Then $z(t_0) = 0$ and

$$z^{\Delta}(t) = \left| \Delta_{t_0}^{\alpha-1} f(t) \right|^q, \quad t \in [t_0, a).$$

Hence,

$$|B(t)| \left| \Delta_{t_0}^{\alpha-1} f(t) \right|^q$$

$$\leq \frac{1}{p} \left| \Delta_{t_0}^{\alpha-1} f(t) \right|^q (G(\alpha, \alpha - 2, p, a, t_0))^p$$

$$+ \frac{1}{q} \left| \Delta_{t_0}^{\alpha-1} f(t) \right|^q \left(\int_{t_0}^{\sigma(t)} \left| \Delta_{t_0}^{\alpha-1} f(\tau) \right|^q \Delta\tau + \int_{t_0}^{t} \left| \Delta_{t_0}^{\alpha-1} f(\tau) \right|^q \Delta\tau \right)$$

$$= \frac{1}{p} \left| \Delta_{t_0}^{\alpha-1} f(t) \right|^q (G(\alpha, \alpha - 2, p, a, t_0))^p z^\Delta(t)$$

$$+ \frac{1}{q} z^\Delta(t) \left(z^\sigma(t) + z(t) \right)$$

$$= \frac{1}{p} \left| \Delta_{t_0}^{\alpha-1} f(t) \right|^q (G(\alpha, \alpha - 2, p, a, t_0))^p z^\Delta(t)$$

$$+ \frac{1}{q} \left(z^2 \right)^\Delta (t), \quad t \in [t_0, a).$$

We integrate the last inequality from t_0 to t and find

$$\int_{t_0}^{t} |B(\tau)| \left| \Delta_{t_0}^{\alpha-1} f(\tau) \right|^q \Delta\tau$$

$$\leq \frac{1}{p} (G(\alpha, \alpha - 2, p, a, t_0))^p \int_{t_0}^{t} z^\Delta(\tau) \Delta\tau$$

$$+ \frac{1}{q} \int_{t_0}^{t} \left(z^2 \right)^\Delta (\tau) \Delta\tau$$

$$= \frac{1}{p} (G(\alpha, \alpha - 2, p, a, t_0))^p z(t) + \frac{1}{q} (z(t))^2$$

$$= \frac{1}{p} (G(\alpha, \alpha - 2, p, a, t_0))^p \int_{t_0}^{t} \left| \Delta_{t_0}^{\alpha-1} f(\tau) \right|^q \Delta\tau$$

$$+ \frac{1}{q} \left(\int_{t_0}^{t} \left| \Delta_{t_0}^{\alpha-1} f(\tau) \right|^q \Delta\tau \right)^2, \quad t \in [t_0, a),$$

whereupon we get the desired inequality. This completes the proof. $\square$

Remark 6.1. Let $l \in \{1, \ldots, m-1\}$. Then

$$B^{\Delta^l}(t) = I_{\Delta,t_0}^{\alpha-l-1}\left(\Delta_{t_0}^{\alpha-1}f\right)(t)$$

$$= \int_{t_0}^t h_{\alpha-l-2}(t, \sigma(u))\Delta_{t_0}^{\alpha-1}f(u)\Delta u, \quad t \in [t_0, a).$$

Hence, we get the following estimate:

$$\int_{t_0}^t \left|B^{\Delta^l}(\tau)\right|\left|\Delta_{t_0}^{\alpha-1}f(\tau)\right|^q \Delta\tau$$

$$\leq \int_{t_0}^t \left|\Delta_{t_0}^{\alpha-1}f(\tau)\right|^q \Delta\tau \left(\frac{1}{p}\left(G(\alpha, \alpha-l-2, p, a, t_0)\right)^p\right.$$

$$\left. + \frac{1}{q}\int_{t_0}^t \left|\Delta_{t_0}^{\alpha-1}f(\tau)\right|^q \Delta\tau\right), \quad t \in [t_0, a),$$

provided that all conditions of Theorem 6.1 hold.

Theorem 6.2. *Let $a < \infty$, $f \in C_{rd}^m([t_0, a))$, $f^{\Delta^k}(t_0) = 0$, $k \in \{0, 1, \ldots, m-1\}$, $p, q \geq 1$, $\frac{1}{p} + \frac{1}{q} = 1$, and $\left|\Delta_{t_0}^{\alpha-1}f(t)\right| \leq 1$, $t \in [t_0, a)$. Then*

$$\int_{t_0}^t \left|B(\tau)\right|\left|\Delta_{t_0}^{\alpha-1}f(\tau)\right|^q \Delta\tau$$

$$\leq \left(G(\alpha, \alpha-2, p, a, t_0)\right)^{\frac{p}{2}}\left(\left(\frac{1}{p} - r_0\right)\left(G(\alpha, \alpha-2, p, a, t_0)\right)^{\frac{1}{2}}\right.$$

$$\left. + 2r_0(a - t_0)^{\frac{1}{2}}\right)\int_{t_0}^t \left|\Delta_{t_0}^{\alpha_1}f(\tau)\right|^q \Delta\tau$$

$$+ \left(\frac{1}{q} - r_0\right)\left(\int_{t_0}^t \left|\Delta_{t_0}^{\alpha_1}f(\tau)\right|^q \Delta\tau\right)^2, \quad t \in [t_0, a),$$

where $r_0 = \min\left\{\frac{1}{p}, \frac{1}{q}\right\}$.

Proof. Let z be as in the proof of Theorem 6.1. Since

$$\left|\Delta_{t_0}^{\alpha-1} f(t)\right| \leq 1, \quad t \in [t_0, a),$$

we have

$$z^{\Delta}(t) \leq \left(z^{\Delta}(t)\right)^{\frac{1}{2}}, \quad t \in [t_0, a).$$

Now, by the definition of B and applying the refined Hölder inequality (see Theorem A.6 in Appendix A), we get

$$|B(t)| \leq \int_{t_0}^{t} h_{\alpha-2}(t, \sigma(\tau)) \left|\Delta_{t_0}^{\alpha-1} f(\tau)\right| \Delta \tau$$

$$\leq 2r_0 \int_{t_0}^{t} \left(h_{\alpha-2}(t, \sigma(\tau))\right)^{\frac{p}{2}} \left|\Delta_{t_0}^{\alpha-1} f(\tau)\right|^{\frac{q}{2}} \Delta \tau$$

$$+ \left(\frac{1}{p} - r_0\right) \int_{t_0}^{t} \left(h_{\alpha-2}(t, \sigma(\tau))\right)^p \Delta \tau$$

$$+ \left(\frac{1}{q} - r_0\right) \int_{t_0}^{t} \left|\Delta_{t_0}^{\alpha-1} f\tau\right|^q \Delta \tau$$

$$\leq 2r_0 \left(\int_{t_0}^{t} \left(h_{\alpha-2}(t, \sigma(\tau))\right)^p \Delta \tau\right)^{\frac{1}{2}} \left(\int_{t_0}^{t} \left|\Delta_{t_0}^{\alpha-1} f(\tau)\right|^q \Delta \tau\right)^{\frac{1}{2}}$$

$$+ \left(\frac{1}{p} - r_0\right) \int_{t_0}^{t} \left(h_{\alpha-2}(t, \sigma(\tau))\right)^p \Delta \tau$$

$$+ \left(\frac{1}{q} - r_0\right) \int_{t_0}^{t} \left|\Delta_{t_0}^{\alpha-1} f(\tau)\right|^q \Delta \tau$$

$$\leq 2r_0 \left(G(\alpha, \alpha - 2, p, a, t_0)\right)^{\frac{p}{2}} \left(\int_{t_0}^{t} \left|\Delta_{t_0}^{\alpha-1} f(\tau)\right|^q \Delta \tau\right.$$

$$\left. + \int_{t_0}^{\sigma(t)} \left|\Delta_{t_0}^{\alpha-1} f(\tau)\right|^q \Delta \tau\right)^{\frac{1}{2}} + \left(\frac{1}{p} - r_0\right) \left(G(\alpha, \alpha - 2, p, a, t_0)\right)^p$$

$$+ \left(\frac{1}{q} - r_0\right) \left(\int_{t_0}^{t} \left|\Delta_{t_0}^{\alpha-1} f(\tau)\right|^q \Delta \tau + \int_{t_0}^{\sigma(t)} \left|\Delta_{t_0}^{\alpha-1} f(\tau)\right|^q \Delta \tau\right)$$

$$= \left(\frac{1}{p} - r_0\right) \left(G(\alpha, \alpha - 2, p, a, t_0)\right)^{\frac{p}{2}}$$

$$+ 2r_0 \left(G(\alpha, \alpha - 2, p, a, t_0)\right)^{\frac{p}{2}} \left(z^\sigma(t) + z(t)\right)^{\frac{1}{2}}$$

$$+ \left(\frac{1}{q} - r_0\right) \left(z^\sigma(t) + z(t)\right), \quad t \in [t_0, a).$$

Hence,

$$|B(t)| \left|\Delta_{t_0}^{\alpha-1} f(t)\right|^q$$

$$= |Bt)|z^\Delta(t) \le \left(\frac{1}{p} - r_0\right) \left(G(\alpha, \alpha - 2, p, a, t_0)\right)^{\frac{p}{2}} z^\Delta(t)$$

$$+ 2r_0 \left(G(\alpha, \alpha - 2, p, a, t_0)\right)^{\frac{p}{2}} z^\Delta(t) \left(z^\sigma(t) + z(t)\right)^{\frac{1}{2}}$$

$$+ \left(\frac{1}{q} - r_0\right) z^\Delta(t) \left(z^\sigma(t) + z(t)\right)$$

$$\le \left(\frac{1}{p} - r_0\right) \left(G(\alpha, \alpha - 2, p, a, t_0)\right)^{\frac{p}{2}} z^\Delta(t)$$

$$+ 2r_0 \left(G(\alpha, \alpha - 2, p, a, t_0)\right)^{\frac{p}{2}} \left(z^\Delta(t) \left(z^\sigma(t) + z(t)\right)\right)^{\frac{1}{2}}$$

$$+ \left(\frac{1}{q} - r_0\right) \left(z^2\right)^\Delta (t)$$

$$= \left(\frac{1}{p} - r_0\right) \left(G(\alpha, \alpha - 2, p, a, t_0)\right)^{\frac{p}{2}} z^\Delta(t)$$

$$+ 2r_0 \left(G(\alpha, \alpha - 2, p, a, t_0)\right)^{\frac{p}{2}} \left(\left(z^2\right)^\Delta (t)\right)^{\frac{1}{2}}$$

$$+ \left(\frac{1}{q} - r_0\right) \left(z^2\right)^\Delta (t), \quad t \in [t_0, a).$$

We integrate the last inequality form t_0 to t and find

$$\int_{t_0}^t |B(\tau)| \left|\Delta_{t_0}^{\alpha-1} f(\tau)\right|^q \Delta\tau$$

$$\le \left(\frac{1}{p} - r_0\right) \left(G(\alpha, \alpha - 2, p, a, t_0)\right)^{\frac{p}{2}} \int_{t_0}^t z^\Delta(\tau)\Delta\tau$$

$$+ 2r_0 \left(G(\alpha, \alpha - 2, p, a, t_0)\right)^{\frac{p}{2}} \int_{t_0}^t \left(\left(z^2\right)^\Delta (\tau)\right)^{\frac{1}{2}} \Delta\tau$$

$$+ \left(\frac{1}{q} - r_0 \right) \int_{t_0}^t \left(z^2 \right)^\Delta (\tau) \Delta \tau$$

$$\leq \left(\frac{1}{p} - r_0 \right) (G(\alpha, \alpha - 2, p, a, t_0))^{\frac{p}{2}} z(t)$$

$$+ 2r_0 \left(G(\alpha, \alpha - 2, p, a, t_0) \right)^{\frac{p}{2}} (t - t_0)^{\frac{1}{2}}$$

$$\times \left(\int_{t_0}^t \left(z^2 \right)^\Delta (\tau) \Delta \tau \right)^{\frac{1}{2}} + \left(\frac{1}{q} - r_0 \right) (z(t))^2$$

$$\leq \left(\frac{1}{p} - r_0 \right) (G(\alpha, \alpha - 2, p, a, t_0))^{\frac{p}{2}} z(t)$$

$$+ 2r_0 \left(G(\alpha, \alpha - 2, p, a, t_0) \right)^{\frac{p}{2}} (a - t_0)^{\frac{1}{2}} z(t) + \left(\frac{1}{q} - r_0 \right) (z(t))^2$$

$$= (G(\alpha, \alpha - 2, p, a, t_0))^{\frac{p}{2}} \left(\left(\frac{1}{p} - r_0 \right) (G(\alpha, \alpha - 2, p, a, t_0))^{\frac{1}{2}} \right.$$

$$\left. + 2r_0 (a - t_0)^{\frac{1}{2}} \right) z(t) + \left(\frac{1}{q} - r_0 \right) (z(t))^2$$

$$= (G(\alpha, \alpha - 2, p, a, t_0))^{\frac{p}{2}} \left(\left(\frac{1}{p} - r_0 \right) (G(\alpha, \alpha - 2, p, a, t_0))^{\frac{1}{2}} \right.$$

$$\left. + 2r_0 (a - t_0)^{\frac{1}{2}} \right) \int_{t_0}^t \left| \Delta_{t_0}^{\alpha_1} f(\tau) \right|^q \Delta \tau$$

$$+ \left(\frac{1}{q} - r_0 \right) \left(\int_{t_0}^t \left| \Delta_{t_0}^{\alpha_1} f(\tau) \right|^q \Delta \tau \right)^2, \quad t \in [t_0, a).$$

This completes the proof. $\qquad \square$

Remark 6.2. Let l be as in Remark 6.1. Suppose that all conditions of Theorem 6.2 hold. Then we have the following estimate:

$$\int_{t_0}^t \left| B^{\Delta^l}(\tau) \right| \left| \Delta_{t_0}^{\alpha - 1} f(\tau) \right|^q \Delta \tau$$

$$\leq (G(\alpha, \alpha - l - 2, p, a, t_0))^{\frac{p}{2}} \left(\left(\frac{1}{p} - r_0 \right) (G(\alpha, \alpha - 2, p, a, t_0))^{\frac{1}{2}} \right.$$

$$+ 2r_0(a - t_0)^{\frac{1}{2}}\right) \int_{t_0}^t \left|\Delta_{t_0}^{\alpha_1} f(\tau)\right|^q \Delta\tau$$

$$+ \left(\frac{1}{q} - r_0\right) \left(\int_{t_0}^t \left|\Delta_{t_0}^{\alpha_1} f(\tau)\right|^q \Delta\tau\right)^2, \quad t \in [t_0, a),$$

Theorem 6.3. *Let* $f \in C_{rd}^m([t_0, a))$, $f^{\Delta^k}(t_0) = 0$, $k \in \{0, 1, \ldots, m - 1\}$,

$$h_{\alpha-2}(t, s) \geq \left|\Delta_{t_0}^{\alpha-1} f(t)\right| > 0, \quad t, s \in [t_0, a), \quad s \leq t, \qquad (6.1)$$

and $p, q \geq 1$, $\frac{1}{p} + \frac{1}{q} = 1$. *Then*

$$\int_{t_0}^t B(\tau) \left|\Delta_{t_0}^{\alpha-1} f(\tau)\right|^q \Delta\tau$$

$$\leq \frac{1}{p} (G(\alpha, \alpha - 2, p, a, t_0))^p \int_{t_0}^t \left|\Delta_{t_0}^{\alpha-1} f(\tau)\right|^q \Delta\tau$$

$$+ \frac{1}{q} \left(\int_{t_0}^t \left|\Delta_{t_0}^{\alpha-1} f(\tau)\right| \Delta\tau\right)^2, \quad t \in [t_0, a).$$

Proof. Let $r = \min\left\{\frac{1}{p}, \frac{1}{q}\right\}$ and S be the Specht ratio. Since $S(1) = 1$, S is increasing on $[1, \infty)$, using (6.1), we get

$$\frac{h_{\alpha-2}(t, s)}{\left|\Delta_{t_0}^{\alpha-1} f(\tau)\right|} \leq \left(\frac{h_{\alpha-2}(t, s)}{\left|\Delta_{t_0}^{\alpha-1} f(\tau)\right|}\right)^r, \quad t, s \in [t_0, a), \quad s \leq t.$$

Let also z be as in the proof of Theorem 6.1. Then, using the definition of B and the refined Young inequality (see Theorem A.7 in Appendix A), we find

$$|B(t)| \leq \int_{t_0}^t h_{\alpha-2}(t, \sigma(\tau)) \left|\Delta_{t_0}^{\alpha} f(\tau)\right| \Delta\tau$$

$$\leq \int_{t_0}^t S\left(\left(\frac{h_{\alpha-2}(t, \sigma(\tau))}{\left|\Delta_{t_0}^{\alpha} f(\tau)\right|}\right)^r\right) h_{\alpha-2}(t, \sigma(\tau)) \left|\Delta_{t_0}^{\alpha} f(\tau)\right| \Delta\tau$$

$$\leq \frac{1}{p} \int_{t_0}^{t} (h_{\alpha-2}(t,\sigma(\tau)))^p \, \Delta\tau + \frac{1}{q} \int_{t_0}^{t} \left| \Delta_{t_0}^\alpha f(\tau) \right|^q \Delta\tau$$

$$\leq \frac{1}{p} \int_{t_0}^{t} (h_{\alpha-2}(t,\sigma(\tau)))^p \, \Delta\tau$$

$$+ \frac{1}{q} \left(\int_{t_0}^{t} \left| \Delta_{t_0}^\alpha f(\tau) \right|^q \Delta\tau + \int_{t_0}^{\sigma(t)} \left| \Delta_{t_0}^\alpha f(\tau) \right|^q \Delta\tau \right)$$

$$= \frac{1}{p} \int_{t_0}^{t} (h_{\alpha-2}(t,\sigma\tau))^p \, \Delta\tau + \frac{1}{q} (z(t) + z^\sigma(t))$$

$$= \frac{1}{p} (G(\alpha, \alpha-2, p, t, t_0))^p + \frac{1}{q} (z(t) + z^\sigma(t))$$

$$\leq \frac{1}{p} (G(\alpha, \alpha-2, p, a, t_0))^p + \frac{1}{q} (z(t) + z^\sigma(t)), \quad t \in [t_0, a).$$

Hence,

$$B(t) \left| \Delta_{t_0}^{\alpha-1} f(t) \right|^p$$

$$= B(t) z^\Delta(t) \leq \frac{1}{p} (G(\alpha, \alpha-2, p, a, t_0))^p z^\Delta(t)$$

$$+ \frac{1}{q} z^\Delta(t) (z(t) + z^\sigma(t))$$

$$\leq \frac{1}{p} (G(\alpha, \alpha-2, p, a, t_0))^p z^\Delta(t) + \frac{1}{q} ((z^2)^\Delta(t)), \quad t \in [t_0, a).$$

Now, we integrate the last inequality from t_0 to t and find

$$\int_{t_0}^{t} B(\tau) \left| \Delta_{t_0}^{\alpha-1} f(\tau) \right|^p \Delta\tau$$

$$\leq \frac{1}{p} (G(\alpha, \alpha-2, p, a, t_0))^p \int_{t_0}^{t} z^\Delta(\tau) \Delta\tau + \frac{1}{q} \int_{t_0}^{t} ((z^2)^\Delta(\tau)) \, \Delta\tau$$

$$= \frac{1}{p} (G(\alpha, \alpha-2, p, a, t_0))^p z(t) + \frac{1}{q} (z(t))^2$$

$$= \frac{1}{p} \left(G(\alpha, \alpha - 2, p, a, t_0) \right)^p \int_{t_0}^t \left| \Delta_{t_0}^\alpha f(\tau) \right|^q \Delta\tau$$

$$+ \frac{1}{q} \left(\int_{t_0}^t \left| \Delta_{t_0}^\alpha f(\tau) \right|^q \Delta\tau \right)^2, \quad t \in [t_0, a).$$

This completes the proof. □

Remark 6.3. Let l be as in Remark 6.1. Suppose that all conditions of Theorem 6.3 hold. Then we have the following estimate:

$$\int_{t_0}^t B^{\Delta^l}(\tau) \left| \Delta_{t_0}^{\alpha-1} f(\tau) \right|^q \Delta\tau$$

$$\leq \frac{1}{p} \left(G(\alpha, \alpha - l - 2, p, a, t_0) \right)^p \int_{t_0}^t \left| \Delta_{t_0}^{\alpha-1} f(\tau) \right|^q \Delta\tau$$

$$+ \frac{1}{q} \left(\int_{t_0}^t \left| \Delta_{t_0}^{\alpha-1} f(\tau) \right| \Delta\tau \right)^2, \quad t \in [t_0, a).$$

6.2   Opial-Type Inequalities for Riemann–Liouville Fractional Derivatives-II

In the previous section, we have deducted some opial-type inequalities for Riemann–Liouville fractional derivatives using the Young inequality and its refinements. In this section, we use the Hölder inequality to deduct some opial-type inequalities for Riemann–Liouville fractional derivatives.

Theorem 6.4. Let $f \in C_{rd}^m([t_0, a))$, $f^{\Delta^k}(t_0) = 0$, $k \in \{0, 1, \ldots, m - 1\}$, $p, q \geq 1$, $\frac{1}{p} + \frac{1}{q} = 1$. Then

$$\int_{t_0}^t \left| B(\tau) \right| \left| \Delta_{t_0}^{\alpha-1} f(\tau) \right| \Delta\tau$$

$$\leq \left(\int_{t_0}^t \left(G(\alpha, \alpha - 2, p, \tau, t_0) \right)^p \Delta\tau \right)^{\frac{1}{p}} \left(\int_{t_0}^t \left| \Delta_{t_0}^{\alpha-1} f(\tau) \right|^q \Delta\tau \right)^{\frac{1}{q}},$$

$$t \in [t_0, a).$$

Proof. Suppose that z is as in the proof of Theorem 6.1. Now, using the definition for B and the classical Hölder inequality, we get

$$
\begin{aligned}
|B(t)| &= \left| \int_{t_0}^{t} h_{\alpha-2}(t, \sigma(\tau)) \Delta_{t_0}^{\alpha-1} f(\tau) \Delta \tau \right| \\
&\leq \int_{t_0}^{t} h_{\alpha-2}(t, \sigma(\tau)) \left| \Delta_{t_0}^{\alpha-1} f(\tau) \right| \Delta \tau \\
&\leq \left(\int_{t_0}^{t} (h_{\alpha-2}(t, \sigma(\tau))) \, \Delta \tau \right)^{\frac{1}{p}} \left(\int_{t_0}^{t} \left| \Delta_{t_0}^{\alpha-1} f(\tau) \right|^q \Delta \tau \right)^{\frac{1}{q}} \\
&= G(\alpha, \alpha - 2, p, t, t_0) \left(\int_{t_0}^{t} \left| \Delta_{t_0}^{\alpha-1} f(\tau) \right|^q \Delta \tau \right)^{\frac{1}{q}} \\
&\leq G(\alpha, \alpha - 2, p, t, t_0) \left(\int_{t_0}^{t} \left| \Delta_{t_0}^{\alpha-1} f(\tau) \right|^q \Delta \tau \right. \\
&\qquad\qquad \left. + \int_{t_0}^{\sigma(t)} \left| \Delta_{t_0}^{\alpha-1} f(\tau) \right|^q \Delta \tau \right)^{\frac{1}{q}} \\
&= G(\alpha, \alpha - 2, p, t, t_0) \left(z^{\sigma}(t) + z(t) \right)^{\frac{1}{q}}, \quad t \in [t_0, a).
\end{aligned}
$$

Hence,

$$
\begin{aligned}
|B(t)| \left| \Delta_{t_0}^{\alpha-1} f(t) \right| &= |B(t)| \left(z^{\Delta}(t) \right)^{\frac{1}{q}} \\
&\leq G(\alpha, \alpha - 2, p, t, t_0) \left(z^{\Delta}(t) \left(z^{\sigma}(t) + z(t) \right) \right)^{\frac{1}{q}}, \\
&\qquad t \in [t_0, a).
\end{aligned}
$$

We integrate the last inequality from t_0 to t and arrive at the inequality

$$
\begin{aligned}
&\int_{t_0}^{t} |B(\tau)| \left| \Delta_{t_0}^{\alpha-1} f(\tau) \right| \Delta \tau \\
&\qquad \leq \int_{t_0}^{t} G(\alpha, \alpha - 2, p, \tau, t_0) \left((z^2)^{\Delta}(\tau) \right)^{\frac{1}{q}} \Delta \tau
\end{aligned}
$$

$$\leq \left(\int_{t_0}^{t} (G(\alpha, \alpha - 2, p, \tau, t_0))^p \, \Delta\tau \right)^{\frac{1}{p}} \left(\int_{t_0}^{t} (z^2)^{\Delta}(\tau) \Delta\tau \right)^{\frac{1}{q}}$$

$$= \left(\int_{t_0}^{t} (G(\alpha, \alpha - 2, p, \tau, t_0))^p \, \Delta\tau \right)^{\frac{1}{p}} (z(t))^{\frac{2}{q}}$$

$$= \left(\int_{t_0}^{t} (G(\alpha, \alpha - 2, p, \tau, t_0))^p \, \Delta\tau \right)^{\frac{1}{p}} \left(\int_{t_0}^{t} |\Delta_{t_0}^{\alpha-1} f(\tau)|^q \, \Delta\tau \right)^{\frac{2}{q}},$$

$$t \in [t_0, a).$$

This completes the proof.   □

Remark 6.4. Let l be as in Remark 6.1. Suppose that all conditions of Theorem 6.4 hold. Then we have the following estimate:

$$\int_{t_0}^{t} |B^{\Delta^l}(\tau)| \, |\Delta_{t_0}^{\alpha-1} f(\tau)| \, \Delta\tau$$

$$\leq \left(\int_{t_0}^{t} (G(\alpha, \alpha - l - 2, p, \tau, t_0))^p \, \Delta\tau \right)^{\frac{1}{p}}$$

$$\times \left(\int_{t_0}^{t} |\Delta_{t_0}^{\alpha-1} f(\tau)|^q \, \Delta\tau \right)^{\frac{1}{q}}, \quad t \in [t_0, a).$$

Theorem 6.5. *Let* $f \in C_{rd}^m([t_0, a))$, $f^{\Delta^k}(t_0) = 0$, $k \in \{0, 1, \ldots, m-1\}$, $p = 1$, $q = \infty$. *Then*

$$\int_{t_0}^{t} |B(\tau)| \, |\Delta_{t_0}^{\alpha-1} f(\tau)| \, \Delta\tau$$

$$\leq \|\Delta_{t_0}^{\alpha-1} f\|_\infty \int_{t_0}^{t} h_{\alpha-1}(t, \sigma(u)) \Delta u, \quad t \in [t_0, a).$$

Proof. By the definition of B, we get

$$|B(t)| = \left| \int_{t_0}^{t} h_{\alpha-2}(t, \sigma(u)) \Delta_{t_0}^{\alpha-1} f(u) \Delta u \right|$$

$$\leq \int_{t_0}^{t} h_{\alpha-2}(t, \sigma(u)) \, |\Delta_{t_0}^{\alpha-1} f(u)| \, \Delta u$$

$$\leq \|\Delta_{t_0}^{\alpha-1} f\|_\infty \left(\int_{t_0}^{t} h_{\alpha-2}(t, \sigma(u)) \Delta u \right), \quad t \in [t_0, a).$$

Hence,

$$|B(t)| \left|\Delta_{t_0}^{\alpha-1} f(t)\right| \le \left\|\Delta_{t_0}^{\alpha-1} f\right\|_\infty \left(\int_{t_0}^t h_{\alpha-2}(t, \sigma(u)) \Delta u\right) \left|\Delta_{t_0}^{\alpha-1} f(t)\right|,$$

$$t \in [t_0, a).$$

Now, we integrate the last inequality from t_0 to t and find

$$\int_{t_0}^t |B(\tau)| \left|\Delta_{t_0}^{\alpha-1} f(\tau)\right| \Delta\tau$$

$$\le \left\|\Delta_{t_0}^{\alpha-1} f\right\|_\infty \int_{t_0}^t \left(\int_{t_0}^\tau h_{\alpha-2}(\tau, \sigma(u)) \Delta u\right) \left|\Delta_{t_0}^{\alpha-1} f(\tau)\right| \Delta\tau$$

$$\le \left\|\Delta_{t_0}^{\alpha-1} f\right\|_\infty^2 \int_{t_0}^t \int_{t_0}^\tau h_{\alpha-2}(\tau, \sigma(u)) \Delta u \Delta\tau$$

$$= \left\|\Delta_{t_0}^{\alpha-1} f\right\|_\infty^2 \int_{t_0}^t \int_{\sigma(u)}^t h_{\alpha-2}(\tau, \sigma(u)) \Delta\tau \Delta u$$

$$= \left\|\Delta_{t_0}^{\alpha-1} f\right\|_\infty^2 \int_{t_0}^t h_{\alpha-1}(t, \sigma(u)) \Delta u, \quad t \in [t_0, a).$$

This completes the proof. □

Remark 6.5. Let l be as in Remark 6.1. Suppose that all conditions of Theorem 6.5 hold. Then we have the following estimate:

$$\int_{t_0}^t \left|B^{\Delta^l}(\tau)\right| \left|\Delta_{t_0}^{\alpha-1} f(\tau)\right| \Delta\tau$$

$$\le \left\|\Delta_{t_0}^{\alpha-1} f\right\|_\infty \int_{t_0}^t h_{\alpha-l-1}(t, \sigma(u)) \Delta u, \quad t \in [t_0, a).$$

6.3 Poincaré-Type Inequalities

In this section, we deduct some Poincaré-type inequalities using the Young and Hölder inequalities.

Theorem 6.6. *Let* $f \in \mathcal{C}_{rd}^m([t_0, a))$, $f^{\Delta^k}(t_0) = 0$, $k \in \{0, 1, \ldots, m-1\}$, $p, q \geq 1$, $\frac{1}{p} + \frac{1}{q} = 1$. *Then*

$$\int_{t_0}^t |B(\tau)|^q \Delta\tau \leq \frac{2^q}{p} \int_{t_0}^t (G(\alpha, \alpha - 2, p, t_1, t_0))^{pq} \, \Delta t_1$$

$$+ \frac{2^q}{q} \int_{t_0}^t \left(\int_{t_0}^{t_1} |\Delta_{t_0}^{\alpha-1} f(\tau)|^q \, \Delta\tau \right)^q \Delta t_1, \quad t \in [t_0, a).$$

Proof. As in the proof of Theorem 6.1, we have

$$|B(t)| \leq \frac{1}{p} \int_{t_0}^t (h_{\alpha-2}(t, \sigma(\tau)))^p \, \Delta\tau + \frac{1}{q} \int_{t_0}^t |\Delta_{t_0}^{\alpha-1} f(\tau)|^q \, \Delta\tau, \quad t \in [t_0, a),$$

whereupon

$$|B(t)|^q \leq \left(\frac{1}{p} (G(\alpha, \alpha - l - 2, p, t, t_0))^p + \frac{1}{q} \int_{t_0}^t |\Delta_{t_0}^{\alpha-1} f(\tau)|^q \, \Delta\tau \right)^q$$

$$\leq \frac{2^q}{p} (G(\alpha, \alpha - 2, p, t_1, t_0))^{pq} + \frac{2^q}{q} \left(\int_{t_0}^t |\Delta_{t_0}^{\alpha-1} f(\tau)|^q \, \Delta\tau \right)^q,$$

$$t \in [t_0, a),$$

and

$$\int_{t_0}^t |B(\tau)|^q \Delta\tau \leq \frac{2^q}{p} \int_{t_0}^t (G(\alpha, \alpha - 2, p, t_1, t_0))^{pq} \, \Delta t_1$$

$$+ \frac{2^q}{q} \int_{t_0}^t \left(\int_{t_0}^{t_1} |\Delta_{t_0}^{\alpha-1} f(\tau)|^q \, \Delta\tau \right)^q \Delta t_1, \quad t \in [t_0, a).$$

This completes the proof. □

Remark 6.6. Let l be as in Remark 6.1. Then, by Theorem 6.1 , we get the following estimate:

$$\int_{t_0}^t |B^{\Delta^l}(\tau)|^q \Delta\tau \leq \frac{2^q}{p} \int_{t_0}^t (G(\alpha, \alpha - l - 2, p, t_1, t_0))^{pq} \, \Delta t_1$$

$$+ \frac{2^q}{q} \int_{t_0}^t \left(\int_{t_0}^{t_1} |\Delta_{t_0}^{\alpha-1} f(\tau)|^q \, \Delta\tau \right)^q \Delta t_1, \quad t \in [t_0, a),$$

provided that all conditions of Theorem 6.1 hold.

Theorem 6.7. *Let* $a < \infty$, $f \in C_{rd}^m([t_0, a))$, $f^{\Delta^k}(t_0) = 0$, $k \in \{0, 1, \ldots, m-1\}$, $p, q \geq 1$, $\frac{1}{p} + \frac{1}{q} = 1$, *and* $\left|\Delta_{t_0}^{\alpha-1} f(t)\right| \leq 1$, $t \in [t_0, a)$. *Then*

$$\int_{t_0}^t |B(t_1)|^q \Delta t_1$$

$$\leq 2^{q+1} r_0^q \left(G(\alpha, \alpha - 2, p, a, t_0)\right)^{\frac{pq}{2}} \int_{t_0}^t \left(\int_{t_0}^{t_1} \left|\Delta_{t_0}^{\alpha-1} f(\tau)\right|^q \Delta \tau\right)^{\frac{q}{2}} \Delta t_1$$

$$+ 2^{2q} \left(\frac{1}{p} - r_0\right)^q \int_{t_0}^t \left(G(\alpha, \alpha - 2, p, t_1, t_0)\right)^{pq} \Delta t_1$$

$$+ 2^{2q} \left(\frac{1}{q} - r_0\right)^q \int_{t_0}^t \left(\int_{t_0}^{t_1} \left|\Delta_{t_0}^{\alpha-1} f(\tau)\right|^q \Delta \tau\right)^q \Delta t_1, \quad t \in [t_0, a),$$

where $r_0 = \min\left\{\frac{1}{p}, \frac{1}{q}\right\}$.

Proof. As in the proof of Theorem 6.2, we get

$$|B(t)| \leq 2r_0 \left(G(\alpha, \alpha - 2, p, a, t_0)\right)^{\frac{p}{2}} \left(\int_{t_0}^t \left|\Delta_{t_0}^{\alpha-1} f(\tau)\right|^q \Delta \tau\right)^{\frac{1}{2}}$$

$$+ \left(\frac{1}{p} - r_0\right) \left(G(\alpha, \alpha - 2, p, t, t_0)\right)^p$$

$$+ \left(\frac{1}{q} - r_0\right) \left(\int_{t_0}^t \left|\Delta_{t_0}^{\alpha-1} f(\tau)\right|^q \Delta \tau\right), \quad t \in [t_0, a).$$

Hence,

$$|B(t)|^q \leq \left(2r_0 \left(G(\alpha, \alpha - 2, p, a, t_0)\right)^{\frac{p}{2}} \left(\int_{t_0}^t \left|\Delta_{t_0}^{\alpha-1} f(\tau)\right|^q \Delta \tau\right)^{\frac{1}{2}}\right.$$

$$+ \left(\frac{1}{p} - r_0\right) \left(G(\alpha, \alpha - 2, p, t, t_0)\right)^p$$

$$\left.+ \left(\frac{1}{q} - r_0\right) \left(\int_{t_0}^t \left|\Delta_{t_0}^{\alpha-1} f(\tau)\right|^q \Delta \tau\right)\right)^q$$

$$\leq 2^{q+1} r_0^q \left(G(\alpha, \alpha - 2, p, a, t_0)\right)^{\frac{pq}{2}} \left(\int_{t_0}^t |\Delta_{t_0}^{\alpha-1} f(\tau)|^q \, \Delta\tau\right)^{\frac{q}{2}}$$

$$+ 2^q \left(\left(\frac{1}{p} - r_0\right) (G(\alpha, \alpha - 2, p, t, t_0))^p\right.$$

$$+ \left(\frac{1}{q} - r_0\right) \left(\int_{t_0}^t |\Delta_{t_0}^{\alpha-1} f(\tau)|^q \, \Delta\tau\right)\right)^q$$

$$\leq 2^{q+1} r_0^q \left(G(\alpha, \alpha - 2, p, a, t_0)\right)^{\frac{pq}{2}} \left(\int_{t_0}^t |\Delta_{t_0}^{\alpha-1} f(\tau)|^q \, \Delta\tau\right)^{\frac{q}{2}}$$

$$+ 2^{2q} \left(\frac{1}{p} - r_0\right)^q (G(\alpha, \alpha - 2, p, t, t_0))^{pq}$$

$$+ 2^{2q} \left(\frac{1}{q} - r_0\right)^q \left(\int_{t_0}^t |\Delta_{t_0}^{\alpha-1} f(\tau)|^q \, \Delta\tau\right)^q, \quad t \in [t_0, a).$$

Therefore,

$$\int_{t_0}^t |B(t_1)|^q \Delta t_1$$

$$\leq 2^{q+1} r_0^q \left(G(\alpha, \alpha - 2, p, a, t_0)\right)^{\frac{pq}{2}} \int_{t_0}^t \left(\int_{t_0}^{t_1} |\Delta_{t_0}^{\alpha-1} f(\tau)|^q \, \Delta\tau\right)^{\frac{q}{2}} \Delta t_1$$

$$+ 2^{2q} \left(\frac{1}{p} - r_0\right)^q \int_{t_0}^t (G(\alpha, \alpha - 2, p, t_1, t_0))^{pq} \, \Delta t_1$$

$$+ 2^{2q} \left(\frac{1}{q} - r_0\right)^q \int_{t_0}^t \left(\int_{t_0}^{t_1} |\Delta_{t_0}^{\alpha-1} f(\tau)|^q \, \Delta\tau\right)^q \Delta t_1, \quad t \in [t_0, a).$$

This completes the proof. $\qquad\square$

Remark 6.7. Let l be as in Remark 6.1. Suppose that all conditions of Theorem 6.7 hold. Then we have the following estimate:

$$\int_{t_0}^t |B^{\Delta^l}(t_1)|^q \Delta t_1$$

$$\leq 2^{q+1} r_0^q \left(G(\alpha, \alpha - l - 2, p, a, t_0)\right)^{\frac{pq}{2}}$$

$$\times \int_{t_0}^{t} \left(\int_{t_0}^{t_1} \left| \Delta_{t_0}^{\alpha-1} f(\tau) \right|^q \Delta\tau \right)^{\frac{q}{2}} \Delta t_1$$

$$+ 2^{2q} \left(\frac{1}{p} - r_0 \right)^q \int_{t_0}^{t} (G(\alpha, \alpha - l - 2, p, t_1, t_0))^{pq} \, \Delta t_1$$

$$+ 2^{2q} \left(\frac{1}{q} - r_0 \right)^q \int_{t_0}^{t} \left(\int_{t_0}^{t_1} \left| \Delta_{t_0}^{\alpha-1} f(\tau) \right|^q \Delta\tau \right)^q \Delta t_1, \quad t \in [t_0, a).$$

Theorem 6.8. *Let* $f \in C_{rd}^m([t_0, a))$, $f^{\Delta^k}(t_0) = 0$, $k \in \{0, 1, \dots, m-1\}$, $p, q \geq 1$, $\frac{1}{p} + \frac{1}{q} = 1$. *Then*

$$\int_{t_0}^{t} |B(t_1)|^q \Delta t_1$$

$$\leq (G(\alpha, \alpha - 2, p, a, t_0))^q \int_{t_0}^{t} \int_{t_0}^{t_1} \left| \Delta_{t_0}^{\alpha-1} f(\tau) \right|^q \Delta\tau \Delta t_1$$

$$= (G(\alpha, \alpha - 2, p, a, t_0))^q \int_{t_0}^{t} (t - \sigma(\tau)) \left| \Delta_{t_0}^{\alpha-1} f(\tau) \right|^q \Delta\tau, \quad t \in [t_0, a).$$

Proof. As in the proof of Theorem 6.4, we get

$$|B(t)| \leq G(\alpha, \alpha - 2, p, a, t_0) \left(\int_{t_0}^{t} \left| \Delta_{t_0}^{\alpha-1} f(\tau) \right|^q \Delta\tau \right)^{\frac{1}{q}}, \quad t \in [t_0, a).$$

Hence,

$$|B(t)|^q \leq (G(\alpha, \alpha - 2, p, a, t_0))^q \int_{t_0}^{t} \left| \Delta_{t_0}^{\alpha-1} f(\tau) \right|^q \Delta\tau, \quad t \in [t_0, a),$$

and

$$\int_{t_0}^{t} |B(t_1)|^q \Delta t_1$$

$$\leq (G(\alpha, \alpha - 2, p, a, t_0))^q \int_{t_0}^{t} \int_{t_0}^{t_1} \left| \Delta_{t_0}^{\alpha-1} f(\tau) \right|^q \Delta\tau \Delta t_1$$

$$= (G(\alpha, \alpha - 2, p, a, t_0))^q \int_{t_0}^{t} (t - \sigma(\tau)) \left| \Delta_{t_0}^{\alpha-1} f(\tau) \right|^q \Delta\tau, \quad t \in [t_0, a).$$

This completes the proof. $\qquad\qquad\qquad\qquad\qquad\qquad\qquad\qquad \square$

Remark 6.8. Let l be as in Remark 6.1. Suppose that all conditions of Theorem 6.8 hold. Then we have the following estimate:

$$\int_{t_0}^t |B^{\Delta^l}(t_1)|^q \Delta t_1$$

$$\le (G(\alpha, \alpha - l - 2, p, a, t_0))^q \int_{t_0}^t \int_{t_0}^{t_1} |\Delta_{t_0}^{\alpha-1} f(\tau)|^q \, \Delta\tau \Delta t_1$$

$$= (G(\alpha, \alpha - l - 2, p, a, t_0))^q \int_{t_0}^t (t - \sigma(\tau)) |\Delta_{t_0}^{\alpha-1} f(\tau)|^q \, \Delta\tau, \quad t \in [t_0, a).$$

Theorem 6.9. *Let* $f \in C_{rd}^m([t_0, a))$, $f^{\Delta^k}(t_0) = 0$, $k \in \{0, 1, \ldots, m-1\}$, $p = 1$, $q = \infty$. *Then*

$$\|B\|_\infty \le G(\alpha, \alpha - 2, 1, a, t_0) \left\| \Delta_{t_0}^{\alpha-1} f \right\|_\infty.$$

Proof. As in the proof of Theorem 6.5, we find

$$|B(t)| \le \left\| \Delta_{t_0}^{\alpha-1} f \right\|_\infty \left(\int_{t_0}^t h_{\alpha-2}(t, \sigma(u)) \Delta u \right)$$

$$= G(\alpha, \alpha - 2, 1, t, t_0) \left\| \Delta_{t_0}^{\alpha-1} f \right\|_\infty, \quad t \in [t_0, a).$$

Hence,

$$\|B\|_\infty \le G(\alpha, \alpha - 2, 1, a, t_0) \left\| \Delta_{t_0}^{\alpha-1} f \right\|_\infty.$$

This completes the proof. □

Remark 6.9. Let l be as in Remark 6.1. Suppose that all conditions of Theorem 6.9 hold. Then we have the following estimate:

$$\left\| B^{\Delta^l} \right\|_\infty \le G(\alpha, \alpha - l - 2, 1, a, t_0) \left\| \Delta_{t_0}^{\alpha-1} f \right\|_\infty.$$

6.4 Ostrowski-Type Inequalities

In this section, we deduct some Ostrowski-type inequalities using the Young and Hölder inequalities.

Theorem 6.10. *Let* $f \in \mathcal{C}_{rd}^m([t_0, a))$, $f^{\Delta^k}(t_0) = 0$, $k \in \{1, \ldots, m-1\}$, $p, q \geq 1$, $\frac{1}{p} + \frac{1}{q} = 1$. *Then*

$$\left| \frac{1}{a - t_0} \int_{t_0}^a B(t) \Delta t - f(t_0) \right|$$

$$\leq \frac{1}{p(a - t_0)} \int_{t_0}^a (G(\alpha, \alpha - 2, p, t_1, t_0))^p \Delta t_1$$

$$+ \frac{1}{q(a - t_0)} \int_{t_0}^a (a - \sigma(\tau)) \left| \Delta_{t_0}^{\alpha-1} f(\tau) \right|^q \Delta \tau, \quad t \in [t_0, a).$$

Proof. As in the proof of Theorem 6.1, we have

$$|B(t) - f(t_0)|$$

$$\leq \frac{1}{p} \int_{t_0}^t (h_{\alpha-2}(t, \sigma(\tau)))^p \Delta \tau + \frac{1}{q} \int_{t_0}^t \left| \Delta_{t_0}^{\alpha-1} f(\tau) \right|^q \Delta \tau$$

$$= \frac{1}{p} (G(\alpha, \alpha - 2, p, t, t_0))^p + \frac{1}{q} \int_{t_0}^t \left| \Delta_{t_0}^{\alpha-1} f(\tau) \right|^q \Delta \tau, \quad t \in [t_0, a).$$

Hence,

$$\left| \frac{1}{a - t_0} \int_{t_0}^a B(\tau) \Delta \tau - f(t_0) \right|$$

$$= \frac{1}{a - t_0} \left| \int_{t_0}^a (B(t) - f(t_0)) \Delta t \right|$$

$$\leq \frac{1}{a - t_0} \int_{t_0}^a |B(t) - f(t_0)| \Delta t$$

$$\leq \frac{1}{p(a - t_0)} \int_{t_0}^a (G(\alpha, \alpha - 2, p, t_1, t_0))^p \Delta t_1$$

$$+ \frac{1}{q(a - t_0)} \int_{t_0}^a \int_{t_0}^t \left| \Delta_{t_0}^{\alpha-1} f(\tau) \right|^q \Delta \tau \Delta t$$

$$= \frac{1}{p(a-t_0)} \int_{t_0}^{a} (G(\alpha, \alpha - 2, p, t_1, t_0))^p \Delta t_1$$

$$+ \frac{1}{q(a-t_0)} \int_{t_0}^{a} (a - \sigma(\tau)) \left| \Delta_{t_0}^{\alpha-1} f(\tau) \right|^q \Delta \tau, \quad t \in [t_0, a).$$

This completes the proof. □

Theorem 6.11. *Let* $a < \infty$, $f \in C_{rd}^m([t_0, a))$, $f^{\Delta^k}(t_0) = 0$, $k \in \{1, \ldots, m-1\}$, $p, q \geq 1$, $\frac{1}{p} + \frac{1}{q} = 1$, *and* $\left| \Delta_{t_0}^{\alpha-1} f(t) \right| \leq 1$, $t \in [t_0, a)$. *Then*

$$\left| \frac{1}{a - t_0} \int_{t_0}^{a} B(\tau) \Delta \tau - f(t_0) \right|$$

$$\leq \frac{2r_0}{(a - t_0)^{\frac{1}{2}}} (G(\alpha, \alpha - 2, p, a, t_0))^{\frac{p}{2}}$$

$$\times \left(\int_{t_0}^{a} (a - \sigma(\tau)) \left| \Delta_{t_0}^{\alpha-1} f(\tau) \right|^q \Delta \tau \right)^{\frac{1}{2}}$$

$$+ \frac{1}{a - t_0} \left(\frac{1}{p} - r_0 \right) \int_{a}^{t_0} (G(\alpha, \alpha - 2, p, t, t_0))^p \Delta t$$

$$+ \frac{1}{a - t_0} \left(\frac{1}{q} - r_0 \right) \left(\int_{t_0}^{a} (a - \sigma(\tau)) \left| \Delta_{t_0}^{\alpha-1} f(\tau) \right|^q \Delta \tau \right),$$

where $r_0 = \min \left\{ \frac{1}{p}, \frac{1}{q} \right\}$.

Proof. As in the proof of Theorem 6.2, we get

$$|B(t) - f(t_0)| \leq 2r_0 (G(\alpha, \alpha - 2, p, a, t_0))^{\frac{p}{2}} \left(\int_{t_0}^{t} \left| \Delta_{t_0}^{\alpha-1} f(\tau) \right|^q \Delta \tau \right)^{\frac{1}{2}}$$

$$+ \left(\frac{1}{p} - r_0 \right) (G(\alpha, \alpha - 2, p, t, t_0))^p$$

$$+ \left(\frac{1}{q} - r_0 \right) \left(\int_{t_0}^{t} \left| \Delta_{t_0}^{\alpha-1} f(\tau) \right|^q \Delta \tau \right), \quad t \in [t_0, a).$$

Hence,

$$\left| \frac{1}{a - t_0} \int_{t_0}^a B(\tau) \Delta \tau - f(t_0) \right|$$

$$= \frac{1}{a - t_0} \left| \int_{t_0}^a (B(t) - f(t_0)) \Delta t \right|$$

$$\leq \frac{1}{a - t_0} \int_{t_0}^a |B(t) - f(t_0)| \Delta t$$

$$\leq \frac{2r_0}{a - t_0} \left(G(\alpha, \alpha - 2, p, a, t_0) \right)^{\frac{p}{2}} \int_{t_0}^a \left(\int_{t_0}^t \left| \Delta_{t_0}^{\alpha - 1} f(\tau) \right|^q \Delta \tau \right)^{\frac{1}{2}} \Delta t$$

$$+ \frac{1}{a - t_0} \left(\frac{1}{p} - r_0 \right) \int_a^{t_0} \left(G(\alpha, \alpha - 2, p, t, t_0) \right)^p \Delta t$$

$$+ \frac{1}{a - t_0} \left(\frac{1}{q} - r_0 \right) \left(\int_{t_0}^a \int_{t_0}^t \left| \Delta_{t_0}^{\alpha - 1} f(\tau) \right|^q \Delta \tau \Delta t \right)$$

$$\leq \frac{2r_0}{a - t_0} \left(G(\alpha, \alpha - 2, p, a, t_0) \right)^{\frac{p}{2}} (a - t_0)^{\frac{1}{2}}$$

$$\times \left(\int_{t_0}^a \int_{t_0}^t \left| \Delta_{t_0}^{\alpha - 1} f(\tau) \right|^q \Delta \tau \Delta t \right)^{\frac{1}{2}}$$

$$+ \frac{1}{a - t_0} \left(\frac{1}{p} - r_0 \right) \int_a^{t_0} \left(G(\alpha, \alpha - 2, p, t, t_0) \right)^p \Delta t$$

$$+ \frac{1}{a - t_0} \left(\frac{1}{q} - r_0 \right) \left(\int_{t_0}^a (a - \sigma(\tau)) \left| \Delta_{t_0}^{\alpha - 1} f(\tau) \right|^q \Delta \tau \right)$$

$$= \frac{2r_0}{(a - t_0)^{\frac{1}{2}}} \left(G(\alpha, \alpha - 2, p, a, t_0) \right)^{\frac{p}{2}}$$

$$\times \left(\int_{t_0}^a (a - \sigma(\tau)) \left| \Delta_{t_0}^{\alpha - 1} f(\tau) \right|^q \Delta \tau \right)^{\frac{1}{2}}$$

$$+ \frac{1}{a - t_0} \left(\frac{1}{p} - r_0 \right) \int_a^{t_0} \left(G(\alpha, \alpha - 2, p, t, t_0) \right)^p \Delta t$$

$$+ \frac{1}{a - t_0} \left(\frac{1}{q} - r_0 \right) \left(\int_{t_0}^a (a - \sigma(\tau)) \left| \Delta_{t_0}^{\alpha - 1} f(\tau) \right|^q \Delta \tau \right).$$

This completes the proof. $\qquad \square$

Theorem 6.12. *Let* $f \in C_{rd}^m([t_0, a))$, $f^{\Delta^k}(t_0) = 0$, $k \in \{0, 1, \ldots,$ $m-1\}$, $p, q \geq 1$, $\frac{1}{p} + \frac{1}{q} = 1$. *Then*

$$\left| \frac{1}{a-t_0} \int_{t_0}^{a} B(\tau) \Delta \tau - f(t_0) \right|$$

$$\leq \frac{1}{(a-t_0)^{\frac{1}{q}}} G(\alpha, \alpha - 2, p, a, t_0) \left(\int_{t_0}^{a} (a - \sigma(\tau)) \left| \Delta_{t_0}^{\alpha-1} f(\tau) \right|^q \Delta \tau \right)^{\frac{1}{q}}.$$

Proof.   As in the proof of Theorem 6.4, we get

$$|B(t)| \leq G(\alpha, \alpha - 2, p, a, t_0) \left(\int_{t_0}^{t} \left| \Delta_{t_0}^{\alpha-1} f(\tau) \right|^q \Delta \tau \right)^{\frac{1}{q}}, \quad t \in [t_0, a).$$

Hence,

$$\left| \frac{1}{a-t_0} \int_{t_0}^{a} B(\tau) \Delta \tau - f(t_0) \right|$$

$$= \frac{1}{a-t_0} \left| \int_{t_0}^{a} (B(t) - f(t_0)) \Delta t \right|$$

$$\leq \frac{1}{a-t_0} \int_{t_0}^{a} |B(t) - f(t_0)| \Delta t$$

$$\leq \frac{1}{a-t_0} G(\alpha, \alpha - 2, p, a, t_0) \int_{t_0}^{a} \left(\int_{t_0}^{t} \left| \Delta_{t_0}^{\alpha-1} f(\tau) \right|^q \Delta \tau \right)^{\frac{1}{q}} \Delta t$$

$$\leq \frac{1}{a-t_0} G(\alpha, \alpha - 2, p, a, t_0)(a-t_0)^{\frac{1}{p}} \left(\int_{t_0}^{a} \int_{t_0}^{t} \left| \Delta_{t_0}^{\alpha-1} f(\tau) \right|^q \Delta \tau \Delta t \right)^{\frac{1}{q}}$$

$$= \frac{1}{(a-t_0)^{\frac{1}{q}}} G(\alpha, \alpha - 2, p, a, t_0) \left(\int_{t_0}^{a} (a - \sigma(\tau)) \left| \Delta_{t_0}^{\alpha-1} f(\tau) \right|^q \Delta \tau \right)^{\frac{1}{q}}.$$

This completes the proof.   □

Theorem 6.13. *Let* $f \in C_{rd}^m([t_0, a))$, $f^{\Delta^k}(t_0) = 0$, $k \in \{0, 1, \ldots,$ $m-1\}$, $p = 1$, $q = \infty$. *Then*

$$\left| \frac{1}{a-t_0} \int_{t_0}^{a} B(\tau) \Delta \tau - f(t_0) \right|$$

$$\leq \frac{1}{a-t_0} \left\| \Delta_{t_0}^{\alpha-1} f \right\|_\infty \int_{t_0}^{a} G(\alpha, \alpha - 2, 1, t, t_0) \Delta t.$$

Proof. As in the proof of Theorem 6.5, we find

$$|B(t) - f(t_0)| \le \left\|\Delta_{t_0}^{\alpha-1} f\right\|_\infty \left(\int_{t_0}^t h_{\alpha-2}(t, \sigma(u)) \Delta u \right)$$

$$= G(\alpha, \alpha - 2, 1, t, t_0) \left\|\Delta_{t_0}^{\alpha-1} f\right\|_\infty, \quad t \in [t_0, a).$$

Hence,

$$\left| \frac{1}{a - t_0} \int_{t_0}^a B(\tau) \Delta \tau - f(t_0) \right|$$

$$= \frac{1}{a - t_0} \left| \int_{t_0}^a (B(t) - f(t_0)) \Delta t \right|$$

$$\le \frac{1}{a - t_0} \int_{t_0}^a |B(t) - f(t_0)| \Delta t$$

$$\le \frac{1}{a - t_0} \left\|\Delta_{t_0}^{\alpha-1} f\right\|_\infty \int_{t_0}^a G(\alpha, \alpha - 2, 1, t, t_0) \Delta t.$$

This completes the proof. □

Chapter 7

Chebyshev-Type Inequalities

In this chapter, we deduct some Chebyshev-type inequalities for two or more functions. Some of the results in this chapter can be found in Ref. [15].

Let $\mathbb{T}$ be a time scale with forward jump operator and delta differentiation operator σ and Δ, respectively. Let also $t_0, a \in \mathbb{T}$, $t_0 < a$. Suppose that $\alpha, \beta \geq 1$.

7.1 Chebyshev-Type Inequalities for Two Functions

Definition 7.1. Let f and g be two integrable functions on $[t_0, a]$. We say that f and g are synchronous functions on $[t_0, a]$ if

$$(f(t_1) - f(t_2))(g(t_1) - g(t_2)) \geq 0$$

for any $t_1, t_2 \in [t_0, a]$.

We start our investigations in this section with the following result.

Theorem 7.1. *Let f and g be two synchronous functions on $[t_0, a]$. Then*

$$h_\alpha(t, t_0) I_{\Delta, t_0}^\alpha (fg)(t) \geq I_{\Delta, t_0}^\alpha f(t) I_{\Delta, t_0}^\alpha g(t), \quad t \in [t_0, a].$$

Proof. Let $t \in [t_0, a]$ be arbitrarily chosen and fixed. Let also $t_1, t_2 \in [t_0, a]$ be such that $t_1, t_2 \leq t$. Since f and g are two synchronous functions on $[t_0, a]$, we have

$$(f(t_1) - f(t_2))(g(t_1) - g(t_2)) \geq 0$$

or

$$f(t_1)g(t_1) - f(t_1)g(t_2) - f(t_2)g(t_1) + f(t_2)g(t_2) \geq 0,$$

or

$$f(t_1)g(t_1) + f(t_2)g(t_2) \geq f(t_1)g(t_2) + f(t_2)g(t_1).$$

Now, we multiply both sides of the last inequality with $h_{\alpha-1}(t, \sigma(t_1))$ and find

$$h_{\alpha-1}(t, \sigma(t_1))f(t_1)g(t_1) + h_{\alpha-1}(t, \sigma(t_1))f(t_2)g(t_2)$$
$$\geq h_{\alpha-1}(t, \sigma(t_1))f(t_1)g(t_2) + h_{\alpha-1}(t, \sigma(t_1))f(t_2)g(t_1).$$

We integrate the last inequality with respect to t_1 from t_0 to t and find

$$\int_{t_0}^{t} h_{\alpha-1}(t, \sigma(t_1))f(t_1)g(t_1)\Delta t_1 + f(t_2)g(t_2) \int_{t_0}^{t} h_{\alpha-1}(t, \sigma(t_1))$$

$$\geq g(t_2) \int_{t_0}^{t} h_{\alpha-1}(t, \sigma(t_1))f(t_1)\Delta t_1$$

$$+ f(t_2) \int_{t_0}^{t} h_{\alpha-1}(t, \sigma(t_1))g(t_1)\Delta t_1.$$

or

$$I_{\Delta,t_0}^{\alpha}(fg)(t) + f(t_2)g(t_2)h_{\alpha}(t, t_0) \geq g(t_2)I_{\Delta,t_0}^{\alpha}f(t) + f(t_2)I_{\Delta,t_0}^{\alpha}g(t).$$

We multiply both sides of the last inequality with $h_{\alpha-1}(t, \sigma(t_2))$ and obtain

$$h_{\alpha-1}(t, \sigma(t_2))I_{\Delta,t_0}^{\alpha}(fg)(t) + h_{\alpha-1}(t, \sigma(t_2))f(t_2)g(t_2)h_{\alpha}(t, t_0)$$
$$\geq h_{\alpha-1}(t, \sigma(t_2))g(t_2)I_{\Delta,t_0}^{\alpha}f(t) + h_{\alpha-1}(t, \sigma(t_2))f(t_2)I_{\Delta,t_0}^{\alpha}g(t),$$

which we integrate with respect to t_2 from t_0 to t and arrive at

$$I_{\Delta,t_0}^{\alpha}(fg)(t)\int_{t_0}^{t}h_{\alpha-1}(t,\sigma(t_2))\Delta t_2$$

$$+h_{\alpha}(t,t_0)\int_{t_0}^{t}h_{\alpha-1}(t,\sigma(t_2))f(t_2)g(t_2)\Delta t_2$$

$$\geq I_{\Delta,t_0}^{\alpha}f(t)\int_{t_0}^{t}h_{\alpha-1}(t,\sigma(t_2))g(t_2)\Delta t_2$$

$$+I_{\Delta,t_0}^{\alpha}g(t)\int_{t_0}^{t}h_{\alpha-1}(t,\sigma(t_2))f(t_2)\Delta t_2$$

or

$$h_{\alpha}(t,t_0)I_{\Delta,t_0}^{\alpha}(fg)(t)+h_{\alpha}(t,t_0)I_{\Delta,t_0}^{\alpha}(fg)(t)$$
$$\geq I_{\Delta,t_0}^{\alpha}f(t)I_{\Delta,t_0}^{\alpha}g(t)+I_{\Delta,t_0}^{\alpha}f(t)I_{\Delta,t_0}^{\alpha}g(t),$$

or

$$2h_{\alpha}(t,t_0)I_{\Delta,t_0}^{\alpha}(fg)(t)\geq I_{\Delta,t_0}^{\alpha}f(t)I_{\Delta,t_0}^{\alpha}g(t),$$

or

$$h_{\alpha}(t,t_0)I_{\Delta,t_0}^{\alpha}(fg)(t)\geq I_{\Delta,t_0}^{\alpha}f(t)I_{\Delta,t_0}^{\alpha}g(t).$$

This completes the proof. □

Theorem 7.2. *Let f and g be two synchronous functions on $[t_0,a]$. Then*

$$h_{\alpha}(t,t_0)I_{\Delta,t_0}^{\beta}(fg)(t)+h_{\beta}(t,t_0)I_{\Delta,t_0}^{\alpha}(fg)(t)$$
$$\geq I_{\Delta,t_0}^{\alpha}f(t)I_{\Delta,t_0}^{\beta}g(t)+I_{\Delta,t_0}^{\beta}f(t)I_{\Delta,t_0}^{\alpha}g(t),\quad t\in[t_0,a].$$

Proof. Let $t\in[t_0,a]$ be arbitrarily chosen and fixed. Let also $t_1,t_2\in[t_0,a]$ be such that $t_1,t_2\leq t$. As in the proof of Theorem 7.2, we get

$$I_{\Delta,t_0}^{\alpha}(fg)(t)+f(t_2)g(t_2)h_{\alpha}(t,t_0)\geq g(t_2)I_{\Delta,t_0}^{\alpha}f(t)+f(t_2)I_{\Delta,t_0}^{\alpha}g(t).$$

We multiply both sides of the last inequality with $h_{\beta-1}(t, \sigma(t_2))$ and obtain

$$h_{\beta-1}(t, \sigma(t_2))I^\alpha_{\Delta,t_0}(fg)(t) + h_{\beta-1}(t, \sigma(t_2))f(t_2)g(t_2)h_\alpha(t, t_0)$$
$$\geq h_{\beta-1}(t, \sigma(t_2))g(t_2)I^\alpha_{\Delta,t_0}f(t) + h_{\beta-1}(t, \sigma(t_2))f(t_2)I^\alpha_{\Delta,t_0}g(t),$$

which we integrate with respect to t_2 from t_0 to t and arrive at

$$I^\alpha_{\Delta,t_0}(fg)(t) \int_{t_0}^t h_{\beta-1}(t, \sigma(t_2))\Delta t_2$$

$$+ h_\alpha(t, t_0) \int_{t_0}^t h_{\beta-1}(t, \sigma(t_2))f(t_2)g(t_2)\Delta t_2$$

$$\geq I^\alpha_{\Delta,t_0}f(t) \int_{t_0}^t h_{\beta-1}(t, \sigma(t_2))g(t_2)\Delta t_2$$

$$+ I^\alpha_{\Delta,t_0}g(t) \int_{t_0}^t h_{\beta-1}(t, \sigma(t_2))f(t_2)\Delta t_2$$

or

$$h_\alpha(t, t_0)I^\beta_{\Delta,t_0}(fg)(t) + h_\beta(t, t_0)I^\alpha_{\Delta,t_0}(fg)(t)$$
$$\geq I^\alpha_{\Delta,t_0}f(t)I^\beta_{\Delta,t_0}g(t) + I^\beta_{\Delta,t_0}f(t)I^\alpha_{\Delta,t_0}g(t).$$

This completes the proof. $\qquad\qquad\qquad\qquad\qquad\qquad\qquad\quad\square$

Theorem 7.3. *Let f and g be two functions defined on $[t_0, a]$ that are Δ-differentiable on $[t_0, a]$ and f be increasing on $[t_0, a]$. Let also there exist a real constant m such that*

$$g^\Delta(t) \geq m, \quad t \in [t_0, a].$$

Then

$$h_\alpha(t, t_0)I^\alpha_{\Delta,t_0}(fg)(t) \geq mh_\alpha(t, t_0)I^\alpha_{\Delta,t_0}(tf(t)) + I^\alpha_{\Delta,t_0}f(t)I^\alpha_{\Delta,t_0}g(t)$$
$$- mI^\alpha_{\Delta,t_0}tI^\alpha_{\Delta,t_0}f(t), \quad t \in [t_0, a].$$

Proof. Let

$$k(t) = g(t) - mt, \quad t \in [t_0, a].$$

Hence,

$$k^\Delta(t) = g^\Delta(t) - m$$
$$\geq 0, \quad t \in [t_0, a].$$

Thus, g is an increasing function on $[t_0, a]$. Since k and f are increasing functions on $[t_0, a]$, it follows that k and f are synchronous functions on $[t_0, a]$. Now, we apply Theorem 7.1 and get

$$h_\alpha(t, t_0) I^\alpha_{\Delta, t_0}(kf)(t) \geq I^\alpha_{\Delta, t_0} k(t) I^\alpha_{\Delta, t_0} f(t), \quad t \in [t_0, a],$$

or

$$h_\alpha(t, t_0) I^\alpha_{\Delta, t_0}\left((g(t) - mt)f(t)\right) \geq I^\alpha_{\Delta, t_0}(g(t) - mt) I^\alpha_{\Delta, t_0} f(t), \quad t \in [t_0, a],$$

or

$$h_\alpha(t, t_0) I^\alpha_{\Delta, t_0}(fg)(t) - h_\alpha(t, t_0) I^\alpha_{\Delta, t_0}(mtf(t))$$
$$\geq I^\alpha_{\Delta, t_0} f(t) I^\alpha_{\Delta, t_0} g(t) - m I^\alpha_{\Delta, t_0} t I^\alpha_{\Delta, t_0} f(t), \quad t \in [t_0, a],$$

or

$$h_\alpha(t, t_0) I^\alpha_{\Delta, t_0}(fg)(t) \geq m h_\alpha(t, t_0) I^\alpha_{\Delta, t_0}(tf(t)) + I^\alpha_{\Delta, t_0} f(t) I^\alpha_{\Delta, t_0} g(t)$$
$$- m I^\alpha_{\Delta, t_0} t I^\alpha_{\Delta, t_0} f(t), \quad t \in [t_0, a].$$

This completes the proof. □

Corollary 7.1. *Let f and g be two functions defined on $[t_0, a]$ that are Δ-differentiable on $[t_0, a]$ and f be decreasing on $[t_0, a]$. Let also there exist a real constant m such that*

$$g^\Delta(t) \leq m, \quad t \in [t_0, a].$$

Then

$$h_\alpha(t,t_0)I^\alpha_{\Delta,t_0}(fg)(t) \geq m h_\alpha(t,t_0)I^\alpha_{\Delta,t_0}(tf(t)) + I^\alpha_{\Delta,t_0}f(t)I^\alpha_{\Delta,t_0}g(t)$$
$$- m I^\alpha_{\Delta,t_0}t I^\alpha_{\Delta,t_0}f(t), \quad t \in [t_0,a].$$

Proof. Let

$$k(t) = g(t) - mt, \quad t \in [t_0,a].$$

Then

$$k^\Delta(t) = g^\Delta(t) - m$$
$$\leq 0, \quad t \in [t_0,a].$$

So, k is a decreasing function on $[t_0,a]$. Since k and f are decreasing functions on $[t_0,a]$, we conclude that k and f are synchronous functions. Now, the proof follows the proof of Theorem 7.3. This completes the proof. □

Theorem 7.4. *Let f and g be two functions defined on $[t_0,a]$ that are Δ-differentiable on $[t_0,a]$. Let also there exist real constants m_1 and m_2 such that*

$$f^\Delta(t) \geq m_1, \quad g^\Delta(t) \geq m_2, \quad t \in [t_0,a].$$

Then

$$h_\alpha(t,t_0) \left(I^\alpha_{\Delta,t_0}(fg)(t) + m_1 m_2 I^\alpha_{\Delta,t_0}t^2 \right)$$
$$+ I^\alpha_{\Delta,t_0}t \left(m_1 I^\alpha_{\Delta,t_0}g(t) + m_2 I^\alpha_{\Delta,t_0}f(t) \right)$$
$$\geq I^\alpha_{\Delta,t_0}f(t)I^\alpha_{\Delta,t_0}g(t) + m_1 m_2 \left(I^\alpha_{\Delta,t_0}t \right)^2$$
$$+ h_\alpha(t,t_0)I^\alpha_{\Delta,t_0}t \left(m_1 g(t) + m_2 f(t) \right),$$
$$t \in [t_0,a].$$

Proof. Let

$$k_1(t) = f(t) - m_1 t,$$
$$k_2(t) = g(t) - m_2 t, \quad t \in [t_0,a].$$

Then

$$k_1^\Delta(t) = f^\Delta(t) - m_1$$
$$\geq 0,$$
$$k_2^\Delta(t) = g^\Delta(t) - m_2$$
$$\geq 0, \quad t \in [t_0, a].$$

Therefore, k_1 and k_2 are increasing functions on $[t_0, a]$. Hence, they are synchronous on $[t_0, a]$. Now, we apply Theorem 7.1 for the functions k_1 and k_2 and get

$$h_\alpha(t, t_0) I_{\Delta, t_0}^\alpha (k_1 k_2)(t) \geq I_{\Delta, t_0}^\alpha k_1(t) I_{\Delta, t_0}^\alpha k_2(t), \quad t \in [t_0, a],$$

or

$$h_\alpha(t, t_0) \, I_{\Delta, t_0}^\alpha \left((f(t) - m_1 t)(f_2(t) - m_2 t) \right)$$
$$\geq I_{\Delta, t_0}^\alpha (f(t) - m_1 t) I_{\Delta, t_0}^\alpha (g(t) - m_2 t), \quad t \in [t_0, a],$$

or

$$h_\alpha(t, t_0) I_{\Delta, t_0}^\alpha \left(f(t)g(t) - t(m_1 g(t) + m_2 f(t)) + m_1 m_2 t^2 \right)$$
$$\geq h_\alpha(t, t_0) \left(I_{\Delta, t_0}^\alpha f(t) - m_1 I_{\Delta, t_0}^\alpha t \right) \left(I_{\Delta, t_0}^\alpha g(t) - m_2 I_{\Delta, t_0}^\alpha t \right),$$
$$t \in [t_0, a],$$

or

$$h_\alpha(t, t_0) \left(I_{\Delta, t_0}^\alpha (fg)(t) + m_1 m_2 I_{\Delta, t_0}^\alpha t^2 \right)$$
$$- h_\alpha(t, t_0) \left(m_2 f(t) I_{\Delta, t_0}^\alpha t + m_1 g(t) I_{\Delta, t_0}^\alpha t \right)$$
$$\geq I_{\Delta, t_0}^\alpha f(t) I_{\Delta, t_0}^\alpha g(t) - m_1 I_{\Delta, t_0}^\alpha t I_{\Delta, t_0}^\alpha g(t)$$
$$- m_2 I_{\Delta, t_0}^\alpha t I_{\Delta, t_0}^\alpha f(t) + m_1 m_2 \left(I_{\Delta, t_0}^\alpha t \right)^2,$$
$$t \in [t_0, a],$$

or

$$h_\alpha(t, t_0) \left(I_{\Delta, t_0}^\alpha (fg)(t) + m_1 m_2 I_{\Delta, t_0}^\alpha t^2 \right)$$
$$+ I_{\Delta, t_0}^\alpha t \left(m_1 I_{\Delta, t_0}^\alpha g(t) + m_2 I_{\Delta, t_0}^\alpha f(t) \right)$$
$$\geq I_{\Delta, t_0}^\alpha f(t) I_{\Delta, t_0}^\alpha g(t) + m_1 m_2 \left(I_{\Delta, t_0}^\alpha t \right)^2$$
$$+ h_\alpha(t, t_0) I_{\Delta, t_0}^\alpha t \left(m_1 g(t) + m_2 f(t) \right),$$
$$t \in [t_0, a].$$

This completes the proof. $\qquad \square$

Corollary 7.2. *Let f and g be two functions defined on $[t_0, a]$ that are Δ-differentiable on $[t_0, a]$. Let also there exist real constants m_1 and m_2 such that*

$$f^\Delta(t) \le m_1, \quad g^\Delta(t) \le m_2, \quad t \in [t_0, a].$$

Then

$$h_\alpha(t, t_0) \left(I^\alpha_{\Delta, t_0}(fg)(t) + m_1 m_2 I^\alpha_{\Delta, t_0} t^2 \right)$$

$$+ I^\alpha_{\Delta, t_0} t \left(m_1 I^\alpha_{\Delta, t_0} g(t) + m_2 I^\alpha_{\Delta, t_0} f(t) \right)$$

$$\ge I^\alpha_{\Delta, t_0} f(t) I^\alpha_{\Delta, t_0} g(t) + m_1 m_2 \left(I^\alpha_{\Delta, t_0} t \right)^2$$

$$+ h_\alpha(t, t_0) I^\alpha_{\Delta, t_0} t \left(m_1 g(t) + m_2 f(t) \right),$$

$$t \in [t_0, a].$$

Proof.   Let

$$k_1(t) = f(t) - m_1 t,$$

$$k_2(t) = g(t) - m_2 t, \quad t \in [t_0, a].$$

Then

$$k_1^\Delta(t) = f^\Delta(t) - m_1$$

$$\le 0,$$

$$k_2^\Delta(t) = g^\Delta(t) - m_2$$

$$\le 0, \quad t \in [t_0, a].$$

Therefore, k_1 and k_2 are decreasing functions on $[t_0, a]$. Hence, they are synchronous on $[t_0, a]$. Now, the proof follows the proof of Theorem 7.4. This completes the proof. $\square$

7.2   Chebyshev-Type Inequalities for n Functions

In this section, we generalize Theorem 7.1 for n functions.

Theorem 7.5. *Let f_j, $j \in \{1, \dots, n\}$, be positive and increasing functions on $[t_0, a]$. Then*

$$(h_\alpha(t, t_0))^{n-1} I^\alpha_{\Delta, t_0} \left(\prod_{j=1}^{n} f_j \right)(t) \ge \prod_{j=1}^{n} I^\alpha_{\Delta, t_0} f_j(t), \quad t \in [t_0, a].$$

Proof. We will use the principle of mathematical induction. For $n = 2$, the assertion follows directly by Theorem 7.1. Assume that the assertion is valid for $n - 1$, i.e., we assume that

$$(h_\alpha(t, t_0))^{n-2} I_{\Delta, t_0}^\alpha \left(\prod_{j=1}^{n-1} f_j \right)(t) \geq \prod_{j=1}^{n-1} I_{\Delta, t_0}^\alpha f_j(t), \quad t \in [t_0, a].$$

We will prove the assertion for n. Set

$$f(t) = \prod_{j=1}^{n-1} f_j(t), \quad t \in [t_0, a].$$

Since f_j, $j \in \{1, \ldots, n-1\}$, are positive and increasing functions on $[t_0, a]$, we have that f is a positive and increasing function on $[t_0, a]$. Since f and f_n are increasing functions on $[t_0, a]$, we obtain that f and f_n are synchronous on $[t_0, a]$. Now, we apply Theorem 7.1 for the functions f and f_n and get

$$h_\alpha(t, t_0) I_{\Delta, t_0}^\alpha (f f_n)(t) \geq I_{\Delta, t_0}^\alpha f(t) I_{\Delta, t_0}^\alpha f_n(t), \quad t \in [t_0, a].$$

Hence,

$$(h_\alpha(t, t_0))^n I_{\Delta, t_0}^\alpha \left(\prod_{j=1}^n f_j \right)(t)$$

$$\geq (h_\alpha(t, t_0))^{n-1} I_{\Delta, t_0}^\alpha \left(\prod_{j=1}^{n-1} f_j \right)(t) I_{\Delta, t_0}^\alpha f_n(t)$$

$$\geq \prod_{j=1}^{n-1} I_{\Delta, t_0}^\alpha f_j(t) I_{\Delta, t_0}^\alpha f_n(t)$$

$$= \prod_{j=1}^n I_{\Delta, t_0}^\alpha f_j(t), \quad t \in [t_0, a].$$

This completes the proof. $\qquad\qquad\qquad\qquad\qquad\qquad\qquad\qquad\square$

Chapter 8

Hardy-Type Fractional Inequalities

In this chapter, we introduce some Hardy-type inequalities on time scales. Some fractional analogues of the Copson, Leindler and Bennett inequalities are deducted. Some of the results in this chapter can be found in Ref. [13].

Let $\mathbb{T}$ be a time scale with forward jump operator and delta differentiation operator σ and Δ, respectively, and also, $\alpha \in (0, 1]$ and $a \in \mathbb{T}$.

Note that when $\alpha = 1$, we obtain the Copson, Leindler and Bennett inequalities on time scales, which can be found in Ref. [1]. Thus, the results in this chapter can be considered as a generalization of the results in Ref. [1].

8.1 Copson-Type Fractional Inequalities

In this section, we deduct some Copson-type inequalities. Throughout this section, suppose that $[a, \infty) \subset \mathbb{T}$. We start with the following result.

Theorem 8.1. *Assume that* $1 < c < k$, $\lambda, g : [a, \infty) \to (0, \infty)$ *are integrable functions and define*

$$f_1(x) = \int_a^x \lambda(s) \Delta_\alpha s,$$

$$f_2(x) = \int_a^x \lambda(s) g(s) \Delta_\alpha s, \quad x \in [a, \infty).$$

If $f_1(\infty) < \infty$ and

$$\int_a^\infty \frac{\lambda(s)}{(f_1^\sigma(s))^{c-\alpha+1}} \Delta_\alpha s < \infty,$$

then

$$\int_a^\infty \frac{\lambda(x)}{(f_1^\sigma(s))^{c-\alpha+1}} (f_2^\sigma(x))^k \Delta_\alpha x$$

$$\leq \left(\frac{k}{c-\alpha}\right)^k \int_a^\infty \frac{\lambda(x)(f_1(x))^{k(\alpha-c)}}{(f_1^\sigma(x))^{(c-\alpha+1)(1-k)}} (g(x))^k \Delta_\alpha x.$$

Proof. Let

$$v(x) = -\int_x^\infty \frac{\lambda(s)}{(f_1^\sigma(s))^{c-\alpha+1}} \Delta_\alpha s, \quad x \in [a, \infty).$$

Then

$$v(x) = -\int_x^\infty \frac{f_1^{\Delta_\alpha}(s)}{(f_1^\sigma(s))^{c-\alpha+1}} \Delta_\alpha s, \quad x \in [a, \infty).$$

Hence, integrating by parts, we arrive at

$$\int_a^\infty \frac{\lambda(x)}{(f_1^\sigma(x))^{c-\alpha+1}} (f_2^\sigma(x))^k \Delta_\alpha x = v(x) \left(f_2(x)\right)^k \Big|_{x=a}^{x=\infty}$$

$$- \int_a^\infty v(x) \left(f_2^k\right)^{\Delta_\alpha} (x) \Delta_\alpha x.$$

$$(8.1)$$

For $x \in [a, \infty)$, by the chain rule, there exists a $\xi \in [x, \sigma(x)]$ such that

$$- \left(f_1^{\alpha-c}\right)^{\Delta_\alpha} (x) = -(\alpha - c) f_1^{\alpha-c-1}(\xi) f_1^{\Delta_\alpha}(x)$$

$$= \frac{(c-\alpha) f_1^{\Delta_\alpha}(x)}{(f_1(x))^{c-\alpha+1}}$$

$$\geq \frac{(c-\alpha) f_1^{\Delta_\alpha}(x)}{(f_1^\sigma(x))^{c-\alpha+1}},$$

whereupon

$$\left(f_1^{\alpha-c}\right)^{\Delta\alpha}(x) \leq -\frac{(c-\alpha)f_1^{\Delta\alpha}(x)}{(f_1^\sigma(x))^{c-\alpha+1}}, \quad x \in [a,\infty).$$

Hence,

$$(f_1^\sigma(x))^{\alpha-c-1} f_1^{\Delta\alpha}(x) \leq -\frac{1}{c-\alpha} \left(f_1^{\alpha-c}\right)^{\Delta\alpha}(x), \quad x \in [a,\infty).$$

Thus,

$$
\begin{aligned}
-v(x) &= \int_x^\infty \frac{\lambda(s)}{(f_1^\sigma(s))^{c-\alpha+1}} \Delta_\alpha s \\
&= \int_x^\infty f_1^{\Delta\alpha}(s) \, (f_1^\sigma(s))^{\alpha-c-1} \Delta_\alpha s \\
&\leq -\frac{1}{c-\alpha} \int_x^\infty \left(f_1^{\alpha-c}\right)^{\Delta\alpha}(s) \Delta_\alpha s \\
&\leq \frac{(f_1(x))^{\alpha-c}}{c-\alpha}, \quad x \in [a,\infty).
\end{aligned}
$$

Again, by the chain rule, for $x \in [a,\infty)$, there exists $\xi_1 \in [x,\sigma(x)]$ such that

$$
\begin{aligned}
\left(f_2^k\right)^{\Delta\alpha}(x) &= k \, (f_2(\xi_1))^{k-1} f_2^{\Delta\alpha}(x) \\
&\leq k \, (f_2^\sigma(x))^{k-1} f_2^{\Delta\alpha}(x) \\
&= k\lambda(x)g(x) \, (f_2^\sigma(x))^{k-1}, \quad x \in [a,\infty).
\end{aligned}
$$

Note that $f_2(a) = 0$, $v(\infty) = 0$. Then, using (3.19) and the Hölder inequality, we find

$$
\begin{aligned}
\int_a^\infty \frac{\lambda(x)}{(f_1^\sigma(x))^{c-\alpha+1}} (f_2^\sigma(x))^k \Delta_\alpha x &= -\int_a^\infty v(x) \left(f_2^k\right)^{\Delta\alpha}(x)\Delta_\alpha x \\
&\leq \int_a^\infty \frac{(f_1(x))^{\alpha-c}}{c-\alpha} \left(f_2^k\right)^{\Delta\alpha}(x)\Delta_\alpha x
\end{aligned}
$$

$$\leq \frac{k}{c-\alpha} \int_a^\infty (f_1(x))^{\alpha-c} \lambda(x)g(x) (f_2^\sigma(x))^{\Delta_\alpha} (x)\Delta_\alpha x$$

$$\leq \frac{k}{c-\alpha} \int_a^\infty (f_1(x))^{\alpha-c} \lambda(x)g(x) (f_2^\sigma(x))^{k-1} \Delta_\alpha x$$

$$= \frac{k}{c-\alpha} \int_a^\infty \frac{\lambda(x)(f_1(x))^{\alpha-c} g(x)}{\left(\lambda(x)(f_1^\sigma(x))^{\alpha-c-1}\right)^{\frac{k-1}{k}}} \left(\frac{\lambda(x)(f_2^\sigma(x))^k}{(f_1^\sigma(x))^{c-\alpha+1}}\right)^{\frac{k-1}{k}} \Delta_\alpha x$$

$$\leq \frac{k}{c-\alpha} \left(\int_a^\infty \left(\frac{\lambda(x)(f_1(x))^{\alpha-c} g(x)}{\left(\lambda(x)(f_1^\sigma(x))^{\alpha-c-1}\right)^{\frac{k-1}{k}}}\right)^k \Delta_\alpha x\right)^{\frac{1}{k}}$$

$$\times \left(\int_a^\infty \left(\frac{\lambda(x)(f_2^\sigma(x))^k}{(f_1^\sigma(x))^{c-\alpha+1}}\right)^{\frac{k-1}{k}} \Delta_\alpha x\right)^{\frac{k-1}{k}},$$

whereupon

$$\left(\int_a^\infty \left(\frac{\lambda(x)(f_2^\sigma(x))^k}{(f_1^\sigma(x))^{c-\alpha+1}}\right)^{\frac{k-1}{k}} \Delta_\alpha x\right)^{\frac{1}{k}}$$

$$\leq \frac{k}{c-\alpha} \left(\int_a^\infty \left(\frac{\lambda(x)(f_1(x))^{\alpha-c} g(x)}{\left(\lambda(x)(f_1^\sigma(x))^{\alpha-c-1}\right)^{\frac{k-1}{k}}}\right)^k \Delta_\alpha x\right)^{\frac{1}{k}}$$

or

$$\int_a^\infty \left(\frac{\lambda(x)(f_2^\sigma(x))^k}{(f_1^\sigma(x))^{c-\alpha+1}}\right)^{\frac{k-1}{k}} \Delta_\alpha x$$

$$\leq \left(\frac{k}{c-\alpha}\right)^k \int_a^\infty \left(\frac{\lambda(x)(f_1(x))^{\alpha-c} g(x)}{\left(\lambda(x)(f_1^\sigma(x))^{\alpha-c-1}\right)^{\frac{k-1}{k}}}\right)^k \Delta_\alpha x.$$

This completes the proof. $\qquad\qquad\qquad\qquad\qquad\qquad\qquad\qquad$ □

Theorem 8.2. *Let* $0 \le c < 1$, $k > 1$, $\lambda, g : [a, \infty) \to (0, \infty)$ *be integrable functions and define*

$$f_1(x) = \int_a^x \lambda(s) \Delta_\alpha s,$$

$$f_2(x) = \int_x^\infty \lambda(s) g(s) \Delta_\alpha s, \quad x \in [a, \infty).$$

If $f_2(a) < \infty$ *and*

$$\int_a^\infty \frac{\lambda(s)}{(f_1^\sigma(s))^{c-\alpha+1}} \Delta_\alpha s < \infty,$$

then

$$\int_a^\infty \frac{\lambda(x)}{(f_1^\sigma(x))^{c-\alpha+1}} \, (f_2(x))^k \, \Delta_\alpha x$$

$$\le \left(\frac{k}{\alpha - c} \right)^k \int_a^\infty (f_1^\sigma(x))^{k-c+\alpha-1} \lambda(x) (g(x))^k \Delta_\alpha x.$$

Proof. Let

$$u(x) = \int_a^x \frac{\lambda(s)}{(f_1^\sigma(s))^{c-\alpha+1}} \Delta_\alpha s, \quad x \in [a, \infty).$$

Then

$$u(x) = \int_a^x f_1^{\Delta_\alpha}(s) \, (f_1^\sigma(s))^{\alpha-c-1} \, \Delta_\alpha s, \quad x \in [a, \infty).$$

Note that

$$f_2^{\Delta_\alpha}(x) = -\lambda(x) g(x)$$

$$\le 0, \quad x \in [a, \infty).$$

Therefore, f_2 is a decreasing function on $[a, \infty)$. Next,

$$\int_a^\infty \frac{\lambda(x)}{(f_1^\sigma(x))^{c-\alpha+1}} \, (f_2(x))^k \, \Delta_\alpha x$$

$$= u(x) \, (f_2(x))^k \Big|_{x=a}^{x=\infty} + \int_a^\infty u^\sigma(x) \left(- \left(f_2^k \right)^{\Delta_\alpha} (x) \right) \Delta_\alpha x. \quad (8.2)$$

For $x \in [a, \infty)$, by the chain rule, there exists a $\xi \in [x, \sigma(x)]$ such that

$$\left(f_1^{\alpha-c}\right)^{\Delta_\alpha}(x) = (\alpha - c)\left(f_1(\xi)\right)^{\alpha-c-1} f_1^{\Delta_\alpha}(x)$$

$$= \frac{(\alpha - c)f_1^{\Delta_\alpha}(x)}{(f_1(\xi))^{c-\alpha+1}}$$

$$\geq \frac{(\alpha - c)f_1^{\Delta_\alpha}(x)}{(f_1^\sigma(x))^{c-\alpha+1}}.$$

Thus,

$$f_1^{\Delta_\alpha}(x)\left(f_1^\sigma(x)\right)^{\alpha-c-1} \leq \frac{1}{\alpha - c}\left(f_1^{\alpha-c}\right)^{\Delta_\alpha}(x), \quad x \in [a, \infty),$$

and

$$u^\sigma(x) = \int_a^{\sigma(x)} f_1^{\Delta_\alpha}(s)\left(f_1^\sigma(s)\right)^{\alpha-c-1} \Delta_\alpha s$$

$$\leq \frac{1}{\alpha - c} \int_a^{\sigma(x)} \left(f_1^{\alpha-c}\right)^{\Delta_\alpha}(s) \Delta_\alpha s$$

$$= \frac{1}{\alpha - c}\left(f_1^\sigma(x)\right)^{\alpha-c}, \quad x \in [a, \infty).$$

For $x \in [a, \infty)$, by the chain rule, there exists $\xi_2 \in [x, \sigma(x)]$ such that

$$-\left(f_2^k\right)^{\Delta_\alpha}(x) = -k\left(f_2(\xi_2)\right)^{k-1} f_2^{\Delta_\alpha}(x)$$

$$= bk\left(f_2(\xi_2)\right)^{k-1} \lambda(x)g(x)$$

$$\leq k\left(f_2(x)\right)^{k-1} \lambda(x)g(x).$$

Hence, using (8.2) and the Hölder inequality, we find

$$\int_a^\infty \frac{\lambda(x)}{(f_1^\sigma(x))^{c-\alpha+1}}(f_2(x))^k \Delta_\alpha x$$

$$\leq \frac{k}{\alpha - c} \int_a^\infty \left(f_1^\sigma(x)\right)^{\alpha-1} \lambda(x)g(x)\left(f_2(x)\right)^{k-1} \Delta_\alpha x$$

$$= \frac{k}{\alpha - c} \int_a^\infty \frac{(f_1^\sigma(x))^{\alpha-1} \lambda(x) g(x)}{\left(\lambda(x) (f_1^\sigma(x))^{\alpha-c-1}\right)^{\frac{k-1}{k}}} \left(\frac{\lambda(x) (f_2(x))^k}{(f_1^\sigma(x))^{c-\alpha+1}}\right)^{\frac{k-1}{k}} \Delta_\alpha x$$

$$\leq \frac{k}{\alpha - c} \left(\int_a^\infty \frac{\left((f_1^\sigma(x))^{\alpha-c} \lambda(x) g(x)\right)^k}{\left(\lambda(x) (f_1^\sigma(x))^{\alpha-c-1}\right)^{k-1}} \Delta_\alpha x\right)^{\frac{1}{k}}$$

$$\times \left(\int_a^\infty \frac{\lambda(x) (f_2(x))^k}{(f_1^\sigma(x))^{c-\alpha+1}} \Delta_\alpha x\right)^{\frac{k-1}{k}},$$

whereupon

$$\left(\int_a^\infty \frac{\lambda(x)}{(f_1^\sigma(x))^{c-\alpha+1}} (f_2(x))^k \Delta_\alpha x\right)^{\frac{1}{k}}$$

$$\leq \frac{k}{\alpha - c} \left(\int_a^\infty \frac{\left((f_1^\sigma(x))^{\alpha-c} \lambda(x) g(x)\right)^k}{\left(\lambda(x) (f_1^\sigma(x))^{\alpha-c-1}\right)^{k-1}} \Delta_\alpha x\right)^{\frac{1}{k}}$$

and

$$\int_a^\infty \frac{\lambda(x)}{(f_1^\sigma(x))^{c-\alpha+1}} (f_2(x))^k \Delta_\alpha x$$

$$\leq \left(\frac{k}{\alpha - c}\right)^k \left(\int_a^\infty \frac{\left((f_1^\sigma(x))^{\alpha-c} \lambda(x) g(x)\right)^k}{\left(\lambda(x) (f_1^\sigma(x))^{\alpha-c-1}\right)^{k-1}} \Delta_\alpha x\right)^{\frac{1}{k}}.$$

This completes the proof. □

8.2 The Leindler Fractional Inequality

In this section, we deduct a fractional analogue of the Leindler inequality.

Theorem 8.3 (The Leindler Inequality). *Assume that* $0 \leq c < 1 < k$, $\lambda, g : [a, \infty) \to (0, \infty)$ *are integrable functions and define*

$$f_1(x) = \int_x^\infty \lambda(s) \Delta_\alpha s,$$

$$f_2(x) = \int_a^x \lambda(s) g(s) \Delta_\alpha s.$$

If $f_2(\infty) < \infty$ *and*

$$\int_a^\infty \frac{\lambda(s)}{(f_1(s))^{c-\alpha+1}} \Delta_\alpha s < \infty,$$

then

$$\int_a^\infty \frac{\lambda(x)}{(f_1(x))^{c-\alpha+1}} (f_2^\sigma(x))^k \, \Delta_\alpha x$$

$$\leq \left(\frac{k}{\alpha - c}\right)^k \int_a^\infty (f_1(x))^{k-c+\alpha-1} \lambda(x) (g(x))^k \Delta_\alpha x.$$

Proof.   Let

$$v(x) = -\int_x^\infty \frac{\lambda(s)}{(f_1(s))^{c-\alpha+1}} \Delta_\alpha s, \quad x \in [a, \infty).$$

Then

$$v^{\Delta_\alpha}(x) = \frac{\lambda(x)}{(f_1(x))^{c-\alpha+1}}, \quad x \in [a, \infty),$$

and

$$\int_a^\infty \frac{\lambda(x)}{(f_1(x))^{c-\alpha+1}} (f_2^\sigma(x))^k \, \Delta_\alpha x = v(x) \, (f_2(x))^k \bigg|_{x=a}^{x=\infty}$$

$$- \int_a^\infty v(x) \left(f_2^k\right)^{\Delta_\alpha}(x) \Delta_\alpha x. \tag{8.3}$$

For $x \in [a, \infty)$, by the chain rule, it follows that there exists a $\xi_1 \in [x, \sigma(x)]$ such that

$$- \left(f_1^{\alpha-c}\right)^{\Delta_\alpha}(x) = -(\alpha - c)\left(f_1(\xi_1)\right)^{\alpha-c-1} f_1^{\Delta_\alpha}(x)$$

$$= \frac{(\alpha - c)\lambda(x)}{\left(f_1(\xi_1)\right)^{c-\alpha-1}}$$

$$\geq \frac{(\alpha - c)\lambda(x)}{\left(f_1(x)\right)^{c-\alpha+1}}.$$

Hence,

$$-v(x) = \int_x^\infty \frac{\lambda(s)}{\left(f_1(s)\right)^{c-\alpha+1}} \Delta_\alpha s$$

$$\leq -\frac{1}{\alpha - c} \int_x^\infty \left(f_1^{\alpha-c}\right)^{\Delta_\alpha}(s) \Delta_\alpha s$$

$$\leq \frac{\left(f_1(x)\right)^{\alpha-c}}{\alpha - c}, \quad x \in [a, \infty).$$

For $x \in [a, \infty)$, by the chain rule, it follows that there exists a $\xi_2 \in [x, \sigma(x)]$ so that

$$\left(f_2^k\right)^{\Delta_\alpha}(x) = k\left(f_2(\xi_2)\right)^{k-1} f_2^{\Delta_\alpha}(x)$$

$$= k\left(f_2(\xi_2)\right)^{k-1} \lambda(x)g(x)$$

$$\leq k\left(f_2^\sigma(x)\right)^{k-1} \lambda(x)g(x).$$

Now, using $f_2(a) = 0$, $v(\infty) = 0$, (8.3) and the Hölder inequality, we find

$$\int_a^\infty \frac{\lambda(x)}{\left(f_1(x)\right)^{c-\alpha+1}} \left(f_2^\sigma(x)\right)^k \Delta_\alpha x = -\int_a^\infty v(x) \left(f_2^k\right)^{\Delta_\alpha}(x) \Delta_\alpha x$$

$$\leq \frac{k}{\alpha - c} \int_a^\infty \left(f_1(x)\right)^{\alpha-c} \left(f_2^\sigma(x)\right)^{k-1} \lambda(x)g(x) \Delta_\alpha x$$

$$= \frac{k}{\alpha - c} \int_a^\infty \frac{(f_1(x))^{(c-\alpha+1)\frac{k-1}{k}}}{(\lambda(x))^{\frac{k-1}{k}} (f_1(x))^{c-\alpha}} \lambda(x) g(x)$$

$$\times \frac{(\lambda(x))^{\frac{k-1}{k}} (f_2^\sigma(x))^{k-1}}{(f_1(x))^{(c-\alpha+1)\frac{k-1}{k}}} \Delta_\alpha x$$

$$\leq \frac{k}{\alpha - c} \left(\int_a^\infty \left(\frac{(f_1(x))^{(c-\alpha+1)\frac{k-1}{k}}}{(\lambda(x))^{\frac{k-1}{k}} (f_1(x))^{c-\alpha}} \lambda(x) g(x) \right)^k \Delta_\alpha x \right)^{\frac{1}{k}}$$

$$\times \left(\int_a^\infty \left(\frac{(\lambda(x))^{\frac{k-1}{k}} (f_2^\sigma(x))^{k-1}}{(f_1(x))^{(c-\alpha+1)\frac{k-1}{k}}} \right)^{\frac{k}{k-1}} \Delta_\alpha x \right)^{\frac{k-1}{k}}$$

$$= \frac{k}{\alpha - c} \left(\int_a^\infty \frac{(f_1(x))^{(c-\alpha+1)(k-1)}}{(\lambda(x))^{k-1} (f_1(x))^{(c-\alpha)k}} (\lambda(x))^k (g(x))^k \Delta_\alpha x \right)^{\frac{1}{k}}$$

$$\times \left(\int_a^\infty \frac{\lambda(x) (f_2^\sigma(x))^k}{(f_1(x))^{c-\alpha+1}} \Delta_\alpha x \right)^{\frac{k-1}{k}},$$

whereupon

$$\left(\int_a^\infty \frac{\lambda(x) (f_2^\sigma(x))^k}{(f_1(x))^{c-\alpha+1}} \Delta_\alpha x \right)^{\frac{1}{k}}$$

$$\leq \frac{k}{\alpha - c} \left(\int_a^\infty \frac{(f_1(x))^{(c-\alpha+1)(k-1)}}{(\lambda(x))^{k-1} (f_1(x))^{(c-\alpha)k}} (\lambda(x))^k (g(x))^k \Delta_\alpha x \right)^{\frac{1}{k}}$$

and

$$\int_a^\infty \frac{\lambda(x) (f_2^\sigma(x))^k}{(f_1(x))^{c-\alpha+1}} \Delta_\alpha x \leq \left(\frac{k}{\alpha - c} \right)^k \int_a^\infty \frac{(f_1(x))^{(c-\alpha+1)(k-1)}}{(\lambda(x))^{k-1} (f_1(x))^{(c-\alpha)k}}$$

$$\times (\lambda(x))^k (g(x))^k \Delta_\alpha x.$$

This completes the proof.   □

8.3 The Bennett Fractional Inequality

In this section, we deduct the fractional analogue of the Bennett inequality.

Theorem 8.4 (The Bennett Inequality). *Let* $0 < \alpha < 1$, $1 < c \leq \alpha$, $\lambda, g : [a, \infty) \to (0, \infty)$ *are integrable functions. Define*

$$f_1(x) = \int_x^\infty \lambda(s) \Delta_\alpha s,$$

$$f_2(x) = \int_x^\infty \lambda(s) g(s) \Delta_\alpha s, \quad x \in [a, \infty).$$

If $f_2(a) < \infty$ *and*

$$\int_a^\infty \frac{\lambda(s)}{(f_1(s))^{c-\alpha+1}} \Delta_\alpha s < \infty,$$

then

$$\int_a^\infty \frac{\lambda(x)}{(f_1(x))^{c-\alpha+1}} (f_2(x))^k \Delta_\alpha x$$

$$\leq \left(\frac{k}{c-\alpha}\right)^k \int_a^\infty (f_1(x))^{k-c+\alpha-1} \lambda(x)(g(x))^k \Delta_\alpha x.$$

Proof. Let

$$u(x) = \int_a^x \frac{\lambda(s)}{(f_1(s))^{c-\alpha+1}} \Delta_\alpha s, \quad x \in [a, \infty).$$

Then

$$u^{\Delta_\alpha}(x) = \frac{\lambda(x)}{(f_1(x))^{c-\alpha+1}}, \quad x \in [a, \infty),$$

and

$$\int_a^\infty \frac{\lambda(x)}{(f_1(x))^{c-\alpha+1}} (f_2(x))^k \Delta_\alpha = u(x) (f_2(x))^k \Big|_{x=a}^{x=\infty}$$

$$+ \int_a^\infty u^\sigma(x) \left(-\left(f_2^k\right)^{\Delta_\alpha}(x)\right) \Delta_\alpha x.$$

$$(8.4)$$

For $x \in [a, \infty)$, by the chain rule, it follows that there exists a $\xi_1 \in [x, \sigma(x)]$ so that

$$\left(f_1^{\alpha-c}\right)^{\Delta_\alpha}(x) = (\alpha - c)\left(f_1(x)\right)^{\alpha-c-1} f_1^{\Delta_\alpha}(x)$$

$$= \frac{(c-\alpha)\lambda(x)}{(f_1(\xi_1))^{c-\alpha+1}}$$

$$\geq \frac{(c-\alpha)\lambda(x)}{(f_1(x))^{c-\alpha+1}}.$$

Thus,

$$u^\sigma(x) = \int_a^{\sigma(x)} \frac{\lambda(s)}{(f_1(s))^{c-\alpha+1}} \Delta_\alpha s$$

$$\leq \frac{1}{c-\alpha} \int_a^{\sigma(x)} \left(f_1^{\alpha-c}\right)^{\Delta_\alpha}(s) \Delta_\alpha s$$

$$= \frac{(f_1^\sigma(x))^{\alpha-c}}{c-\alpha} - \frac{(f_1(a))^{\alpha-c}}{c-\alpha}$$

$$\leq \frac{(f_1^\sigma(x))^{\alpha-c}}{c-\alpha}$$

$$\leq \frac{(f_1(x))^{\alpha-c}}{c-\alpha}, \quad x \in [a, \infty).$$

For $x \in [a, \infty)$, by the chain rule, it follows that there exists a $\xi_2 \in [x, \sigma(x)]$ such that

$$\left(f_2^k\right)^{\Delta_\alpha}(x) = k\left(f_2(\xi_2)\right)^{k-1} f_2^{\Delta_\alpha}(x)$$

$$= -k\left(f_2(\xi_2)\right)^{k-1} \lambda(x)g(x)$$

$$\geq -k\left(f_2(x)\right)^{k-1} \lambda(x)g(x).$$

Now, using (8.4), $f_2(\infty) = 0$, $u(a) = 0$ and the Hölder inequality, we get

$$\int_a^\infty \frac{\lambda(x)}{(f_1(x))^{c-\alpha+1}} (f_2(x))^k \, \Delta_\alpha x \leq \int_a^\infty \frac{\lambda(x)}{(f_1(x))^{c-\alpha+1}} (f_2^\sigma(x))^k \, \Delta_\alpha x$$

$$\leq \frac{k}{c-\alpha} \int_a^\infty (f_1(x))^{\alpha-c} \lambda(x) g(x) (f_2(x))^{k-1} \, \Delta_\alpha x$$

$$= \frac{k}{c-\alpha} \int_a^\infty \frac{(f_1(x))^{(c-\alpha+1)\frac{k-1}{k}}}{(\lambda(x))^{\frac{k-1}{k}} (f_1(x))^{c-\alpha}} \lambda(x) g(x) (\lambda(x))^{\frac{k-1}{k}}$$

$$\times \frac{(f_2(x))^{k-1}}{(f_1(x))^{(c-\alpha+1)\frac{k-1}{k}}} \Delta_\alpha x$$

$$\leq \frac{k}{c-\alpha} \left(\int_a^\infty \left(\frac{k}{c-\alpha} \int_a^\infty \frac{(f_1(x))^{(c-\alpha+1)\frac{k-1}{k}}}{(\lambda(x))^{\frac{k-1}{k}} (f_1(x))^{c-\alpha}} \lambda(x) g(x) \right)^k \Delta_\alpha x \right)^{\frac{1}{k}}$$

$$\times \left(\int_a^\infty \left((\lambda(x))^{\frac{k-1}{k}} \frac{(f_2(x))^{k-1}}{(f_1(x))^{(c-\alpha+1)\frac{k-1}{k}}} \right)^{\frac{k}{k-1}} \Delta_\alpha x \right)^{\frac{k-1}{k}}$$

$$= \frac{k}{c-\alpha} \left(\int_a^\infty \frac{(f_1(x))^{(c-\alpha+1)(k-1)}}{(\lambda(x))^{k-1} (f_1(x))^{(c-\alpha)k}} (\lambda(x))^k (g(x))^k \, \Delta_\alpha x \right)^{\frac{1}{k}}$$

$$\times \left(\int_a^\infty \lambda(x) \frac{(f_2(x))^k}{(f_1(x))^{c-\alpha+1}} \, \Delta_\alpha x \right)^{\frac{k-1}{k}}$$

$$= \frac{k}{c-\alpha} \left(\int_a^\infty \lambda(x) (f_1(x))^{k-c+\alpha-1} (g(x))^k \, \Delta_\alpha x \right)^{\frac{1}{k}}$$

$$\times \left(\int_a^\infty \lambda(x) \frac{(f_2(x))^k}{(f_1(x))^{c-\alpha+1}} \, \Delta_\alpha x \right)^{\frac{k-1}{k}},$$

whereupon

$$\left(\int_a^\infty \lambda(x) \frac{(f_2(x))^k}{(f_1(x))^{c-\alpha+1}} \Delta_\alpha x \right)^{\frac{1}{k}}$$

$$\leq \frac{k}{c-\alpha} \left(\int_a^\infty \lambda(x) (f_1(x))^{k-c+\alpha-1} (g(x))^k \Delta_\alpha x \right)^{\frac{1}{k}}$$

and

$$\int_a^\infty \lambda(x) \frac{(f_2(x))^k}{(f_1(x))^{c-\alpha+1}} \Delta_\alpha x$$

$$\leq \left(\frac{k}{c-\alpha} \right)^k \int_a^\infty \lambda(x) (f_1(x))^{k-c+\alpha-1} (g(x))^k \Delta_\alpha x.$$

This completes the proof.   □

Chapter 9

Reverse Hardy-Type Fractional Inequalities

In this chapter, we investigate some reverse Copson-type, Leindler-type and Bennett-type fractional inequalities on arbitrary time scales.

Let $\mathbb{T}$ be a time scale with forward jump operator and delta differentiation operator σ and Δ, respectively. Let also $a \in \mathbb{T}$, $[a, \infty) \subset \mathbb{T}$ and $\alpha \in (0, 1]$.

Note that when $\alpha = 1$, we get the reverse Copson, Leindler and Bennett inequalities on time scales, which can be found in Ref. [1]. Thus, the results in this chapter can be considered as a generalization of the results in Ref. [1].

9.1 Reverse Copson-Type Fractional Inequalities

In this section, we deduct some reverse Copson-type fractional inequalities.

Theorem 9.1 (Reverse Copson-Type Inequality). *Let* $k \le 0 < h < 1$, $\lambda, g : [a, \infty) \to (0, \infty)$ *be integrable functions. Define*

$$f_1(x) = \int_x^\infty \lambda(s)\Delta_\alpha s,$$

$$f_2(x) = \int_a^x \lambda(s)g(s)\Delta_\alpha s, \quad x \in [a, \infty).$$

Then

$$\int_a^\infty \frac{\lambda(x)}{(f_1(x))^{k-\alpha+1}} \left(f_2^\sigma(x)\right)^h \Delta_\alpha x$$

$$\geq \left(\frac{h}{\alpha-k}\right)^h \int_a^\infty \lambda(x)(g(x))^h \left(f_1(x)\right)^{k-k+\alpha-1}(x)\Delta_\alpha x.$$

Proof. Let

$$v(x) = -\int_{+}^{x\infty} \frac{\lambda(s)}{(f_1(s))^{k-\alpha+1}} \Delta_\alpha s, \quad x \in [a,\infty).$$

Then

$$v^{\Delta_\alpha}(x) = \frac{\lambda(x)}{(f_1(x))^{k-\alpha+1}}, \quad x \in [a,\infty).$$

Hence,

$$\int_a^\infty \frac{\lambda(x)}{(f_1(x))^{k-\alpha+1}} \left(f_2^\sigma(x)\right)^h \Delta_\alpha x = v(x)\left(f_2(x)\right)^h \Big|_{x=a}^{x=\infty}$$

$$- \int_a^\infty v(x) \left(f_2^h\right)^{\Delta_\alpha}(x)\Delta_\alpha x$$

$$= -\int_a^\infty v(x) \left(f_2^h\right)^{\Delta_\alpha}(x)\Delta_\alpha x,$$

$$(9.1)$$

where we have used that $v(\infty) = 0$ and $f_2(a) = 0$. For $x \in [a,\infty)$, by the chain rule, we note that there exists a $\xi_1 \in [x,\sigma(x)]$ so that

$$\left(f_2^h\right)^{\Delta_\alpha}(x) = h\left(f_2(\xi)\right)^{h-1} f_2^{\Delta_\alpha}(x)$$

$$= \frac{h f_2^{\Delta_\alpha}(x)}{(f_2(\xi_1))^{1-h}}$$

$$\geq \frac{h f_2^{\Delta_\alpha}(x)}{(f_2^\sigma(x))^{1-h}}.$$

Next, for $x \in [a, \infty)$, by the chain rule, it follows that there exists a $\xi_2 \in [x, \sigma(x)]$ such that

$$\left(f_1^{\alpha-k}\right)^{\Delta_\alpha}(x) = (\alpha - k)\left(f_1(\xi_2)\right)^{\alpha-k-1} f_1^{\Delta_\alpha}(x)$$

$$= \frac{(\alpha - k) f_1^{\Delta_\alpha}(x)}{(f_1(\xi_2))^{k+1-\alpha}}$$

$$\leq -\frac{(\alpha - k)\lambda(x)}{(f_1(x))^{k+1-\alpha}},$$

whereupon

$$(f_1(x))^{-k+\alpha-1} \lambda(x) \geq -\frac{1}{\alpha - k}\left(f_1^{\alpha-k}\right)^{\Delta_\alpha}(x).$$

Therefore,

$$-v(x) = b\int_x^\infty \frac{\lambda(s)}{(f_1(s))^{k-\alpha+1}}\Delta_\alpha s$$

$$\geq -\frac{1}{\alpha - k}\int_x^\infty \left(f_1^{\alpha-k}\right)^{\Delta_\alpha}(s)\Delta_\alpha s$$

$$= \frac{1}{(\alpha - k)\left(f_1(x)\right)^{\alpha-k}}, \quad x \in [a, \infty).$$

Hence, from (9.1), we arrive at the inequality

$$\int_a^\infty \frac{\lambda(x)}{(f_1(x))^{k-\alpha+1}}\left(f_2^\sigma(x)\right)^h \Delta_\alpha x$$

$$\geq \frac{h}{\alpha - k}\int_a^\infty \frac{\lambda(x)g(x)}{(f_1(x))^{k-\alpha}\left(f_2^\sigma(x)\right)^{1-h}}\Delta_\alpha x. \tag{9.2}$$

By the Hölder inequality, we find

$$\int_a^\infty \frac{(g(x))^h(\lambda(x))^h}{(f_1(x))^{h(k-\alpha)}\left(f_2^\sigma(x)\right)^{h(1-h)}}$$

$$\times \left(\frac{\lambda(x)}{(f_1(x))^{k-\alpha+1}}\right)^{1-h}\left(f_2^\sigma(x)\right)^{h(1-h)}\Delta_\alpha x$$

$$\leq \left(\int_a^\infty \left(\frac{(g(x))^h (\lambda(x))^h}{(f_1(x))^{h(k-\alpha)} (f_2^\sigma(x))^{h(1-h)}} \right)^{\frac{1}{h}} \Delta_\alpha x \right)^h$$

$$\times \left(\int_a^\infty \left(\frac{\lambda(x)}{(f_1(x))^{k-\alpha+1}} \right)^{1-h} (f_2^\sigma(x))^{h(1-h)} \Delta_\alpha x \right)^{1-h},$$

whereupon

$$\left(\int_a^\infty \left(\frac{(g(x))^h (\lambda(x))^h}{(f_1(x))^{h(k-\alpha)} (f_2^\sigma(x))^{h(1-h)}} \right)^{\frac{1}{h}} \Delta_\alpha x \right)^h$$

$$\geq \int_a^\infty \frac{(g(x))^h (\lambda(x))^h}{(f_1(x))^{h(k-\alpha)} (f_2^\sigma(x))^{h(1-h)}}$$

$$\times \left(\frac{\lambda(x)}{(f_1(x))^{k-\alpha+1}} \right)^{1-h} (f_2^\sigma(x))^{h(1-h)} \Delta_\alpha x$$

$$\times \left(\int_a^\infty \left(\frac{\lambda(x)}{(f_1(x))^{k-\alpha+1}} \right)^{1-h} (f_2^\sigma(x))^{h(1-h)} \Delta_\alpha x \right)^{h-1}$$

$$= \int_a^\infty \frac{(g(x))^h \lambda(x)}{(f_1(x))^{k-\alpha+1-h}} \Delta_\alpha x$$

$$\times \left(\int_a^\infty \left(\frac{\lambda(x)}{(f_1(x))^{k-\alpha+1}} \right)^{1-h} (f_2^\sigma(x))^{h(1-h)} \Delta_\alpha x \right)^{h-1}$$

$$= \frac{\int_a^\infty \frac{(g(x))^h \lambda(x)}{(f_1(x))^{k-\alpha+1-h}} \Delta_\alpha x}{\left(\int_a^\infty \left(\frac{\lambda(x)}{(f_1(x))^{k-\alpha+1}} \right)^{1-h} (f_2^\sigma(x))^{h(1-h)} \Delta_\alpha x \right)^{1-h}}.$$

Now, we apply the inequality (9.2) and find

$$\int_a^\infty \frac{\lambda(x)}{(f_1(x))^{k-\alpha+1}} \left(f_2^\sigma(x)\right)^h \Delta_\alpha x$$

$$\geq \frac{h}{\alpha-k} \int_a^\infty \frac{\lambda(x)g(x)}{(f_1(x))^{k-\alpha} \left(f_2^\sigma(x)\right)^{1-h}} \Delta_\alpha x$$

$$\geq \left(\frac{h}{\alpha-k}\right)^h \frac{\int_a^\infty \frac{(g(x))^h \lambda(x)}{(f_1(x))^{k-\alpha+1-h}} \Delta_\alpha x}{\left(\int_a^\infty \left(\frac{\lambda(x)}{(f_1(x))^{k-\alpha+1}}\right)^{1-h} \left(f_2^\sigma(x)\right)^{h(1-h)} \Delta_\alpha x\right)^{1-h}}$$

or

$$\int_a^\infty \frac{\lambda(x)}{(f_1(x))^{k-\alpha+1}} \left(f_2^\sigma(x)\right)^h \Delta_\alpha x$$

$$\geq \left(\frac{h}{\alpha-k}\right)^h \int_a^\infty \lambda(x)(g(x))^h \left(f_1(x)\right)^{k-k+\alpha-1} (x) \Delta_\alpha x.$$

This completes the proof. $\qquad \square$

Theorem 9.2 (Reverse Copson Fractional Inequality). *Suppose that $0 < h < 1 < k$, $\lambda, g : [a, \infty) \to (0, \infty)$. Let f_1 be as in Theorem 9.1 and define*

$$f_2(x) = \int_x^\infty \lambda(s)g(s)\Delta_\alpha s, \quad x \in [a, \infty).$$

If

$$M = \inf_{x \in \mathbb{T}} \frac{f_1^\sigma(x)}{f_1(x)} > 0,$$

then

$$\int_a^\infty \frac{\lambda(x)}{(f_1(x))^{k-\alpha+1}} \left(f_2(x)\right)^h \Delta_\alpha x$$

$$\geq \left(\frac{hM^{k-\alpha+1}}{k-\alpha}\right)^h \int_a^\infty \lambda(x)(g(x))^h \left(f_1(x)\right)^{h-k+\alpha-1} \Delta_\alpha x.$$

Proof. Let

$$u(x) = \int_a^x \frac{\lambda(s)}{(f_1(s))^{k-\alpha+1}} \Delta_\alpha s, \quad x \in [a, \infty).$$

Then

$$u^{\Delta_\alpha}(x) = \frac{\lambda(x)}{(f_1(x))^{k-\alpha+1}}, \quad x \in [a, \infty).$$

Using $u(a) = 0$ and $f_2(\infty) = 0$, we get

$$\int_a^\infty \frac{\lambda(x)}{(f_1(x))^{k-\alpha+1}} \left(f_2(x)\right)^h \Delta_\alpha x = u(x) \left(f_2(x)\right)^h \Big|_{x=a}^{x=\infty}$$

$$- \int_a^\infty u^\sigma(x) \left(f_2^h\right)^{\Delta_\alpha}(x)\Delta_\alpha x$$

$$= - \int_a^\infty u^\sigma(x) \left(f_2^h\right)^{\Delta_\alpha}(x)\Delta_\alpha x.$$

$$(9.3)$$

For $x \in [a, \infty)$, by the chain rule, it follows that there exists a $\xi_1 \in [x, \sigma(x)]$ such that

$$- \left(f_2^h\right)^{-\int_a^\infty u^\sigma(x)(f_2^h)^{\Delta_\alpha}(x)\Delta_\alpha x}(x) = -h \left(f_2(\xi_1)\right)^{h-1} f_2^{\Delta_\alpha}(x)$$

$$= \frac{h\lambda(x)g(x)}{(f_2(\xi_1))^{1-h}}$$

$$\geq \frac{h\lambda(x)g(x)}{(f_2(x))^{1-h}}.$$

Now, using (9.3), we find

$$\int_a^\infty \frac{\lambda(x)}{(f_1(x))^{k-\alpha+1}}\left(f_2(x)\right)^h \Delta_\alpha x \geq h\int_a^\infty \frac{\lambda(x)g(x)}{(f_2(x))^{1-h}}u^\sigma(x)\Delta_\alpha x.$$

$$(9.4)$$

Next, for $x \in [a, \infty)$, by the chain rule, it follows that there exists a $\xi_2 \in [x, \sigma(x)]$ so that

$$\left(f_1^{\alpha-k}\right)^{\Delta_\alpha}(x) = (\alpha - k)\left(f_1(\xi_2)\right)^{\alpha-k-1}f_1^{\Delta_\alpha}(x)$$

$$= \frac{(\alpha - k)f_1^{\Delta_\alpha}(x)}{(f_1(\xi_2))^{k-\alpha+1}}$$

$$\leq \frac{(\alpha - k)f_1^{\Delta_\alpha}(x)}{(f_1^\sigma(x))^{k-\alpha+1}}$$

$$= -\frac{(\alpha - k)\lambda(x)}{(f_1^\sigma(x))^{k-\alpha+1}}$$

$$= (k - \alpha)\frac{\lambda(x)}{(f_1^\sigma(x))^{k-\alpha+1}}\frac{(f_1(x))^{k+1-\alpha}}{(f_1(x))^{k+1-0\alpha}}$$

$$\leq (k - \alpha)\frac{\lambda(x)}{M^{k-\alpha+1}(f_1(x))^{k+1-\alpha}}.$$

Hence,

$$u(x) = \int_a^x \frac{\lambda(s)}{(f_1(s))^{k+1-\alpha}}\Delta_\alpha s$$

$$\geq \frac{M^{k-\alpha+1}}{k-\alpha}\int_a^x \left(f_1^{\alpha-k}\right)^{\Delta_\alpha}(x)\Delta_\alpha s$$

$$= \frac{M^{k-\alpha+1}}{k-\alpha}\left(f_1(x)\right)^{\alpha-k}, \quad x \in [a, \infty).$$

Now, we apply the last inequality to (9.4) and find

$$\int_a^\infty \frac{\lambda(x)}{(f_1(x))^{k-\alpha+1}} \left(f_2(x)\right)^h \Delta_\alpha x$$

$$\geq h \int_a^\infty \frac{\lambda(x)g(x)}{(f_1(x))^{1-h}} u^\sigma(x)\Delta_\alpha x$$

$$\geq h \int_a^\infty \frac{\lambda(x)g(x)}{(f_2(x))^{1-h}} u(x)\Delta_\alpha x$$

$$\geq \frac{hM^{k-\alpha+1}}{k-\alpha} \int_a^\infty \frac{\lambda(x)g(x)}{(f_2(x))^{1-h}} \left(f_1(x)\right)^{\alpha-k} \Delta_\alpha x.$$

Hence, using the rest of the proof of Theorem 9.1, we obtain

$$\left(\int_a^\infty \frac{\lambda(x)}{(f_1(x))^{k-\alpha+1}} \left(f_2(x)\right)^h \Delta_\alpha x\right)^h$$

$$\geq \left(\frac{hM^{k-\alpha+1}}{k-\alpha}\right)^h \left(\int_a^\infty \left(\frac{(\lambda(x))^h (g(x))^h}{(f_1(x))^{(k-\alpha)h} (f_2(x))^{h(1-h)}}\right)^{\frac{1}{h}} \Delta_\alpha x\right)^h$$

$$\geq \frac{\int_a^\infty \frac{(g(x))^h \lambda(x)}{(f_1(x))^{k-\alpha+1-h}}\Delta_\alpha x}{\left(\int_a^\infty \frac{\lambda(x)(f_2(x))^h}{(f_1(x))^{k-\alpha+1}}\Delta_\alpha x\right)^{1-h}}$$

and

$$\int_a^\infty \frac{\lambda(x)}{(f_1(x))^{k-\alpha+1}} \left(f_2(x)\right)^h \Delta_\alpha x$$

$$\geq \left(\frac{hM^{k-\alpha+1}}{k-\alpha}\right)^h \int_a^\infty \lambda(x)(g(x))^h \left(f_1(x)\right)^{h-k+\alpha-1} \Delta_\alpha x.$$

This completes the proof. □

9.2 The Reverse Leindler Fractional Inequality

In this section, we deduct the reverse Leindler fractional inequality.

Theorem 9.3 (The Reverse Leindler Fractional Inequality).
Let $k \geq 0 < h < 1$, $\lambda, g : [a, \infty) \to (0, \infty)$ be integrable functions and define

$$f_1(x) = \int_a^x \lambda(s)\Delta_\alpha s,$$

$$f_2(x) = \int_x^\infty \lambda(s)g(s)\Delta_\alpha s, \quad x \in [a, \infty).$$

Then

$$\int_a^\infty \frac{\lambda(x)}{(f_1^\sigma(x))^{k-\alpha+1}} (f_2(x))^h \, \Delta_\alpha x$$

$$\geq \left(\frac{h}{\alpha - k}\right)^h \int_a^\infty \lambda(x)(g(x))^h (f_2^\sigma(x))^{h-k+\alpha-1} \, \Delta_\alpha x.$$

Proof. Let

$$u(x) = \int_a^x \frac{\lambda(s)}{(f_1(s))^{k-\alpha+1}} \Delta_\alpha s, \quad x \in [a, \infty).$$

Then

$$u^{\Delta_\alpha}(x) = \frac{\lambda(x)}{(f_1^\sigma(x))^{k-\alpha+1}}, \quad x \in [a, \infty),$$

$u(a) = 0$. Now, using $f_2(\infty) = 0$, we get

$$\int_a^\infty \frac{\lambda(x)}{(f_1^\sigma(x))^{k-\alpha+1}} (f_2(x))^h \, \Delta_\alpha x$$

$$= u(x) (f_2(x))^h \Big|_{x=a}^{x=\infty} - \int_a^\infty u^\sigma(x) \left(f_2^h\right)^{\Delta_\alpha}(x)\Delta_\alpha x$$

$$= -\int_a^\infty u^\sigma(x) \left(f_2^h\right)^{\Delta_\alpha}(x)\Delta_\alpha x. \tag{9.5}$$

For $x \in [a, \infty)$, by the chain rule, it follows that there exists a $\xi_1 \in [x, \sigma(x)]$ such that

$$-\left(f_2^h\right)^{\Delta_\alpha}(x) = -h (f_2(\xi_1))^{h-1} f_2^{\Delta_\alpha}(x)$$

$$= h (f_2(\xi_1))^{h-1} \lambda(x)g(x)$$

$$= \frac{h\lambda(x)g(x)}{(f_2(\xi_1))^{1-h}}$$

$$\geq \frac{h\lambda(x)g(x)}{(f_2(x))^{1-h}}.$$

Hence, by (9.5), we get

$$\int_a^\infty \frac{\lambda(x)}{(f_1^\sigma(x))^{k-\alpha+1}} (f_2(x))^h \, \Delta_\alpha x \geq h \int_a^\infty u^\sigma(x) \frac{\lambda(x)g(x)}{(f_2(x))^{1-h}} \Delta_\alpha x.$$

$$(9.6)$$

Next, for $x \in [a, \infty)$, by the chain rule, it follows that there exists a $\xi_2 \in [x, \sigma(x)]$ so that

$$\left(f_1^{\alpha-k}\right)^{\Delta_\alpha}(x) = b(\alpha - k) (f_1(\xi_2))^{\alpha-k-1} f_1^{\Delta_\alpha}(x)$$

$$= (\alpha - k) \frac{\lambda(x)}{(f_1(\xi_2))^{k-\alpha+}}$$

$$\leq \frac{(\alpha - k)\lambda(x)}{(f_1(x))^{k-\alpha+1}}.$$

This implies that

$$u^\sigma(x) = \int_a^{\sigma(x)} \frac{\lambda(s)}{(f_1(s))^{k-\alpha+1}} \Delta_\alpha s$$

$$\geq \frac{1}{\alpha - k} \int_a^{\sigma(x)} \left(f_1^{\alpha-k}\right)^{\Delta_\alpha}(s) \Delta_\alpha s$$

$$= \frac{(f_1^\sigma(x))^{\alpha-k}}{\alpha - k}, \quad x \in [a, \infty).$$

By the last inequality and (9.6), we arrive at the inequality

$$\int_a^\infty \frac{\lambda(x)}{(f_1^\sigma(x))^{k-\alpha+1}} (f_2(x))^h \, \Delta_\alpha x$$

$$\geq \frac{h}{\alpha - k} \int_a^\infty (f_1^\sigma(x))^{\alpha-k} \frac{\lambda(x)g(x)}{(f_2(x))^{1-h}} \Delta_\alpha x.$$

Now, using the rest of the proof of Theorem 9.1, by the last inequality, we find

$$\left(\int_a^\infty \frac{\lambda(x)}{(f_1^\sigma(x))^{k-\alpha+1}} \left(f_2(x)\right)^h \Delta_\alpha x \right)^{\frac{1}{h}}$$

$$\geq \left(\frac{h}{\alpha-k} \right)^h \left(\int_a^\infty \left((f_1^\sigma(x))^{h(\alpha-k)} \frac{(\lambda(x))((g(x))^h}{(f_2(x))^{h(1-h)}} \right)^{\frac{1}{h}} \Delta_\alpha x \right)^h$$

$$= \left(\frac{h}{\alpha-k} \right)^h \left(\int_a^\infty \left(\frac{(\lambda(x))((g(x))^h}{(f_1^\sigma(x))^{h(\alpha-k)}(f_2(x))^{h(1-h)}} \right)^{\frac{1}{h}} \Delta_\alpha x \right)^h$$

$$\geq \left(\frac{h}{\alpha-k} \right)^h \frac{\int_a^\infty \frac{(g(x))^h\lambda(x)}{(f_1^\sigma(x))^{k-\alpha+1-h}} \Delta_\alpha x}{\left(\int_a^\infty \frac{\lambda(x)(f_2(x))^h}{(f_1^\sigma(x))^{k-\alpha+1}} \Delta_\alpha x \right)^{1-h}},$$

whereupon

$$\int_a^\infty \frac{\lambda(x)}{(f_1^\sigma(x))^{k-\alpha+1}} \left(f_2(x)\right)^h \Delta_\alpha x$$

$$\geq \left(\frac{h}{\alpha-k} \right)^h \int_a^\infty \frac{(g(x))^h\lambda(x)}{(f_1^\sigma(x))^{k-\alpha+1-h}} \Delta_\alpha x.$$

This completes the proof. □

9.3 The Reverse Bennett Fractional Inequality

In this section, we deduct the reverse Bennett inequality.

Theorem 9.4 (The Reverse Bennett Fractional Inequality).
Let $0 < h \leq 1 < k$, $\lambda, g : [a, \infty) \to (0, \infty)$ be integrable functions and define

$$f_1(x) = \int_a^x \lambda(s)\Delta_\alpha s,$$

$$f_2(x) = \int_a^x \lambda(s)g(s)\Delta_\alpha s, \quad x \in [a, \infty).$$

If

$$L = \inf_{x \in \mathbb{T}} \frac{f_1(x)}{f_1^\sigma(x)} > 0,$$

then

$$\int_a^\infty \frac{\lambda(x)}{(f_1^\sigma(x))^{k-\alpha+1}} \left(f_2^\sigma(x)\right)^h \Delta_\alpha x$$

$$\geq \left(\frac{hL^{\alpha-k}}{k-\alpha}\right)^h \int_a^\infty \lambda(x)(g(x))^h \left(f_1^\sigma(x)\right)^{h-k+\alpha-1} \Delta_\alpha x.$$

Proof. Let

$$v(x) = \int_a^x \frac{\lambda(s)}{(f_1^\sigma(s))^{k-\alpha+1}} \Delta_\alpha s, \quad x \in [a, \infty).$$

Then $v(0) = 0$ and

$$v^{\Delta_\alpha}(x) = \frac{\lambda(x)}{(f_1^\sigma(x))^{k-\alpha+1}}, \quad x \in [a, \infty).$$

Hence,

$$\int_a^\infty \frac{\lambda(x)}{(f_1^\sigma(x))^{k-\alpha+1}} \left(f_2^\sigma(x)\right)^h \Delta_\alpha x$$

$$= v(x) \left(f_2(x)\right)^h \Big|_{x=a}^{x=\infty} - \int_a^\infty v_2(x) \left(f_2^h\right)^{\Delta_\alpha}(x) \Delta_\alpha x$$

$$= - \int_a^\infty v(x) \left(f_2^h\right)^{\Delta_\alpha}(x) \Delta_\alpha x. \tag{9.7}$$

For $x \in [a, \infty)$, by the chain rule, it follows that there exists a $\xi_1 \in [x, \sigma(x)]$ such that

$$\left(f_2^h\right)^{\Delta_\alpha}(x) = h \left(f_2(\xi_1)\right)^{h-1} f_2^{\Delta_\alpha}(x)$$

$$= h \frac{\lambda(x)g(x)}{(f_2(\xi_1))^{1-h}}$$

$$\geq h \frac{\lambda(x)g(x)}{(f_2^\sigma(x))^{1-h}}.$$

Hence, from (9.7), we get

$$\int_a^\infty \frac{\lambda(x)}{(f_1^\sigma(x))^{k-\alpha+1}} (f_2^\sigma(x))^h \,\Delta_\alpha x \geq h \int_a^\infty (-v(x)) \frac{\lambda(x)g(x)}{(f_2^\sigma(x))^{1-h}} \Delta_\alpha x.$$

$$(9.8)$$

Next, for $x \in [a, \infty)$, by the chain rule, it follows that there exists a $\xi_2 \in [x, \sigma(x)]$ so that

$$\left(f_1^{\alpha-k}\right)^{\Delta_\alpha} (x) = (\alpha - k) \left(f_1(\xi_2)\right)^{\alpha-k-1} f_1^{\Delta_\alpha}(x)$$

$$= (\alpha - k) \frac{f_1^{\Delta_\alpha}(x)}{(f_1(\xi_2))^{k-\alpha+1}}$$

$$= (\alpha - k) \frac{\lambda(x)}{(f_1(\xi_2))^{k-\alpha+1}}$$

$$\leq (\alpha - k) \frac{\lambda(x)}{(f_1^\sigma(x))^{k-\alpha+1}}$$

and

$$-v(x) = -\int_a^x \frac{\lambda(s)}{(f_1^\sigma(s))^{k-\alpha+1}} \Delta_\alpha s$$

$$\geq \frac{1}{k-\alpha} \int_a^x \left(f_1^{\alpha-k}\right)^{\Delta_\alpha} (s) \Delta_\alpha s$$

$$\geq \frac{1}{k-\alpha} (f_1(x))^{\alpha-k}$$

$$= \frac{1}{k-\alpha} \left(\frac{f_1(x)}{f_1^\sigma(x)}\right)^{\alpha-k} (f_1^\sigma(x))^{\alpha-k}$$

$$\geq \frac{L^{\alpha-k}}{k-\alpha} (f_1^\sigma(x))^{\alpha-k}.$$

Hence, from (9.8) and the rest of the proof of Theorem 9.1, it follows that

$$\int_a^\infty \frac{\lambda(x)}{(f_1^\sigma(x))^{k-\alpha+1}} (f_2^\sigma(x))^h \,\Delta_\alpha x$$

$$\geq \frac{hL^{\alpha-k}}{k-\alpha} \int_a^\infty \frac{\lambda(x)g(x)}{(f_1^\sigma(x))^{k-\alpha} (f_2^\sigma(x))^{1-h}} \Delta_\alpha x$$

and

$$\left(\int_a^\infty \frac{\lambda(x)}{(f_1^\sigma(x))^{k-\alpha+1}} \left(f_2^\sigma(x)\right)^h \Delta_\alpha x \right)^h$$

$$\geq \left(\frac{hL^{\alpha-k}}{k-\alpha} \right)^h \left(\int_a^\infty \left(\frac{(\lambda(x))^h (g(x))^h}{(f_1^\sigma(x))^{h(k-\alpha)} (f_2^\sigma(x))^{h(1-h)}} \right)^{\frac{1}{h}} \Delta_\alpha x \right)^h$$

$$\geq \left(\frac{hL^{\alpha-k}}{k-\alpha} \right)^h \frac{\int_a^\infty \frac{(g(x))^h \lambda(x)}{(f_1^\sigma(x))^{k-\alpha+1-h}} \Delta_\alpha x}{\left(\int_a^\infty \frac{\lambda(x)(f_2(x))^h}{(f_1^\sigma(x))^{k-\alpha+1}} \Delta_\alpha x \right)^{1-h}},$$

whereupon

$$\int_a^\infty \frac{\lambda(x)}{(f_1^\sigma(x))^{k-\alpha+1}} \left(f_2^\sigma(x)\right)^h \Delta_\alpha x$$

$$\geq \left(\frac{hL^{\alpha-k}}{k-\alpha} \right)^h \int_a^\infty \lambda(x)(g(x))^h \left(f_1^\sigma(x)\right)^{h-k+\alpha-1} \Delta_\alpha x.$$

This completes the proof. □

Chapter 10

Inequalities for Generalized Riemann–Liouville Fractional Integrals

In this chapter, we investigate some inequalities for generalized Riemann–Liouville fractional integral operators. Some fractional analogues of Minkowski and Grüss-type inequalities are deducted. Some of the results in this chapter can be found in Ref. [12].

Let $\mathbb{T}$ be a time scale with forward jump operator and delta differentiation operator σ and Δ, respectively. Suppose that $\delta > 1$, t_0, $a \in \mathbb{T}$, $t_0 < a$, $\phi : [t_0, a] \to (0, \infty)$ is a monotone that has a delta derivative ϕ^Δ so that $\phi^\Delta(t) \neq 0$, $t \in [t_0, a]$.

10.1 Minkowski-Type Inequalities

In this section, we deduct some Minkowski-type inequalities.

Theorem 10.1. *Let $p, q \geq 1$, $\frac{1}{p} + \frac{1}{q} = 1$, $f, g : [t_0, a] \to (0, \infty)$ be positive functions such that $_\phi I^\delta_{\Delta, t_0} f(t) < \infty$, $_\phi I^\delta_{\Delta, t_0} g(t) < \infty$, $t \in [t_0, a]$. Let also there exist positive constants m and M such that*

$$0 < m \leq \frac{f(t)}{g(t)} \leq M, \quad t \in [t_0, a].$$

Then

$$\left(\phi I^\delta_{\Delta,t_0} f(t) \right)^{\frac{1}{p}} \left(\phi I^\delta_{\Delta,t_0} g(t) \right)^{\frac{1}{q}}$$

$$\geq \left(\frac{M}{m} \right)^{\frac{1}{pq}} \left(\phi I^\delta_{\Delta,t_0} \left((f(t))^{\frac{1}{p}} (g(t))^{\frac{1}{q}} \right) \right), \quad t \in [t_0, a].$$

Proof.   Since

$$\frac{f(t)}{g(t)} \leq M, \quad t \in [t_0, a],$$

we get

$$M^{\frac{1}{q}} \geq \frac{(f(t))^{\frac{1}{q}}}{(g(t))^{\frac{1}{q}}}, \quad t \in [t_0, a],$$

or

$$(g(t))^{\frac{1}{q}} \geq M^{-\frac{1}{q}} (f(t))^{\frac{1}{q}}, \quad t \in [t_0, a].$$

We multiply the last inequality with $(f(t))^{\frac{1}{p}}$, $t \in [t_0, a]$ and find

$$(f(t))^{\frac{1}{p}} (g(t))^{\frac{1}{q}} \geq M^{-\frac{1}{q}} (f(t))^{\frac{1}{q}} (f(t))^{\frac{1}{p}}$$

$$= M^{-\frac{1}{q}} f(t), \quad t \in [t_0, a],$$

whereupon

$$\frac{(\phi(t) - \phi(s))^{\delta-1}}{\Gamma(\delta)} \phi^\Delta(s) (f(s))^{\frac{1}{p}} (g(s))^{\frac{1}{q}}$$

$$\geq M^{-\frac{1}{q}} \frac{(\phi(t) - \phi(s))^{\delta-1}}{\Gamma(\delta)} \phi^\Delta(s) f(s), \quad t, s \in [t_0, a].$$

We integrate the last inequality with respect to s from t_0 to t and obtain

$$\int_{t_0}^t \frac{(\phi(t) - \phi(s))^{\delta-1}}{\Gamma(\delta)} \phi^\Delta(s) (f(s))^{\frac{1}{p}} (g(s))^{\frac{1}{q}} \Delta s$$

$$\geq M^{-\frac{1}{q}} \int_{t_0}^t \frac{(\phi(t) - \phi(s))^{\delta-1}}{\Gamma(\delta)} \phi^\Delta(s) f(s) \Delta s, \quad t \in [t_0, a],$$

or

$$\phi I_{\Delta,t_0}^{\delta} \left(f^{\frac{1}{p}} g^{\frac{1}{q}} \right) (t) \geq M^{-\frac{1}{q}} {}_{\phi} I_{\Delta,t_0}^{\delta}(f)(t), \quad t \in [t_0, a],$$

whereupon

$$\left({}_{\phi} I_{\Delta,t_0}^{\delta} \left(f^{\frac{1}{p}} g^{\frac{1}{q}} \right) (t) \right)^{\frac{1}{p}} \geq M^{-\frac{1}{pq}} \left({}_{\phi} I_{\Delta,t_0}^{\delta}(f)(t) \right)^{\frac{1}{p}}, \quad t \in [t_0, a].$$

$$(10.1)$$

Now, by the inequality

$$mg(t) \leq f(t), \quad t \in [t_0, a],$$

we find

$$m^{\frac{1}{p}} \left(g(t) \right)^{\frac{1}{p}} \leq \left(f(t) \right)^{\frac{1}{p}}, \quad t \in [t_0, a],$$

whereupon

$$m^{\frac{1}{p}} g(t) = m^{\frac{1}{p}} \left(g(t) \right)^{\frac{1}{p}} \left(g(t) \right)^{\frac{1}{q}}$$

$$\leq \left(f(t) \right)^{\frac{1}{p}} \left(g(t) \right)^{\frac{1}{q}}, \quad t \in [t_0, a],$$

and

$$m^{\frac{1}{p}} \frac{\left(\phi(t) - \phi(s) \right)^{\delta-1}}{\Gamma(\delta)} \phi^{\Delta}(s) g(s)$$

$$\leq \frac{\left(\phi(t) - \phi(s) \right)^{\delta-1}}{\Gamma(\delta)} \phi^{\Delta}(s) \left(g(s) \right)^{\frac{1}{q}}, \quad t, s \in [t_0, a].$$

We integrate the last inequality with respect to s from t_0 to t and arrive at the inequality

$$m^{\frac{1}{p}} \int_{t_0}^{t} \frac{\left(\phi(t) - \phi(s) \right)^{\delta-1}}{\Gamma(\delta)} \phi^{\Delta}(s) g(s) \Delta s$$

$$\leq \int_{t_0}^{t} \frac{\left(\phi(t) - \phi(s) \right)^{\delta-1}}{\Gamma(\delta)} \phi^{\Delta}(s) \left(g(s) \right)^{\frac{1}{q}} \Delta s, \quad t, s \in [t_0, a],$$

or

$$m^{\frac{1}{p}}{}_\phi I^\delta_{\Delta,t_0} g(t) \leq {}_\phi I^\delta_{\Delta,t_0}\left(f^{\frac{1}{p}}g^{\frac{1}{q}}\right)(t), \quad t \in [t_0, a],$$

and

$$m^{\frac{1}{pq}}\left({}_\phi I^\delta_{\Delta,t_0} g(t)\right)^{\frac{1}{q}} \leq \left({}_\phi I^\delta_{\Delta,t_0}\left(f^{\frac{1}{p}}g^{\frac{1}{q}}\right)(t)\right)^{\frac{1}{p}}, \quad t \in [t_0, a],$$

We multiply the last inequality with the inequality (10.1) and find

$${}_\phi I^\delta_{\Delta,t_0}\left(f^{\frac{1}{p}}g^{\frac{1}{q}}\right)(t) \geq \left(\frac{m}{M}\right)^{\frac{1}{pq}}\left({}_\phi I^\delta_{\Delta,t_0} f(t)\right)^{\frac{1}{p}}\left({}_\phi I^\delta_{\Delta,t_0} g(t)\right)^{\frac{1}{q}}, \quad t \in [t_0, a].$$

This completes the proof. □

Theorem 10.2. *Let* $p, q \geq 1$, $\frac{1}{p} + \frac{1}{q} = 1$, $f, g : [t_0, a] \to (0, \infty)$ *be such that*

$$_\phi I^\delta_{\Delta,t_0} f^p(t) < \infty, \quad {}_\phi I^\delta_{\Delta,t_0} g1(t) < \infty, \quad t \in [t_0, a].$$

If there exist positive constants m *and* M *such that*

$$0 < m \leq \frac{(f(t))^p}{(g(t))^q} \leq M < \infty, \quad t \in [t_0, a],$$

then

$$\left({}_\phi I^\delta_{\Delta,t_0} f^p(t)\right)^{\frac{1}{p}}\left({}_\phi I^\delta_{\Delta,t_0} g^q(t)\right)^{\frac{1}{q}} \leq \left(\frac{M}{m}\right)^{\frac{1}{pq}}\left({}_\phi I^\delta_{\Delta,t_0}(fg)(t)\right), \quad t \in [t_0, a].$$

Proof. We replace f and g with f^p and g^q, respectively, in Theorem 10.1 and get the desired result. This completes the proof. □

10.2   Grüss-Type Inequalities

In this section, we deduct Grüss-type inequalities.

Theorem 10.3 (The Grüss-Type Inequality). *Let* $f, \phi_1, \phi_2 : [t_0, a] \to (0, \infty)$ *be integrable functions such that*

$$\phi_1(t) \leq f(t) \leq \phi_2(t), \quad t \in [t_0, a].$$

Then

$$
{}_\phi I^\delta_{\Delta,t_0} \phi_2(t)\, {}_\phi I^\gamma_{\Delta,t_0} f(t) + {}_\phi I^\delta_{\Delta,t_0} f(t)\, {}_\phi I^\gamma_{\Delta,t_0} \phi_1(t)
$$
$$
\geq {}_\phi I^\delta_{\Delta,t_0} \phi_2(t)\, {}_\phi I^\gamma_{\Delta,t_0} \phi_1(t) + {}_\phi I^\delta_{\Delta,t_0} f(t)\, {}_\phi I^\gamma_{\Delta,t_0} f(t), \quad t \in [t_0, a].
$$

Proof.   First, note that for any $s_1, s_2 \in [t_0, a]$, we have

$$
(\phi_2(s_1) - f(s_1))\,(f(s_2) - \phi_1(s_2)) \geq 0.
$$

Then

$$
\phi_2(s_1)f(s_2) + f(s_1)\phi_1(s_2) \geq f(s_1)f(s_2) + \phi_1(s_2)\phi_2(s_1).
$$

Hence,

$$
f(s_2)\frac{(\phi(t) - \phi(s_1))^{\delta-1}\,\phi^\Delta(s_1)}{\Gamma(\delta)}\phi_2(s_1)
$$
$$
+ \phi_1(s_2)\frac{(\phi(t) - \phi(s_1))^{\delta-1}\,\phi^\Delta(s_1)}{\Gamma(\delta)}f(s_1)
$$
$$
\geq f(s_2)\frac{(\phi(t) - \phi(s_1))^{\delta-1}\,\phi^\Delta(s_1)}{\Gamma(\delta)}f(s_1)
$$
$$
+ \phi_1(s_2)\frac{(\phi(t) - \phi(s_1))^{\delta-1}\,\phi^\Delta(s_1)}{\Gamma(\delta)}\phi_2(s_1), \quad t, s_1, s_2 \in [t_0, a].
$$

Now, we integrate the last inequality with respect to s_1 from t_0 to t and find

$$
f(s_2)\int_{t_0}^t \frac{(\phi(t) - \phi(s_1))^{\delta-1}\,\phi^\Delta(s_1)}{\Gamma(\delta)}\phi_2(s_1)\Delta s_1
$$
$$
+ \phi_1(s_2)\int_{t_0}^t \frac{(\phi(t) - \phi(s_1))^{\delta-1}\,\phi^\Delta(s_1)}{\Gamma(\delta)}f(s_1)\Delta s_1
$$
$$
\geq f(s_2)\int_{t_0}^t \frac{(\phi(t) - \phi(s_1))^{\delta-1}\,\phi^\Delta(s_1)}{\Gamma(\delta)}f(s_1)\Delta s_1
$$
$$
+ \phi_1(s_2)\int_{t_0}^t \frac{(\phi(t) - \phi(s_1))^{\delta-1}\,\phi^\Delta(s_1)}{\Gamma(\delta)}\phi_2(s_1)\Delta s_1, \quad t, s_2 \in [t_0, a],
$$

or

$$f(s_2)_\phi I^\delta_{\Delta,t_0}\phi_2(t) + \phi_1(s_2)_\phi I^\delta_{\Delta,t_0}f(t)$$
$$\geq f(s_2)_\phi I^\delta_{\Delta,t_0}f(t) + \phi_1(s_2)_\phi I^\delta_{\Delta,t_0}\phi_2(t), \quad t, s_2 \in [t_0, a].$$

Hence,

$$_\phi I^\delta_{\Delta,t_0}\phi_2(t)\frac{(\phi(t) - \phi(s_2))^{\gamma-1}\phi^\Delta(s_2)}{\Gamma(\delta)}f(s_2)$$

$$+ _\phi I^\delta_{\Delta,t_0}f(t)\frac{(\phi(t) - \phi(s_2))^{\gamma-1}\phi^\Delta(s_2)}{\Gamma(\delta)}\phi_1(s_2)$$

$$\geq _\phi I^\delta_{\Delta,t_0}f(t)\frac{(\phi(t) - \phi(s_2))^{\gamma-1}\phi^\Delta(s_2)}{\Gamma(\delta)}f(s_2)$$

$$+ _\phi I^\delta_{\Delta,t_0}\phi_2(t)\frac{(\phi(t) - \phi(s_2))^{\gamma-1}\phi^\Delta(s_2)}{\Gamma(\delta)}\phi_1(s_2), \quad t \in [t_0, a],$$

which we integrate with respect to s_2 from t_0 to t and get

$$_\phi I^\delta_{\Delta,t_0}\phi_2(t)\int_{t_0}^t \frac{(\phi(t) - \phi(s_2))^{\gamma-1}\phi^\Delta(s_2)}{\Gamma(\delta)}f(s_2)\Delta s_2$$

$$+ _\phi I^\delta_{\Delta,t_0}f(t)\int_{t_0}^t \frac{(\phi(t) - \phi(s_2))^{\gamma-1}\phi^\Delta(s_2)}{\Gamma(\delta)}\phi_1(s_2)\Delta s_2$$

$$\geq _\phi I^\delta_{\Delta,t_0}f(t)\int_{t_0}^t \frac{(\phi(t) - \phi(s_2))^{\gamma-1}\phi^\Delta(s_2)}{\Gamma(\delta)}f(s_2)\Delta s_2$$

$$+ _\phi I^\delta_{\Delta,t_0}\phi_2(t)\int_{t_0}^t \frac{(\phi(t) - \phi(s_2))^{\gamma-1}\phi^\Delta(s_2)}{\Gamma(\delta)}\phi_1(s_2)\Delta s_2, \quad t \in [t_0, a],$$

or

$$_\phi I^\delta_{\Delta,t_0}\phi_2(t)_\phi I^\gamma_{\Delta,t_0}f(t) + _\phi I^\delta_{\Delta,t_0}f(t)_\phi I^\gamma_{\Delta,t_0}\phi_1(t)$$
$$+ _\phi I^\delta_{\Delta,t_0}f(t)_\phi I^\gamma_{\Delta,t_0}{}_\phi I^\gamma_{\Delta,t_0}f(t) + _\phi I^\delta_{\Delta,t_0}\phi_2(t)_\phi I^\gamma_{\Delta,t_0}\phi_1(t), \quad t \in [t_0, a].$$

This completes the proof. $\qquad\square$

Theorem 10.4. *Let* $f, g, \phi_1, \phi_2, w_1, w_2 : [t_0, a] \to (0, \infty)$ *be integrable functions,* $\gamma > 1$, *and*

$$\phi_1(t) \le f(t) \le \phi_2(t),$$
$$w_1(t) \le g(t) \le w_2(t), \quad t \in [t_0, a]. \tag{10.2}$$

Then, we have the following inequalities:

1.

$$_\phi I^\delta_{\Delta,t_0} \phi_2(t) {}_\phi I^\gamma_{\Delta,t_0} g(t) + {}_\phi I^\delta_{\Delta,t_0} f(t) {}_\phi I^\gamma_{\Delta,t_0} w_1(t)$$
$$\ge {}_\phi I^\delta_{\Delta,t_0} \phi_2(t) {}_\phi I^\gamma_{\Delta,t_0} w_1(t) + {}_\phi I^\delta_{\Delta,t_0} f(t) {}_\phi I^\gamma_{\Delta,t_0} g(t), \quad t \in [t_0, a],$$

2.

$$_\phi I^\delta_{\Delta,t_0} \phi_2(t) {}_\phi I^\gamma_{\Delta,t_0} w_2(t) + {}_\phi I^\delta_{\Delta,t_0} f(t) {}_\phi I^\gamma_{\Delta,t_0} g(t)$$
$$\ge {}_\phi I^\delta_{\Delta,t_0} f(t) {}_\phi I^\gamma_{\Delta,t_0} w_2(t) + {}_\phi I^\delta_{\Delta,t_0} \phi_2(t) {}_\phi I^\gamma_{\Delta,t_0} g(t), \quad t \in [t_0, a],$$

3.

$$_\phi I^\delta_{\Delta,t_0} f(t) {}_\phi I^\gamma_{\Delta,t_0} g(t) + {}_\phi I^\delta_{\Delta,t_0} \phi_1(t) {}_\phi I^\gamma_{\Delta,t_0} w_1(t)$$
$$\ge {}_\phi I^\delta_{\Delta,t_0} \phi_1(t) {}_\phi I^\gamma_{\Delta,t_0} g(t) + {}_\phi I^\delta_{\Delta,t_0} f(t) {}_\phi I^\gamma_{\Delta,t_0} w_1(t), \quad t \in [t_0, a],$$

4.

$$_\phi I^\delta_{\Delta,t_0} f(t) {}_\phi I^\gamma_{\Delta,t_0} w_2(t) + {}_\phi I^\delta_{\Delta,t_0} g(t) {}_\phi I^\gamma_{\Delta,t_0} \phi_1(t)$$
$$\ge {}_\phi I^\delta_{\Delta,t_0} \phi_1(t) {}_\phi I^\gamma_{\Delta,t_0} w_2(t) + {}_\phi I^\delta_{\Delta,t_0} f(t) {}_\phi I^\gamma_{\Delta,t_0} g(t), \quad t \in [t_0, a].$$

Proof.

1. By the inequalities (10.2), it follows that

$$(\phi_2(s_1) - f(s_1)) (g(s_2) - w_1(s_2)) \ge 0, \quad s_1, s_2 \in [t_0, a].$$

Hence,

$$\phi_2(s_1)g(s_2) + f(s_1)w_1(s_2)$$
$$\ge f(s_1)g(s_2) + \phi_2(s_1)w_1(s_2), \quad s_1, s_2 \in [t_0, a],$$

and

$$g(s_2)\frac{(\phi(t)-\phi(s_1))^{\delta-1}}{\Gamma(\delta)}\phi^\Delta(s_1)\phi_2(s_1)$$

$$+w_1(s_2)\frac{(\phi(t)-\phi(s_1))^{\delta-1}}{\Gamma(\delta)}\phi^\Delta(s_1)$$

$$\geq g(s_2)\frac{(\phi(t)-\phi(s_1))^{\delta-1}}{\Gamma(\delta)}\phi^\Delta(s_1)f(s_1)$$

$$+w_1(s_2)\frac{(\phi(t)-\phi(s_1))^{\delta-1}}{\Gamma(\delta)}\phi^\Delta(s_1)\phi_)2(s_1),\quad s_1,s_2,t\in[t_0,a].$$

Now, we integrate the last inequality with respect to s_1 from t_0 to t and find

$$g(s_2)\int_{t_0}^t\frac{(\phi(t)-\phi(s_1))^{\delta-1}}{\Gamma(\delta)}\phi^\Delta(s_1)\phi_2(s_1)\Delta s_1$$

$$+w_1(s_2)\int_{t_0}^t\frac{(\phi(t)-\phi(s_1))^{\delta-1}}{\Gamma(\delta)}\phi^\Delta(s_1)\Delta s_1$$

$$\geq g(s_2)\int_{t_0}^t\frac{(\phi(t)-\phi(s_1))^{\delta-1}}{\Gamma(\delta)}\phi^\Delta(s_1)f(s_1)\Delta s_1$$

$$+w_1(s_2)\int_{t_0}^t\frac{(\phi(t)-\phi(s_1))^{\delta-1}}{\Gamma(\delta)}\phi^\Delta(s_1)\phi_)2(s_1)\Delta s_1,\quad s_1,s_2,t\in[t_0,a],$$

or

$$g(s_2)_\phi I_{\Delta,t_0}^\delta\phi_2(t)+w_1(s_2)_\phi I_{\Delta,t_0}^\delta f(t)$$

$$\geq g(s_2)_\phi I_{\Delta,t_0}^\delta f(t)+w_1(s_2)_\phi I_{\Delta,t_0}^\delta,\quad s_2,t\in[t_0,a].$$

We multiply the last inequality with

$$\frac{(\phi(t)-\phi(s_2))^{\gamma-1}}{\Gamma(\gamma)}\phi^\Delta(s_2)$$

and get

$$_\phi I_{\Delta,t_0}^\delta \phi_2(t)\frac{(\phi(t)-\phi(s_2))^{\gamma-1}}{\Gamma(\gamma)}\phi^\Delta(s_2)g(s_2)$$

$$+\,_\phi I_{\Delta,t_0}^\delta f(t)\frac{(\phi(t)-\phi(s_2))^{\gamma-1}}{\Gamma(\gamma)}\phi^\Delta(s_2)w_1(s_2)$$

$$\geq\,_\phi I_{\Delta,t_0}^\delta f(t)\frac{(\phi(t)-\phi(s_2))^{\gamma-1}}{\Gamma(\gamma)}\phi^\Delta(s_2)g(s_2)$$

$$+\,_\phi I_{\Delta,t_0}^\delta \phi_2(t)\frac{(\phi(t)-\phi(s_2))^{\gamma-1}}{\Gamma(\gamma)}\phi^\Delta(s_2)w_1(s_2),\quad t,s_2\in[t_0,a].$$

Now, we integrate the last inequality with respect to s_2 from t_0 to t and arrive at the inequality

$$_\phi I_{\Delta,t_0}^\delta \phi_2(t)\int_{t_0}^t \frac{(\phi(t)-\phi(s_2))^{\gamma-1}}{\Gamma(\gamma)}\phi^\Delta(s_2)g(s_2)\Delta s_2$$

$$+\,_\phi I_{\Delta,t_0}^\delta f(t)\int_{t_0}^t \frac{(\phi(t)-\phi(s_2))^{\gamma-1}}{\Gamma(\gamma)}\phi^\Delta(s_2)w_1(s_2)\Delta s_2$$

$$+\,_\phi I_{\Delta,t_0}^\delta f(t)\int_{t_0}^t \frac{(\phi(t)-\phi(s_2))^{\gamma-1}}{\Gamma(\gamma)}\phi^\Delta(s_2)g(s_2)\Delta s_2$$

$$+\,_\phi I_{\Delta,t_0}^\delta \phi_2(t)\int_{t_0}^t \frac{(\phi(t)-\phi(s_2))^{\gamma-1}}{\Gamma(\gamma)}\phi^\Delta(s_2)w_1(s_2)\Delta s_2,\quad t\in[t_0,a],$$

or

$$_\phi I_{\Delta,t_0}^\delta \phi_2(t)\,_\phi I_{\Delta,t_0}^\gamma g(t)+\,_\phi I_{\Delta,t_0}^\delta f(t)\,_\phi I_{\Delta,t_0}^\gamma w_1(t)$$

$$\geq\,_\phi I_{\Delta,t_0}^\delta \phi_2(t)\,_\phi I_{\Delta,t_0}^\gamma w_1(t)+\,_\phi I_{\Delta,t_0}^\delta f(t)\,_\phi I_{\Delta,t_0}^\gamma g(t),\quad t\in[t_0,a].$$

2. By the inequalities (10.2), we find

$$(\phi_2(s_1)-f(s_1))\,(w_2(s_2)-g(s_2))\geq 0,\quad s_1,s_2\in[t_0,a],$$

whereupon

$$\phi_2(s_1)w_2(s_2)+f(s_1)g(s_2)$$

$$\geq f(s_1)w_2(s_2)+\phi_2(s_1)g(s_2),\quad s_1,s_2\in[a,b].$$

We multiply the last inequality with

$$\frac{(\phi(t) - \phi(s_2))^{\gamma-1}}{\Gamma(\gamma)} \phi^{\Delta}(s_2), \quad t, s_2 \in [t_0, a],$$

and find

$$\phi_2(s_1) \frac{(\phi(t) - \phi(s_2))^{\gamma-1}}{\Gamma(\gamma)} \phi^{\Delta}(s_2) w_2(s_2)$$

$$+ f(s_1) \frac{(\phi(t) - \phi(s_2))^{\gamma-1}}{\Gamma(\gamma)} \phi^{\Delta}(s_2) g(s_2)$$

$$\geq f(s_1) \frac{(\phi(t) - \phi(s_2))^{\gamma-1}}{\Gamma(\gamma)} \phi^{\Delta}(s_2) w_2(s_2)$$

$$+ \phi_2(s_1) \frac{(\phi(t) - \phi(s_2))^{\gamma-1}}{\Gamma(\gamma)} \phi^{\Delta}(s_2) g(s_2), \quad t, s_1, s_2 \in [t_0, a].$$

We integrate the last inequality with respect to s_2 from t_0 to t and obtain

$$\phi_2(s_1) \int_{t_0}^{t} \frac{(\phi(t) - \phi(s_2))^{\gamma-1}}{\Gamma(\gamma)} \phi^{\Delta}(s_2) w_2(s_2) \Delta s_2$$

$$+ f(s_1) \int_{t_0}^{t} \frac{(\phi(t) - \phi(s_2))^{\gamma-1}}{\Gamma(\gamma)} \phi^{\Delta}(s_2) g(s_2) \Delta s_2$$

$$\geq f(s_1) \int_{t_0}^{t} \frac{(\phi(t) - \phi(s_2))^{\gamma-1}}{\Gamma(\gamma)} \phi^{\Delta}(s_2) w_2(s_2) \Delta s_2$$

$$+ \phi_2(s_1) \int_{t_0}^{t} \frac{(\phi(t) - \phi(s_2))^{\gamma-1}}{\Gamma(\gamma)} \phi^{\Delta}(s_2) g(s_2) \Delta s_2, \quad t, s_1 \in [t_0, a],$$

or

$$_\phi I^{\delta}_{\Delta, t_0} w_2(t) \phi_2(s_1) + {_\phi I^{\delta}_{\Delta, t_0}} g(t) f(s_1)$$

$$\geq {_\phi I^{\delta}_{\Delta, t_0}} w_2(t) f(s_1) + {_\phi I^{\delta}_{\Delta, t_0}} g(t) \phi_2(s_1), \quad t, s_1 \in [t_0, a].$$

We multiply the last inequality with

$$\frac{(\phi(t) - \phi(s_1))^{\delta-1}}{\Gamma(\delta)} \phi^{\Delta}(s_1), \quad t, s_1 \in [t_0, a],$$

and arrive at

$$_\phi I_{\Delta,t_0}^\delta w_2(t)\frac{(\phi(t)-\phi(s_1))^{\delta-1}}{\Gamma(\delta)}\phi^\Delta(s_1)\phi_2(s_1)$$

$$+_\phi I_{\Delta,t_0}^\delta g(t)\frac{(\phi(t)-\phi(s_1))^{\delta-1}}{\Gamma(\delta)}\phi^\Delta(s_1)f(s_1)$$

$$\geq _\phi I_{\Delta,t_0}^\delta w_2(t)\frac{(\phi(t)-\phi(s_1))^{\delta-1}}{\Gamma(\delta)}\phi^\Delta(s_1)f(s_1)$$

$$+_\phi I_{\Delta,t_0}^\delta g(t)\frac{(\phi(t)-\phi(s_1))^{\delta-1}}{\Gamma(\delta)}\phi^\Delta(s_1)\phi_2(s_1),\quad t,s_1\in[t_0,a].$$

We integrate the last inequality with respect to s_1 from t_0 to t and get

$$_\phi I_{\Delta,t_0}^\delta w_2(t)\int_{t_0}^t\frac{(\phi(t)-\phi(s_1))^{\delta-1}}{\Gamma(\delta)}\phi^\Delta(s_1)\phi_2(s_1)\Delta s_1$$

$$+_\phi I_{\Delta,t_0}^\delta g(t)\int_{t_0}^t\frac{(\phi(t)-\phi(s_1))^{\delta-1}}{\Gamma(\delta)}\phi^\Delta(s_1)f(s_1)\Delta s_1$$

$$\geq _\phi I_{\Delta,t_0}^\delta w_2(t)\int_{t_0}^t\frac{(\phi(t)-\phi(s_1))^{\delta-1}}{\Gamma(\delta)}\phi^\Delta(s_1)f(s_1)\Delta s_1$$

$$+_\phi I_{\Delta,t_0}^\delta g(t)\int_{t_0}^t\frac{(\phi(t)-\phi(s_1))^{\delta-1}}{\Gamma(\delta)}\phi^\Delta(s_1)\phi_2(s_1)\Delta s_1,\quad t\in[t_0,a],$$

or

$$_\phi I_{\Delta,t_0}^\delta\phi_2(t)_\phi I_{\Delta,t_0}^\gamma w_2(t)+_\phi I_{\Delta,t_0}^\delta f(t)_\phi I_{\Delta,t_0}^\gamma g(t)$$

$$\geq _\phi I_{\Delta,t_0}^\delta f(t)_\phi I_{\Delta,t_0}^\gamma w_2(t)+_\phi I_{\Delta,t_0}^\delta\phi_2(t)_\phi I_{\Delta,t_0}^\gamma g(t),\quad t\in[t_0,a].$$

3. By the inequalities (10.2), we find

$$(f(s_1)-\phi_1(s_1))(g(s_2)-w_1(s_2))\geq 0,\quad s_1,s_2\in[t_0,a],$$

whereupon

$$f(s_1)g(s_2)+\phi_1(s_1)w_1(s_2)$$

$$\geq \phi_1(s_1)g(s_2)+f(s_1)w_1(s_2),\quad s_1,s_2\in[t_0,a].$$

Hence,

$$f(s_1)\frac{(\phi(t) - \phi(s_2))^{\gamma-1}}{\Gamma(\gamma)}\phi^\Delta(s_2)g(s_2)$$

$$+ \phi_1(s_1)\frac{(\phi(t) - \phi(s_2))^{\gamma-1}}{\Gamma(\gamma)}\phi^\Delta(s_2)w_1(s_2)$$

$$\geq \phi_1(s_1)\frac{(\phi(t) - \phi(s_2))^{\gamma-1}}{\Gamma(\gamma)}\phi^\Delta(s_2)g(s_2)$$

$$+ f(s_1)\frac{(\phi(t) - \phi(s_2))^{\gamma-1}}{\Gamma(\gamma)}\phi^\Delta(s_2)w_1(s_2), \quad t, s_1, s_2 \in [t_0, a].$$

We integrate the last inequality with respect to s_2 from t_0 to t and find

$$f(s_1)\int_{t_0}^t \frac{(\phi(t) - \phi(s_2))^{\gamma-1}}{\Gamma(\gamma)}\phi^\Delta(s_2)g(s_2)\Delta s_2$$

$$+ \phi_1(s_1)\int_{t_0}^t \frac{(\phi(t) - \phi(s_2))^{\gamma-1}}{\Gamma(\gamma)}\phi^\Delta(s_2)w_1(s_2)\Delta s_2$$

$$\geq \phi_1(s_1)\int_{t_0}^t \frac{(\phi(t) - \phi(s_2))^{\gamma-1}}{\Gamma(\gamma)}\phi^\Delta(s_2)g(s_2)\Delta s_2$$

$$+ f(s_1)\int_{t_0}^t \frac{(\phi(t) - \phi(s_2))^{\gamma-1}}{\Gamma(\gamma)}\phi^\Delta(s_2)w_1(s_2)\Delta s_2, \quad t, s_1 \in [t_0, a],$$

or

$$\phi I^\gamma_{\Delta,t_0}g(t)f(s_1) + \phi I^\gamma_{\Delta,t_0}w_1(t)\phi_1(s_1)$$

$$\geq \phi I^\gamma_{\Delta,t_0}g(t)\phi_1(s_1) + \phi I^\gamma_{\Delta,t_0}w_1(t)f(s_1), \quad t, s_1 \in [t_0, a].$$

From the last inequality, we arrive at

$$\phi I^\gamma_{\Delta,t_0}g(t)\frac{(\phi(t) - \phi(s_1))^{\delta-1}}{\Gamma(\delta)}\phi^\Delta(s_1)f(s_1)$$

$$+ \phi I^\gamma_{\Delta,t_0}w_1(t)\frac{(\phi(t) - \phi(s_1))^{\delta-1}}{\Gamma(\delta)}\phi^\Delta(s_1)\phi_1(s_1)$$

$$\geq {}_\phi I^\gamma_{\Delta,t_0} g(t) \frac{(\phi(t) - \phi(s_1))^{\delta-1}}{\Gamma(\delta)} \phi^\Delta(s_1)\phi_1(s_1)$$

$$+ {}_\phi I^\gamma_{\Delta,t_0} w_1(t) \frac{(\phi(t) - \phi(s_1))^{\delta-1}}{\Gamma(\delta)} \phi^\Delta(s_1) f(s_1), \quad t, s_1 \in [t_0, a].$$

We integrate the last inequality with respect to s_1 from t_0 to t and find

$${}_\phi I^\gamma_{\Delta,t_0} g(t) \int_{t_0}^t \frac{(\phi(t) - \phi(s_1))^{\delta-1}}{\Gamma(\delta)} \phi^\Delta(s_1) f(s_1) \Delta s_1$$

$$+ {}_\phi I^\gamma_{\Delta,t_0} w_1(t) \int_{t_0}^t \frac{(\phi(t) - \phi(s_1))^{\delta-1}}{\Gamma(\delta)} \phi^\Delta(s_1)\phi_1(s_1) \Delta s_1$$

$$\geq {}_\phi I^\gamma_{\Delta,t_0} g(t) \int_{t_0}^t \frac{(\phi(t) - \phi(s_1))^{\delta-1}}{\Gamma(\delta)} \phi^\Delta(s_1)\phi_1(s_1) \Delta s_1$$

$$+ {}_\phi I^\gamma_{\Delta,t_0} w_1(t) \int_{t_0}^t \frac{(\phi(t) - \phi(s_1))^{\delta-1}}{\Gamma(\delta)} \phi^\Delta(s_1) f(s_1) \Delta s_1,$$

$$t, s_1 \in [t_0, a],$$

or

$${}_\phi I^\delta_{\Delta,t_0} f(t) {}_\phi I^\gamma_{\Delta,t_0} g(t) + {}_\phi I^\delta_{\Delta,t_0} \phi_1(t) {}_\phi I^\gamma_{\Delta,t_0} w_1(t)$$

$$\geq {}_\phi I^\delta_{\Delta,t_0} \phi_1(t) {}_\phi I^\gamma_{\Delta,t_0} g(t) + {}_\phi I^\delta_{\Delta,t_0} f(t) {}_\phi I^\gamma_{\Delta,t_0} w_1(t), \quad t \in [t_0, a].$$

4. By the inequalities (10.2), we find

$$(f(s_1) - \phi_1(s_1))(w_2(s_2) - g(s_2)) \geq 0, \quad s_1, s_2 \in [t_0, a],$$

or

$$f(s_1)w_2(s_2) + \phi_1(s_1)g(s_2) \geq \phi_1(s_1)w_2(s_2) + f(s_1)g(s_2).$$

By the last inequality, we get the inequality

$$f(s_1) \frac{(\phi(t) - \phi(s_2))^{\gamma-1}}{\Gamma(\gamma)} \phi^\Delta(s_2) w_2(s_2)$$

$$+ \phi_1(s_1) \frac{(\phi(t) - \phi(s_2))^{\gamma-1}}{\Gamma(\gamma)} \phi^\Delta(s_2) g(s_2)$$

$$\geq \phi_1(s_1) \frac{(\phi(t) - \phi(s_2))^{\gamma-1}}{\Gamma(\gamma)} \phi^\Delta(s_2) w_2(s_2)$$

$$+ f(s_1) \frac{(\phi(t) - \phi(s_2))^{\gamma-1}}{\Gamma(\gamma)} \phi^\Delta(s_2) g(s_2), \quad s_1, s_2 \in [t_0, a].$$

We integrate the last inequality with respect to s_2 from t_0 to t and find

$$f(s_1) \int_{t_0}^t \frac{(\phi(t) - \phi(s_2))^{\gamma-1}}{\Gamma(\gamma)} \phi^\Delta(s_2) w_2(s_2) \Delta s_2$$

$$+ \phi_1(s_1) \int_{t_0}^t \frac{(\phi(t) - \phi(s_2))^{\gamma-1}}{\Gamma(\gamma)} \phi^\Delta(s_2) g(s_2) \Delta s_2$$

$$\geq \phi_1(s_1) \int_{t_0}^t \frac{(\phi(t) - \phi(s_2))^{\gamma-1}}{\Gamma(\gamma)} \phi^\Delta(s_2) w_2(s_2) \Delta s_2$$

$$+ f(s_1) \int_{t_0}^t \frac{(\phi(t) - \phi(s_2))^{\gamma-1}}{\Gamma(\gamma)} \phi^\Delta(s_2) g(s_2) \Delta s_2,$$

$$s_1, s_2 \in [t_0, a],$$

or

$$_\phi I^\gamma_{\Delta, t_0} w_2(t) f(s_1) + {_\phi I^\gamma_{\Delta, t_0}} g(t) \phi_1(s_1)$$

$$\geq {_\phi I^\gamma_{\Delta, t_0}} w_2(t) \phi_1(s_1) + {_\phi I^\gamma_{\Delta, t_0}} g(t) f(s_1), \quad t, s_1 \in [t_0, a].$$

From here,

$$_\phi I^\gamma_{\Delta, t_0} w_2(t) \frac{(\phi(t) - \phi(s_1))^{\delta-1}}{\Gamma(\delta)} \phi^\Delta(s_1) f(s_1)$$

$$+ {_\phi I^\gamma_{\Delta, t_0}} g(t) \frac{(\phi(t) - \phi(s_1))^{\delta-1}}{\Gamma(\delta)} \phi^\Delta(s_1) \phi_1(s_1)$$

$$\geq {_\phi I^\gamma_{\Delta, t_0}} w_2(t) \frac{(\phi(t) - \phi(s_1))^{\delta-1}}{\Gamma(\delta)} \phi^\Delta(s_1) \phi_1(s_1)$$

$$+ {_\phi I^\gamma_{\Delta, t_0}} g(t) \frac{(\phi(t) - \phi(s_1))^{\delta-1}}{\Gamma(\delta)} \phi^\Delta(s_1) f(s_1), \quad t, s_1 \in [t_0, a].$$

We integrate the last inequality with respect to s_1 from t_0 to t and go to

$$_\phi I_{\Delta,t_0}^\gamma w_2(t) \int_{t_0}^t \frac{(\phi(t) - \phi(s_1))^{\delta-1}}{\Gamma(\delta)} \phi^\Delta(s_1) f(s_1) \Delta s_1$$

$$+ {}_\phi I_{\Delta,t_0}^\gamma g(t) \int_{t_0}^t \frac{(\phi(t) - \phi(s_1))^{\delta-1}}{\Gamma(\delta)} \phi^\Delta(s_1) \phi_1(s_1) \Delta s_1$$

$$\geq {}_\phi I_{\Delta,t_0}^\gamma w_2(t) \int_{t_0}^t \frac{(\phi(t) - \phi(s_1))^{\delta-1}}{\Gamma(\delta)} \phi^\Delta(s_1) \phi_1(s_1) \Delta s_1$$

$$+ {}_\phi I_{\Delta,t_0}^\gamma g(t) \int_{t_0}^t \frac{(\phi(t) - \phi(s_1))^{\delta-1}}{\Gamma(\delta)} \phi^\Delta(s_1) f(s_1), \quad t \in [t_0, a],$$

or

$$_\phi I_{\Delta,t_0}^\delta f(t) {}_\phi I_{\Delta,t_0}^\gamma w_2(t) + {}_\phi I_{\Delta,t_0}^\delta g(t) {}_\phi I_{\Delta,t_0}^\gamma \phi_1(t)$$

$$\geq {}_\phi I_{\Delta,t_0}^\delta \phi_1(t) {}_\phi I_{\Delta,t_0}^\gamma w_2(t) + {}_\phi I_{\Delta,t_0}^\delta f(t) {}_\phi I_{\Delta,t_0}^\gamma g(t), \quad t \in [t_0, a].$$

This completes the proof. □

10.3 Some Other Inequalities

In this section, we deduct some other inequalities for generalized Riemann–Liouville fractional integral using the classical Young inequality.

Theorem 10.5. *Let* $\gamma > 1$, $p, q \geq 1$, $\frac{1}{p} + \frac{1}{q} = 1$, $f, g : [t_0, a] \to (0, \infty)$ *be integrable functions. Then we have the following inequalities:*

1.

$$\frac{1}{p} {}_\phi I_{\Delta,t_0}^\delta f^p(t) {}_\phi I_{\Delta,t_0}^\gamma g^p(t) + \frac{1}{q} {}_\phi I_{\Delta,t_0}^\gamma f^q(t) {}_\phi I_{\Delta,t_0}^\delta g^q(t)$$

$$\geq {}_\phi I_{\Delta,t_0}^\delta (fg)(t) {}_\phi I_{\Delta,t_0}^\gamma (fg)(t), \quad t \in [t_0, a],$$

2.

$$\frac{1}{p} {}_\phi I_{\Delta,t_0}^\delta g^q(t) {}_\phi I_{\Delta,t_0}^\gamma f^p(t) + \frac{1}{q} {}_\phi I_{\Delta,t_0}^\gamma g^q(t) {}_\phi I_{\Delta,t_0}^\delta f^p(t)$$

$$\geq {}_\phi I_{\Delta,t_0}^\delta (fg)^{p-1}(t) {}_\phi I_{\Delta,t_0}^\gamma (fg)(t), \quad t \in [t_0, a],$$

3.

$$\frac{1}{p} {}_\phi I^\delta_{\Delta, t_0} f^p(t) {}_\phi I^\gamma_{\Delta, t_0} g^2(t) + \frac{1}{q} {}_\phi I^\gamma_{\Delta, t_0} f^2(t) {}_\phi I^\delta_{\Delta, t_0} g^q(t)$$

$$\geq {}_\phi I^\delta_{\Delta, t_0}(fg)(t) {}_\phi I^\gamma_{\Delta, t_0} \left(f^{\frac{2}{p}} g^{\frac{2}{q}} \right)(t), \quad t \in [t_0, a],$$

4.

$$\frac{1}{p} {}_\phi I^\delta_{\Delta, t_0} f^2(t) {}_\phi I^\gamma_{\Delta, t_0} f^p(t) + \frac{1}{q} {}_\phi I^\gamma_{\Delta, t_0} g^q(t) {}_\phi I^\delta_{\Delta, t_0} g^2(t)$$

$$\geq {}_\phi I^\delta_{\Delta, t_0} \left(f^{\frac{2}{p}} g^{\frac{2}{q}} \right)(t) {}_\phi I^\gamma_{\Delta, t_0}(fg)(t), \quad t \in [t_0, a].$$

Proof.

1. By the Young inequality, we get

$$\frac{1}{p}(f(s_1)g(s_2))^p + \frac{1}{q}(f(s_2)g(s_1))^q \geq f(s_1)g(s_1)f(s_2)g(s_2)$$

for any $s_1, s_2 \in [t_0, a]$. We multiply both sides of the last inequality with

$$\frac{(\phi(t) - \phi(s_2))^{\gamma-1}}{\Gamma(\gamma)} \phi^\Delta(s_2), \quad t, s_2 \in [t_0, a], \qquad (10.3)$$

and find

$$\frac{1}{p}(f(s_1))^p \frac{(\phi(t) - \phi(s_2))^{\gamma-1}}{\Gamma(\gamma)} \phi^\Delta(s_2) (g(s_2))^p$$

$$+ \frac{1}{q}(g(s_1))^q \frac{(\phi(t) - \phi(s_2))^{\gamma-1}}{\Gamma(\gamma)} \phi^\Delta(s_2) (f(s_2))^q$$

$$\geq f(s_1)g(s_1) \frac{(\phi(t) - \phi(s_2))^{\gamma-1}}{\Gamma(\gamma)} \phi^\Delta(s_2) f(s_2)g(s_2), \quad t, s_1, s_2 \in [t_0, a].$$

We integrate the last inequality with respect to s_2 from t_0 to t and get

$$\frac{1}{p}(f(s_1))^p \int_{t_0}^t \frac{(\phi(t) - \phi(s_2))^{\gamma-1}}{\Gamma(\gamma)} \phi^\Delta(s_2) (g(s_2))^p \Delta s_2$$

$$+ \frac{1}{q}(g(s_1))^q \int_{t_0}^t \frac{(\phi(t) - \phi(s_2))^{\gamma-1}}{\Gamma(\gamma)} \phi^\Delta(s_2) (f(s_2))^q \Delta s_2$$

$$\geq f(s_1)g(s_1) \int_{t_0}^t \frac{(\phi(t) - \phi(s_2))^{\gamma-1}}{\Gamma(\gamma)} \phi^\Delta(s_2) f(s_2)g(s_2) \Delta s_2,$$

$$t, s_1 \in [t_0, a],$$

whereupon

$$\phi I_{\Delta,t_0}^\gamma g^p(t)(f(s_1))^p + \frac{1}{q}\phi I_{\Delta,t_0}^\gamma f^q(t)(g(s_1))^q$$

$$\geq \phi I_{\Delta,t_0}^\gamma (fg)(t)f(s_1)g(s_1), \quad t,s_1 \in [t_0,a].$$

We multiply both sides of the last inequality with

$$\frac{(\phi(t)-\phi(s_1))^{\delta-1}}{\Gamma(\delta)}\phi^\Delta(s_1), \quad t,s_1 \in [t_0,a], \tag{10.4}$$

and obtain

$$\phi I_{\Delta,t_0}^\gamma g^p(t)\frac{(\phi(t)-\phi(s_1))^{\delta-1}}{\Gamma(\delta)}\phi^\Delta(s_1)(f(s_1))^p$$

$$+\frac{1}{q}\phi I_{\Delta,t_0}^\gamma f^q(t)\frac{(\phi(t)-\phi(s_1))^{\delta-1}}{\Gamma(\delta)}\phi^\Delta(s_1)(g(s_1))^q$$

$$\geq \phi I_{\Delta,t_0}^\gamma (fg)(t)\frac{(\phi(t)-\phi(s_1))^{\delta-1}}{\Gamma(\delta)}\phi^\Delta(s_1)f(s_1)g(s_1),$$

$$t,s_1 \in [t_0,a].$$

Now, we integrate both sides of the last inequality with respect to s_1 from t_0 to t and arrive at

$$\phi I_{\Delta,t_0}^\gamma g^p(t)\int_{t_0}^t \frac{(\phi(t)-\phi(s_1))^{\delta-1}}{\Gamma(\delta)}\phi^\Delta(s_1)(f(s_1))^p\Delta s_1$$

$$+\frac{1}{q}\phi I_{\Delta,t_0}^\gamma f^q(t)\int_{t_0}^t \frac{(\phi(t)-\phi(s_1))^{\delta-1}}{\Gamma(\delta)}\phi^\Delta(s_1)(g(s_1))^q\Delta s_1$$

$$\geq \phi I_{\Delta,t_0}^\gamma (fg)(t)\int_{t_0}^t \frac{(\phi(t)-\phi(s_1))^{\delta-1}}{\Gamma(\delta)}\phi^\Delta(s_1)f(s_1)g(s_1)\Delta s_1,$$

$$t \in [t_0,a],$$

or

$$\frac{1}{p}\phi I_{\Delta,t_0}^\delta f^p(t)\phi I_{\Delta,t_0}^\gamma g^p(t) + \frac{1}{q}\phi I_{\Delta,t_0}^\gamma f^q(t)\phi I_{\Delta,t_0}^\delta g^q(t)$$

$$\geq \phi I_{\Delta,t_0}^\delta (fg)(t)\phi I_{\Delta,t_0}^\gamma (fg)(t), \quad t \in [t_0,a].$$

2. We apply the Young inequality and get

$$\frac{1}{p}\frac{(f(s_2))^p}{(f(s_1))^p} + \frac{1}{q}\frac{(g(s_2))^q}{(g(s_1))^q} \geq \frac{f(s_2)g(s_2)}{f(s_1)g(s_1)}, \quad s_1, s_2 \in [t_0, a],$$

whereupon

$$\frac{1}{p}(f(s_2))^p (g(s_1))^q + \frac{1}{q}(g(s_2))^q (f(s_1))^p$$

$$\geq f(s_2)g(s_2)(f(s_1))^{p-1}(g(s_1))^{p-1}, \quad s_1, s_2 \in [t_0, a].$$

We multiply both sides of the last inequality with (10.3) and get

$$\frac{1}{p}(g(s_1))^q \frac{(\phi(t) - \phi(s_2))^{\gamma-1}}{\Gamma(\gamma)} \phi^\Delta(s_2)(f(s_2))^p$$

$$+ \frac{1}{q}(f(s_1))^p \frac{(\phi(t) - \phi(s_2))^{\gamma-1}}{\Gamma(\gamma)} \phi^\Delta(s_2)(g(s_2))^q$$

$$\geq (f(s_1))^{p-1}(g(s_1))^{p-1} \frac{(\phi(t) - \phi(s_2))^{\gamma-1}}{\Gamma(\gamma)} \phi^\Delta(s_2)f(s_2)g(s_2),$$

$$t, s_1, s_2 \in [t_0, a].$$

We integrate the last inequality with respect to s_2 from t_0 to t and find

$$\frac{1}{p}(g(s_1))^q \int_{t_0}^t \frac{(\phi(t) - \phi(s_2))^{\gamma-1}}{\Gamma(\gamma)} \phi^\Delta(s_2)(f(s_2))^p \Delta s_2$$

$$+ \frac{1}{q}(f(s_1))^p \int_{t_0}^t \frac{(\phi(t) - \phi(s_2))^{\gamma-1}}{\Gamma(\gamma)} \phi^\Delta(s_2)(g(s_2))^q \Delta s_2$$

$$\geq (f(s_1))^{p-1}(g(s_1))^{p-1} \int_{t_0}^t \frac{(\phi(t) - \phi(s_2))^{\gamma-1}}{\Gamma(\gamma)}$$

$$\times \phi^\Delta(s_2)f(s_2)g(s_2)\Delta s_2, \quad t, s_1 \in [t_0, a],$$

or

$$\frac{1}{p}\phi I^\gamma_{\Delta,t_0} f^p(t)(g(s_1))^q + \frac{1}{q}\phi I^\gamma_{\Delta,t_0} g^q(t)(f(s_1))^p$$

$$\geq \phi I^\gamma_{\Delta,t_0}(fg)(t)(f(s_1))^{p-1}(g(s_1))^{p-1}, \quad t, s_1 \in [t_0, a].$$

We multiply both sides of the last inequality with (10.4) and arrive at

$$\frac{1}{p} {}_\phi I^\gamma_{\Delta,t_0} f^p(t) \frac{(\phi(t) - \phi(s_1))^{\delta-1}}{\Gamma(\delta)} \phi^\Delta(s_1) \, (g(s_1))^q$$

$$+ \frac{1}{q} {}_\phi I^\gamma_{\Delta,t_0} g^q(t) \frac{(\phi(t) - \phi(s_1))^{\delta-1}}{\Gamma(\delta)} \phi^\Delta(s_1) \, (f(s_1))^p$$

$$\geq {}_\phi I^\gamma_{\Delta,t_0} (fg)(t) \frac{(\phi(t) - \phi(s_1))^{\delta-1}}{\Gamma(\delta)} \phi^\Delta(s_1) \, (f(s_1))^{p-1} \, (g(s_1))^{p-1},$$

$$t, s_1 \in [t_0, a].$$

We integrate the last inequality with respect to s_1 from t_0 to t and get

$$\frac{1}{p} {}_\phi I^\gamma_{\Delta,t_0} f^p(t) \int_{t_0}^t \frac{(\phi(t) - \phi(s_1))^{\delta-1}}{\Gamma(\delta)} \phi^\Delta(s_1) \, (g(s_1))^q \, \Delta s_1$$

$$+ \frac{1}{q} {}_\phi I^\gamma_{\Delta,t_0} g^q(t) \int_{t_0}^t \frac{(\phi(t) - \phi(s_1))^{\delta-1}}{\Gamma(\delta)} \phi^\Delta(s_1) \, (f(s_1))^p \, \Delta s_1$$

$$\geq {}_\phi I^\gamma_{\Delta,t_0} (fg)(t) \int_{t_0}^t \frac{(\phi(t) - \phi(s_1))^{\delta-1}}{\Gamma(\delta)}$$

$$\times \phi^\Delta(s_1) \, (f(s_1))^{p-1} \, (g(s_1))^{p-1} \, \Delta s_1, \quad t, s_1 \in [t_0, a],$$

or

$$\frac{1}{p} {}_\phi I^\delta_{\Delta,t_0} g^q(t) {}_\phi I^\gamma_{\Delta,t_0} f^p(t) + \frac{1}{q} {}_\phi I^\gamma_{\Delta,t_0} g^q(t) {}_\phi I^\delta_{\Delta,t_0} f^p(t)$$

$$\geq {}_\phi I^\delta_{\Delta,t_0} (fg)^{p-1}(t) {}_\phi I^\gamma_{\Delta,t_0} (fg)(t), \quad t \in [t_0, a].$$

3. We apply the Young inequality and find

$$\frac{1}{p} (f(s_1))^p \, (g(s_2))^2 + \frac{1}{q} (f(s_2))^2 \, (g(s_1))^q$$

$$\geq f(s_1) g(s_1) \, (f(s_2))^{\frac{2}{p}} \, (g(s_2))^{\frac{2}{q}}, \quad s_1, s_2 \in [t_0, a].$$

We multiply both sides of the last inequality with (10.3) and get

$$
\frac{1}{p}\left(f(s_1)\right)^p \frac{(\phi(t)-\phi(s_2))^{\gamma-1}}{\Gamma(\gamma)}\phi^\Delta(s_2)\left(g(s_2)\right)^2
$$

$$
+\frac{1}{q}\left(g(s_1)\right)^q \frac{(\phi(t)-\phi(s_2))^{\gamma-1}}{\Gamma(\gamma)}\phi^\Delta(s_2)\left(f(s_2)\right)^2
$$

$$
\geq f(s_1)g(s_1)\frac{(\phi(t)-\phi(s_2))^{\gamma-1}}{\Gamma(\gamma)}\phi^\Delta(s_2)\left(f(s_2)\right)^{\frac{2}{p}}\left(g(s_2)\right)^{\frac{2}{q}},
$$

$$
t,s_1,s_2 \in [t_0,a].
$$

We integrate the last inequality with respect to s_2 from t_0 to t and arrive at

$$
\frac{1}{p}\left(f(s_1)\right)^p \int_{t_0}^t \frac{(\phi(t)-\phi(s_2))^{\gamma-1}}{\Gamma(\gamma)}\phi^\Delta(s_2)\left(g(s_2)\right)^2 \Delta s_2
$$

$$
+\frac{1}{q}\left(g(s_1)\right)^q \int_{t_0}^t \frac{(\phi(t)-\phi(s_2))^{\gamma-1}}{\Gamma(\gamma)}\phi^\Delta(s_2)\left(f(s_2)\right)^2 \Delta s_2
$$

$$
\geq f(s_1)g(s_1)\int_{t_0}^t \frac{(\phi(t)-\phi(s_2))^{\gamma-1}}{\Gamma(\gamma)}
$$

$$
\times \phi^\Delta(s_2)\left(f(s_2)\right)^{\frac{2}{p}}\left(g(s_2)\right)^{\frac{2}{q}} \Delta s_2, \quad t,s_1 \in [t_0,a],
$$

whereupon

$$
\frac{1}{p}{}_\phi I^\gamma_{\Delta,t_0}g^2(t)\left(f(s_1)\right)^p + \frac{1}{q}{}_\phi I^\gamma_{\Delta,t_0}f^2(t)\left(g(s_1)\right)^q
$$

$$
\geq {}_\phi I^\gamma_{\Delta,t_0}\left(f^{\frac{2}{p}}g^{\frac{2}{q}}\right)(t)f(s_1)g(s_1), \quad t,s_1 \in [t_0,a].
$$

Now, we multiply the last inequality with (10.4) and find

$$
\frac{1}{p}{}_\phi I^\gamma_{\Delta,t_0}g^2(t)\frac{(\phi(t)-\phi(s_1))^{\delta-1}}{\Gamma(\delta)}\phi^\Delta(s_1)\left(f(s_1)\right)^p
$$

$$
+\frac{1}{q}{}_\phi I^\gamma_{\Delta,t_0}f^2(t)\frac{(\phi(t)-\phi(s_1))^{\delta-1}}{\Gamma(\delta)}\phi^\Delta(s_1)\left(g(s_1)\right)^q
$$

$$
\geq {}_\phi I^\gamma_{\Delta,t_0}\left(f^{\frac{2}{p}}g^{\frac{2}{q}}\right)(t)\frac{(\phi(t)-\phi(s_1))^{\delta-1}}{\Gamma(\delta)}\phi^\Delta(s_1)f(s_1)g(s_1),
$$

$$
t,s_1 \in [t_0,a].
$$

We integrate the last inequality with respect to s_1 from t_0 to t and obtain

$$\frac{1}{p} {}_\phi I^\gamma_{\Delta,t_0} g^2(t) \int_{t_0}^t \frac{(\phi(t) - \phi(s_1))^{\delta-1}}{\Gamma(\delta)} \phi^\Delta(s_1) \left(f(s_1) \right)^p \Delta s_1$$

$$+ \frac{1}{q} {}_\phi I^\gamma_{\Delta,t_0} f^2(t) \int_{t_0}^t \frac{(\phi(t) - \phi(s_1))^{\delta-1}}{\Gamma(\delta)} \phi^\Delta(s_1) \left(g(s_1) \right)^q \Delta s_1$$

$$\geq {}_\phi I^\gamma_{\Delta,t_0} \left(f^{\frac{2}{p}} g^{\frac{2}{q}} \right)(t) \int_{t_0}^t \frac{(\phi(t) - \phi(s_1))^{\delta-1}}{\Gamma(\delta)} \phi^\Delta(s_1) f(s_1) g(s_1) \Delta s_1,$$

$$t \in [t_0, a],$$

or

$$\frac{1}{p} {}_\phi I^\delta_{\Delta,t_0} f^p(t) {}_\phi I^\gamma_{\Delta,t_0} g^2(t) + \frac{1}{q} {}_\phi I^\gamma_{\Delta,t_0} f^2(t) {}_\phi I^\delta_{\Delta,t_0} g^q(t)$$

$$\geq {}_\phi I^\delta_{\Delta,t_0}(fg)(t) {}_\phi I^\gamma_{\Delta,t_0} \left(f^{\frac{2}{p}} g^{\frac{2}{q}} \right)(t), \quad t \in [t_0, a].$$

4. We apply the Young inequality and get

$$\frac{1}{p} \left(f(s_1) \right)^2 \left(f(s_2) \right)^p + \frac{1}{q} \left(g(s_1) \right)^2 \left(g(s_2) \right)^q$$

$$\geq \left(f(s_1) \right)^{\frac{2}{p}} \left(g(s_1) \right)^{\frac{2}{q}} f(s_2) g(s_2), \quad s_1, s_2 \in [t_0, a].$$

We multiply both sides of the last inequality with (10.3) and arrive at

$$\frac{1}{p} \left(f(s_1) \right)^2 \frac{(\phi(t) - \phi(s_2))^{\gamma-1}}{\Gamma(\gamma)} \phi^\Delta(s_2) \left(f(s_2) \right)^p$$

$$+ \frac{1}{q} \left(g(s_1) \right)^2 \frac{(\phi(t) - \phi(s_2))^{\gamma-1}}{\Gamma(\gamma)} \phi^\Delta(s_2) \left(g(s_2) \right)^q$$

$$\geq \left(f(s_1) \right)^{\frac{2}{p}} \left(g(s_1) \right)^{\frac{2}{q}} \frac{(\phi(t) - \phi(s_2))^{\gamma-1}}{\Gamma(\gamma)} \phi^\Delta(s_2) f(s_2) g(s_2),$$

$$t, s_1, s_2 \in [t_0, a].$$

We integrate the last inequality with respect to s_2 from t_0 to t and get

$$\frac{1}{p} \left(f(s_1) \right)^2 \int_{t_0}^t \frac{(\phi(t) - \phi(s_2))^{\gamma-1}}{\Gamma(\gamma)} \phi^\Delta(s_2) \left(f(s_2) \right)^p \Delta s_2$$

$$+ \frac{1}{q} \left(g(s_1) \right)^2 \int_{t_0}^t \frac{(\phi(t) - \phi(s_2))^{\gamma-1}}{\Gamma(\gamma)} \phi^\Delta(s_2) \left(g(s_2) \right)^q \Delta s_2$$

$$\geq (f(s_1))^{\frac{2}{p}} (g(s_1))^{\frac{2}{q}} \int_{t_0}^{t} \frac{(\phi(t) - \phi(s_2))^{\gamma-1}}{\Gamma(\gamma)}$$

$$\times \phi^{\Delta}(s_2) f(s_2) g(s_2) \Delta s_2, \quad t, s_1 \in [t_0, a],$$

or

$$\frac{1}{p} {}_{\phi}I^{\gamma}_{\Delta,t_0} f^p(t) (f(s_1))^2 + \frac{1}{q} {}_{\phi}I^{\gamma}_{\Delta,t_0} g^2(t) (g(s_1))^2$$

$$\geq {}_{\phi}I^{\gamma}_{\Delta,t_0} (fg)(t) (f(s_1))^{\frac{2}{p}} (g(s_1))^{\frac{2}{q}}, \quad t, s_1 \in [t_0, a].$$

We multiply both sides of the last inequality with (10.4) and get

$$\frac{1}{p} {}_{\phi}I^{\gamma}_{\Delta,t_0} f^p(t) \frac{(\phi(t) - \phi(s_1))^{\delta-1}}{\Gamma(\delta)} \phi^{\Delta}(s_1) (f(s_1))^2$$

$$+ \frac{1}{q} {}_{\phi}I^{\gamma}_{\Delta,t_0} g^2(t) \frac{(\phi(t) - \phi(s_1))^{\delta-1}}{\Gamma(\delta)} \phi^{\Delta}(s_1) (g(s_1))^2$$

$$\geq {}_{\phi}I^{\gamma}_{\Delta,t_0} (fg)(t) \frac{(\phi(t) - \phi(s_1))^{\delta-1}}{\Gamma(\delta)} \phi^{\Delta}(s_1) (f(s_1))^{\frac{2}{p}} (g(s_1))^{\frac{2}{q}},$$

$$t, s_1 \in [t_0, a].$$

We integrate the last inequality with respect to s_1 from t_0 to t and get

$$\frac{1}{p} {}_{\phi}I^{\gamma}_{\Delta,t_0} f^p(t) \int_{t_0}^{t} \frac{(\phi(t) - \phi(s_1))^{\delta-1}}{\Gamma(\delta)} \phi^{\Delta}(s_1) (f(s_1))^2 \, \Delta s_1$$

$$+ \frac{1}{q} {}_{\phi}I^{\gamma}_{\Delta,t_0} g^2(t) \int_{t_0}^{t} \frac{(\phi(t) - \phi(s_1))^{\delta-1}}{\Gamma(\delta)} \phi^{\Delta}(s_1) (g(s_1))^2 \, \Delta s_1$$

$$\geq {}_{\phi}I^{\gamma}_{\Delta,t_0} (fg)(t) \int_{t_0}^{t} \frac{(\phi(t) - \phi(s_1))^{\delta-1}}{\Gamma(\delta)}$$

$$\times \phi^{\Delta}(s_1) (f(s_1))^{\frac{2}{p}} (g(s_1))^{\frac{2}{q}} \, \Delta s_1, \quad t \in [t_0, a],$$

or

$$\frac{1}{p}{}_\phi I_{\Delta,t_0}^\delta f^2(t){}_\phi I_{\Delta,t_0}^\gamma f^p(t) + \frac{1}{q}{}_\phi I_{\Delta,t_0}^\gamma g^q(t){}_\phi I_{\Delta,t_0}^\delta g^2(t)$$

$$\geq {}_\phi I_{\Delta,t_0}^\delta \left(f^{\frac{2}{p}} g^{\frac{2}{q}}\right)(t){}_\phi I_{\Delta,t_0}^\gamma (fg)(t), \quad t \in [t_0, a].$$

This completes the proof. $\qquad\qquad\qquad\qquad\qquad\qquad\qquad\qquad$ □

Young and Hölder Inequalities

In this appendix, we deduct the Young inequalities and their reverse versions using the Specht ratio. Then the Hölder inequality and its reverse versions are also investigated. Some of the results discussed here can be found in Refs. [6, 7, 10, 14].

A.1 Young Inequalities

We will give here some crucial inequalities, which play an important role in applied mathematics.

Notation: Let $1 \leq p \leq \infty$, we denote by q the conjugate of p, i.e., $\frac{1}{p} + \frac{1}{q} = 1$.

Theorem A.1 (The Young Inequality). *Let $a, b > 0$, $p.q \in (1, \infty)$, $\frac{1}{p} + \frac{1}{q} = 1$. Then*

$$ab \leq \frac{a^p}{p} + \frac{b^q}{q}. \tag{A.1}$$

The equality holds if and only if $a^p = b^q$.

Proof.   Since the mapping $x \to e^x$ is convex, we get

$$ab = e^{\log a + \log b}$$

$$= e^{\frac{1}{p} \log a^p + \frac{1}{q} \log b^q}$$

$$\leq \frac{1}{p} e^{\log a^p} + \frac{1}{q} e^{\log b^q}$$

$$= \frac{a^p}{p} + \frac{b^q}{q}.$$

Now, we will prove that the equality holds if and only if $a^p = b^q$. When $a = 0$ or $b = 0$, the assertion is evident. Assume that $a \neq 0$ and $b \neq 0$. We set $t = a b^{-\frac{q}{p}}$ or $a = t b^{\frac{q}{p}}$. Then

$$\frac{a^p}{p} + \frac{b^q}{q} - ab = \frac{t^p b^q}{p} + \frac{b^q}{q} - t b^{\frac{q}{p}+1}$$

$$= \frac{t^p b^q}{p} + \frac{b^q}{q} - t b^{q\left(\frac{1}{p}+\frac{1}{q}\right)}$$

$$= \frac{t^p b^q}{p} + \frac{b^q}{q} - t b^q$$

$$= b^q \left(\frac{t^p}{p} + \frac{1}{q} - t \right).$$

Hence,

$$\frac{a^p}{p} + \frac{b^q}{q} - ab = 0 \iff b^q \left(\frac{t^p}{p} + \frac{1}{q} - t \right) = 0$$

$$\iff \frac{t^p}{p} + \frac{1}{q} - t = 0 \iff t = 1 \iff a^p = b^q$$

(because, if we take $f(t) = \frac{t^p}{p} + \frac{1}{q} - t$, we have $f(1) = 0$, $f'(t) = t^{p-1} - 1$, $f'(t) \geq 0$ for $t \geq 1$ and $f'(t) \leq 0$ for $t \in [0,1]$). This completes the proof. $\qquad\square$

Remark A.1. Let a and b be positive numbers. Then the inequality

$$(1 - \lambda)a + \lambda b \geq a^{1-\lambda}b^{\lambda} \qquad (A.2)$$

holds for any $\lambda \in [0, 1]$. Really, let $\lambda \in (0, 1)$ and

$$a_1 = a^{1-\lambda}, \quad b_1 = b^{\lambda}.$$

Observe that

$$\frac{1}{\frac{1}{1-\lambda}} + \frac{1}{\frac{1}{\lambda}} = 1 - \lambda + \lambda$$

$$= 1.$$

Then, we apply the Young inequality for $\frac{1}{1-\lambda}$, $\frac{1}{\lambda}$, a_1 and b_1, and get

$$a_1 b_1 \leq (1 - \lambda)a_1^{\frac{1}{1-\lambda}} + \lambda b_1^{\frac{1}{\lambda}}$$

or

$$(1 - \lambda)a + \lambda b \geq a^{1-\lambda}b^{\lambda}.$$

If $\lambda = 0$ or $\lambda = 1$, then the inequality (A.2) is evident.

Definition A.1. The inequality (A.2) is also called λ-weighted arithmetic–geometric mean inequality.

Theorem A.2 (The Young Inequality with ϵ). *Let $a, b, \epsilon > 0$, $p, q \in (1, \infty)$, $\frac{1}{p} + \frac{1}{q} = 1$. Then*

$$ab \leq \epsilon a^p + C(\epsilon)b^q, \quad C(\epsilon) = \frac{1}{q(\epsilon p)^{\frac{q}{p}}}.$$

The equality holds if and only if $a^p = \frac{b^q}{(\epsilon p)^q}$.

Proof. Let

$$a_1 = (\epsilon p)^{\frac{1}{p}} a, \quad b_1 = \frac{b}{(\epsilon p)^{\frac{1}{p}}}.$$

We apply the Young inequality (A.1) for a_1 and b_1 and get

$$ab \leq \frac{\epsilon p a^p}{p} + \frac{1}{(\epsilon p)^{\frac{q}{p}} q} b^q$$

$$= \epsilon a^p + C(\epsilon)b^q.$$

The equality holds if and only if $a_1^p = b_1^q$. This completes the proof. $\qquad \square$

A.2   The Specht Ratio

Definition A.2 (The Specht Ratio). For $h > 0$, define the Specht ratio as follows:

$$S(h) = \frac{h^{\frac{1}{h-1}}}{e \log h^{\frac{1}{h-1}}}.$$

In this section, we deduct some of the properties of the Specht ratio.

Theorem A.3. *We have*

$$S(1) = 1.$$

Proof. By the L'Hôpital rule, we get

$$
\begin{aligned}
\lim_{h \to 1} \log S(h) &= \lim_{h \to 1} \log \frac{h^{\frac{1}{h-1}}}{e \log h^{\frac{1}{h-1}}} \\
&= \lim_{h \to 1} \left(\log h^{\frac{1}{h-1}} - \log \left(e \log h^{\frac{1}{h-1}} \right) \right) \\
&= \lim_{h \to 1} \left(\frac{\log h}{h-1} - \log e - \log \left(\frac{\log h}{h-1} \right) \right) \\
&= \lim_{h \to 1} \left(\frac{1}{h} - 1 - \log \frac{1}{h} \right) \\
&= 0.
\end{aligned}
$$

So,

$$S(1) = 1.$$

This completes the proof. □

Theorem A.4. *We have*

$$S\left(\frac{1}{h} \right) = S(h), \quad h > 0.$$

Proof. By the equality

$$
\left(\frac{1}{h} \right)^{\frac{1}{\frac{1}{h}-1}} = \left(\frac{1}{h} \right)^{\frac{h}{1-h}}
$$

$$
= h^{\frac{1}{h-1}}, \quad h > 0,
$$

we conclude that

$$S(h) = S\left(\frac{1}{h}\right), \quad h > 0.$$

This completes the proof. □

Theorem A.5. *S is strictly decreasing on* $(0,1)$ *and strictly increasing on* $(1,\infty)$.

Proof. We have

$$\frac{d}{dh}\log S(h) = \frac{d}{dh}\left(\frac{\log h}{h-1} - 1 - \log\frac{\log h}{h-1}\right)$$

$$= \frac{\frac{1}{h}(h-1) - \log h}{(h-1)^2} - \frac{h-1}{\log h}\cdot\frac{\frac{1}{h}(h-1) - \log h}{(h-1)^2}$$

$$= \frac{1 - \frac{1}{h} - \log h}{(h-1)^2} - \frac{1 - \frac{1}{h} - \log h}{(h-1)\log h}$$

$$= \frac{\left(1 - \frac{1}{h} - \log h\right)\left(\log h - h + 1\right)}{(h-1)^2\log h}, \quad h > 0.$$

By the Klein inequality, we have

$$1 - \frac{1}{h} \le \log h \le h - 1, \quad h > 0.$$

Thus, if $h \in (0,1)$, we have

$$\frac{d}{dh}\log S(h) < 0,$$

and for $h \in (1,\infty)$, we have

$$\frac{d}{dh}\log S(h) > 0.$$

This completes the proof. □

A.3   Refined Young Inequalities

We begin this section with the following refinement of the scalar Young inequality.

Theorem A.6. *If $a, b \geq 0$ and $\lambda \in [0, 1]$, then*

$$a^\lambda b^{1-\lambda} + r_0(\sqrt{a} + \sqrt{b})^2 \leq \lambda a + (1 - \lambda)b,$$

where

$$r_0 = \min\{\lambda, 1 - \lambda\}.$$

Proof.   Let $\lambda = \frac{1}{2}$. Then $r_0 = \frac{1}{2}$ and

$$a^\lambda b^{1-\lambda} + r_0(\sqrt{a} - \sqrt{b})^2 = \sqrt{a}\sqrt{b} + \frac{1}{2}(\sqrt{a} - \sqrt{b})^2$$

$$= \frac{1}{2}(a + b),$$

i.e., in the case when $\lambda = \frac{1}{2}$, we have equality. Let now $\lambda < \frac{1}{2}$. Then $r_0 = \lambda$ and applying the classical Young inequality, we arrive at the inequality

$$\begin{aligned}
\lambda a + (1 - \lambda)b - r_0(\sqrt{a} - \sqrt{b})^2 &= \lambda a + (1 - \lambda)b - \lambda(\sqrt{a} - \sqrt{b})^2 \\
&= \lambda a + (1 - \lambda)b - \lambda a - \lambda b + 2\lambda\sqrt{ab} \\
&= (1 - 2\lambda)b + 2\lambda\sqrt{ab} \\
&\geq b^{1-2\lambda}(ab)^\lambda \\
&= a^\lambda b^{1-\lambda},
\end{aligned}$$

whereupon

$$a^\lambda b^{1-\lambda} + r_0(\sqrt{a} - \sqrt{b})^2 \leq \lambda a + (1 - \lambda)b.$$

Now, suppose that $\lambda > \frac{1}{2}$ and applying the classical Young inequality, we find

$$\begin{aligned}
&\lambda a + (1 - \lambda)b - r_0(\sqrt{a} - \sqrt{b})^2 \\
&= \lambda a + (1 - \lambda)b - (1 - \lambda)(\sqrt{a} - \sqrt{b})^2
\end{aligned}$$

$$= \lambda a + (1 - \lambda)b - (1 - \lambda)a - (1 - \lambda)b + \lambda(1 - \lambda)\sqrt{ab}$$
$$= (2\lambda - 1)a + (2 - 2\lambda)\sqrt{ab}$$
$$\geq a^{2\lambda - 1}(ab)^{1-\lambda}$$
$$= a^{\lambda}b^{1-\lambda},$$

whereupon we get the desired result. This completes the proof. □

Corollary A.1. *Let $a, b \geq 0$ and $\lambda \in [0, 1]$. Then*

$$\left(a^{\lambda}b^{1-\lambda} + a^{1-\lambda}b^{\lambda}\right)^2 + 2r_0(a - b)^2 \leq (a + b)^2,$$

where

$$r_0 = \min\{\lambda, 1 - \lambda\}.$$

Proof. Applying Theorem A.6, we obtain

$$(a + b)^2 - \left(a^{\lambda}b^{1-\lambda} + a^{1-\lambda}b^{\lambda}\right)^2$$
$$= a^2 + 2ab + b^2 - a^{2\lambda}b^{2-2\lambda} - 2ab - a^{2(1-\lambda)}b^{2\lambda}$$
$$= a^2 + b^2 - a^{2\lambda}b^{2-2\lambda} - a^{2-2\lambda}b^{2\lambda}$$
$$= \lambda a^2 + (1 - \lambda)b^2 - a^{2\lambda}b^{2-2\lambda}$$
$$\quad + (1 - \lambda)a^2 + \lambda b^2 - a^{2-2\lambda}b^{2\lambda}$$
$$\geq r_0(a - b)^2 + r_0(a - b)^2$$
$$= 2r_0(a - b)^2.$$

This completes the proof. □

Before proving the next refinement of the classical Young inequality, we will prove the following auxiliary lemmas.

Lemma A.1. *For $x \geq 1$, we have*

$$\frac{2(x - 1)}{x + 1} \leq \log x \leq \frac{x - 1}{\sqrt{x}}.$$

Proof.   First, we will prove

$$\log x \le \frac{x-1}{\sqrt{x}}, \quad x \ge 1. \tag{A.3}$$

Set

$$t = \sqrt{x}, \quad x \ge 1.$$

Then, the inequality (A.3) is equivalent to the inequality

$$2\log t \le \frac{t^2-1}{t}, \quad t \ge 1. \tag{A.4}$$

Let

$$g(t) = \frac{t^2-1}{t} - 2\log t, \quad t \ge 1.$$

Then

$$\begin{aligned}
g'(t) &= \frac{2t^2 - t^2 + 1}{t^2} - \frac{2}{t} \\
&= \frac{t^2+1}{t^2} - \frac{2}{t} \\
&= \frac{t^2 - 2t + 1}{t^2} \\
&= \frac{(t-1)^2}{t^2} \\
&\ge 0, \quad t \ge 1.
\end{aligned}$$

Therefore,

$$g(t) \ge g(1), \quad t \ge 1,$$

whereupon we get (A.4). Now, we will prove the inequality

$$\frac{2(x-1)}{x+1} \le \log x, \quad x \ge 1,$$

which is equivalent to the inequality

$$2(x-1) \le (x+1)\log x, \quad x \ge 1.$$

Set

$$f(x) = (x+1)\log x - 2(x-1), \quad x \ge 1.$$

Then

$$f'(x) = \log x + \frac{x+1}{x} - 2, \quad x \geq 2,$$

$$f'(a) = 0,$$

$$f''(x) = \frac{1}{x} - \frac{1}{x^2}$$

$$= \frac{x-1}{x^2}$$

$$\geq 0, \quad x \geq 1.$$

Consequently,

$$f'(x) \geq f'(1)$$
$$= 0, \quad x \geq 1,$$

and hence,

$$f(x) \geq f(1)$$
$$= 0, \quad x \geq 1.$$

This completes the proof. □

Lemma A.2. *For $t > 0$, we have*

$$e(t^2 + 1) \geq (t+1)t^{\frac{t}{t-1}}. \tag{A.5}$$

Proof.   Let $t \geq 1$. Set

$$f(t) = e(t^2 + 1) - (t+1)t^{\frac{t}{t-1}},$$

$$g(t) = t^{\frac{t}{t-1}}, \quad t \geq 1.$$

Then

$$\log g(t) = \frac{t}{t-1} \log t,$$

$$\frac{g'(t)}{g(t)} = \frac{t-1-t}{(t-1)^2} \log t + \frac{1}{t-1}$$

$$= -\frac{\log t}{(t-1)^2} + \frac{1}{t-1}$$

$$= \frac{t-1-\log t}{(t-1)^2}, \quad t \geq 1.$$

Thus,

$$g'(t) = -\frac{\log t - (t-1)}{(t-1)^2} t^{\frac{t}{t-1}}, \quad t \geq 1.$$

Hence,

$$f'(t) = 2et - t^{\frac{t}{t-1}} - (t+1)g'(t)$$

$$= 2et - t^{\frac{t}{t-1}} + (t+1)\frac{\log t - (t-1)}{(t-1)^2} t^{\frac{t}{t-1}}$$

$$= \frac{2et(t-1)^2 - (t-1)^2 t^{\frac{t}{t-1}} + (t+1)t^{\frac{t}{t-1}}\log t - (t^2-1)t^{\frac{t}{t-1}}}{(t-1)^2}$$

$$= \frac{2et(t-1)^2 + 2t(1-t)t^{\frac{t}{t-1}} + (t+1)t^{\frac{t}{t-1}}\log t}{(t-1)^2}$$

$$= \frac{2et(t-1)^2 + 2t(1-t)t^{\frac{t}{t-1}} + 2(t-1)t^{\frac{t}{t-1}}}{(t-1)^2}$$

$$= \frac{2et(t-1)^2 - 2(t-1)^2 t^{\frac{t}{t-1}}}{(t-1)^2}$$

$$= \frac{2et(t-1)^2 - 2t(t-1)^2 t^{\frac{1}{t-1}}}{(t-1)^2}$$

$$\geq \frac{2t(t-1)^2 t^{\frac{1}{t-1}} - 2t(t-1)^2 t^{\frac{1}{t-1}}}{(t-1)^2}$$

$$= 0, \quad t \geq 1.$$

Consequently,

$$f(t) \geq f(1)$$
$$= 0, \quad t \geq 1.$$

Let now, $t \in (0,1)$. Then, we set

$$s = \frac{1}{t}.$$

Observe that

$$e(t^2 + 1) \geq (t+1)t^{\frac{t}{t-1}}$$

$$\iff e\left(\frac{1}{s^2} + 1\right) \geq \left(\frac{1}{s} + 1\right)\left(\frac{1}{s}\right)^{\frac{\frac{1}{s}}{\frac{1}{s}-1}}$$

$$\iff \frac{e(s^2+1)}{s^2} \geq \frac{s+1}{s}\left(\frac{1}{s}\right)^{\frac{1}{1-s}}$$

$$\iff \frac{e(s^2+1)}{s^2} \geq \frac{s+1}{s}s^{\frac{1}{s-1}}$$

$$\iff \frac{e(s^2+1)}{s^2} \geq \frac{s+1}{s^2}s^{\frac{s}{s-1}}$$

$$\iff e(s^2+1) \geq (s+1)s^{\frac{s}{s-1}},$$

i.e., we have reduced this case to the previous one. This completes the proof. □

The next result is a refinement of the classical Young inequality.

Theorem A.7. *For any* $a, b > 0$ *and* $\lambda \in [0,1]$, *we have*

$$(1-\lambda)a + \lambda b \geq S\left(\left(\frac{b}{a}\right)^r\right)a^{1-\lambda}b^{\lambda},$$

where

$$r = \min\{\lambda, 1-\lambda\}.$$

Proof. First, we will prove the following inequality:

$$\frac{(b-1)\lambda + 1}{b^{\lambda}S(b^{\lambda})} = \frac{e((b-1)\lambda + 1)\log b^{\lambda}}{(b^{\lambda})^{\frac{b^{\lambda}}{b^{\lambda}-1}}(b^{\lambda}-1)}$$

$$\geq 1, \quad \lambda \in \left[0, \frac{1}{2}\right].$$

Suppose that $\lambda \in \left[0, \frac{1}{2}\right]$. By Lemma A.1, we get

$$\frac{\log b^{\lambda}}{b^{\lambda}-1} \geq \frac{2}{b^{\lambda}-1},$$

whereupon

$$\frac{e((b-1)\lambda + 1)\log b^\lambda}{(b^\lambda)^{\frac{b^\lambda}{b^\lambda - 1}}(b^\lambda - 1)} \geq \frac{2e((b-1)\lambda + 1)}{(b^\lambda)^{\frac{b^\lambda}{b^\lambda - 1}}(b^\lambda + 1)}.$$

Set

$$f_b(\lambda) = 2e((b-1)\lambda + 1) - \left(b^\lambda\right)^{\frac{b^\lambda}{b^\lambda - 1}}(b^\lambda + 1),$$

$$g_b(\lambda) = \left(b^\lambda\right)^{\frac{b^\lambda}{b^\lambda - 1}}.$$

Then

$$f_b(\lambda) = 2e((b-1)\lambda + 1) - g_b(\lambda)(b^\lambda + 1).$$

We will compute the first and second derivatives of f_b. For this purpose, we need the first and second derivatives of g_b. We have

$$g_b'(\lambda) = g_b(\lambda)\log b \frac{\left(b^\lambda + \lambda b^\lambda \log b\right)\left(b^\lambda - 1\right) - \lambda \left(b^\lambda\right)^2 \log b}{\left(b^\lambda - 1\right)^2}$$

$$= g_b(\lambda)\log b \frac{b^\lambda\left(b^\lambda - 1\right) + \lambda \left(b^\lambda\right)^2 \log b - \lambda b^\lambda \log b - \lambda \left(b^\lambda\right)^2 \log b}{\left(b^\lambda - 1\right)^2}$$

$$= g_b(\lambda)\log b \frac{\left(b^\lambda\right)^2 - b^\lambda - \lambda b^\lambda \log b}{\left(b^\lambda - 1\right)^2},$$

$$g_b''(\lambda) = g_b'(\lambda)\log b \frac{\left(b^\lambda\right)^2 - b^\lambda - \lambda b^\lambda \log b}{\left(b^\lambda - 1\right)^2}$$

$$+ \frac{g_b(\lambda)\log b}{\left(b^\lambda - 1\right)^4}\left(\left(2\left(b^\lambda\right)^2 \log b - b^\lambda \log b - b^\lambda \log b - \lambda b^\lambda (\log b)^2\right)\right.$$

$$\times \left(b^\lambda - 1\right)^2 - 2\left(b^\lambda - 1\right)b^\lambda \log b\left(\left(b^\lambda\right)^2 - b^\lambda - \lambda b^\lambda \log b\right)\right)$$

$$= g_b(\lambda)(\log b)^2 \frac{\left(b^\lambda\left(b^\lambda - 1\right) - \lambda b^\lambda \log b\right)^2}{\left(b^\lambda - 1\right)^4} + \frac{g_b(\lambda)\log b}{\left(b^\lambda - 1\right)^4}$$

$$\times \left(\left(2\left(b^\lambda\right)^2 \log b - 2b^\lambda \log b - \lambda b^\lambda (\log b)^2\right)\left(b^\lambda - 1\right)^2\right.$$

$$\left. - 2\left(b^\lambda - 1\right)b^\lambda \log b \left(b^\lambda\left(b^\lambda - 1\right) - \lambda b^\lambda \log b\right)\right)$$

$$= g_b(\lambda)(\log b)^2 \frac{\left(b^\lambda\left(b^\lambda - 1\right) - \lambda b^\lambda \log b\right)^2}{\left(b^\lambda - 1\right)^4} + \frac{g_b(\lambda)(\log b)^2}{\left(b^\lambda - 1\right)^4}$$

$$\times \left(\left(2b^\lambda\left(b^\lambda - 1\right) - \lambda b^\lambda \log b\right)\left(b^\lambda - 1\right)^2\right.$$

$$\left. - 2\left(b^\lambda\right)^2\left(b^\lambda - 1\right)^2 + 2\lambda\left(b^\lambda\right)^2\left(b^\lambda - 1\right)\log b\right)$$

$$= \frac{g_b(\lambda)(\log b)^2}{\left(b^\lambda - 1\right)^4} \left(\left(b^\lambda\right)^2\left(b^\lambda - 1\right)^2 - 2\lambda\left(b^\lambda\right)^2\left(b^\lambda - 1\right)\log b\right.$$

$$+ \lambda^2\left(b^\lambda\right)^2 (\log b)^2 + 2b^\lambda\left(b^\lambda - 1\right)^3 - b^\lambda\left(b^\lambda - 1\right)^2 \log b^\lambda$$

$$\left. - 2\left(b^\lambda\right)^2\left(b^\lambda - 1\right)^2 + \lambda\left(b^\lambda\right)^2\left(b^\lambda - 01\right)\log b^\lambda\right)$$

$$= \frac{g_b(\lambda)(\log b)^2}{\left(b^\lambda - 1\right)^4} \left(2b^\lambda\left(b^\lambda - 1\right)^3 - \left(b^\lambda\right)^2\left(b^\lambda - 1\right)^2\right.$$

$$-b^\lambda \left(b^\lambda - 1\right)^2 \log b^\lambda + \left(b^\lambda\right)^2 \left(\log b^\lambda\right)^2\right)$$

$$= \frac{g_b(\lambda)(\log b)^2}{\left(b^\lambda - 1\right)^4} \left(\left(b^\lambda - 1\right)^2 \left(\left(b^\lambda\right)^2 - 2b^\lambda\right)\right.$$

$$\left. -b^\lambda \left(b^\lambda - 1\right) \log b^\lambda + \left(b^\lambda\right)^2 \left(\log b^\lambda\right)^2\right).$$

Hence,

$$f_b'(\lambda) = 2e(b - 1) - g_b'(\lambda)\left(b^\lambda + 1\right) - g_b(\lambda)b^\lambda \log b,$$

$$f_b''(\lambda) = -g_b''(\lambda)\left(b^\lambda + 1\right) - g_b'(\lambda)b^\lambda \log b$$

$$\quad - g_b'(\lambda)b^\lambda \log b - g_b(\lambda)b^\lambda (\log b)^2$$

$$= -g_b''(\lambda)\left(b^\lambda + 1\right) - 2g_b'(\lambda)b^\lambda \log b - g_b(\lambda)b^\lambda (\log b)^2$$

$$= -\frac{g_b(\lambda)(\log b)^2}{\left(b^\lambda - 1\right)^4}\left(\left(b^\lambda + 1\right)\left(b^\lambda - 1\right)^2\left(\left(b^\lambda\right)^2 - 2b^\lambda\right)\right.$$

$$\left. -b^\lambda\left(b^\lambda - 1\right)^2\left(b^\lambda + 1\right)\log b^\lambda + \left(b^\lambda + 1\right)\left(b^\lambda\right)^2\left(\log b^\lambda\right)^2\right)$$

$$\quad - 2g_b(\lambda)b^\lambda(\log b)^2 \frac{b^\lambda\left(b^\lambda - 1\right) - \lambda b^\lambda \log b}{\left(b^\lambda - 1\right)^2} - g_b(\lambda)b^\lambda(\log b)^2$$

$$= -\frac{g_b(\lambda)(\log b)^2}{\left(b^\lambda - 1\right)^4}\left(\left(b^\lambda + 1\right)\left(b^\lambda - 1\right)^2\left(\left(b^\lambda\right)^2 - 2b^\lambda\right)\right.$$

$$\left. -b^\lambda\left(b^\lambda - 1\right)^2\left(b^\lambda + 1\right)\log b^\lambda + \left(b^\lambda + 1\right)\left(b^\lambda\right)^2\left(\log b^\lambda\right)^2\right.$$

$$+ 2\left(b^\lambda - 1\right)^2 \left(\left(b^\lambda\right)^2 - b^\lambda\right)$$

$$- 2\left(b^\lambda - 1\right)^2 \lambda b^\lambda \log b + \left(b^\lambda - 1\right)^4\right)$$

$$= -\frac{g_b(\lambda)(\log b)^2}{\left(b^\lambda - 1\right)^4}\left(\left(b^\lambda - 1\right)^2 \left(\left(b^\lambda\right)^2 - b^\lambda - 2\right)\right.$$

$$- \left(b^\lambda - 1\right)^2 \left(b^\lambda + 1\right)\log b^\lambda + b^\lambda \left(b^\lambda + 1\right)\left(\log b^\lambda\right)^2$$

$$+ \left(b^\lambda - 1\right)^2 \left(2\left(b^\lambda\right)^2 - 2b^\lambda\right)$$

$$- 2\left(b^\lambda - 1\right)^2 b^\lambda \log b^\lambda + \left(b^\lambda - 1\right)^4\right)$$

$$= -\frac{g_b(\lambda)(\log b)^2}{\left(b^\lambda - 1\right)^4}\left(\left(b^\lambda - 1\right)^2 \left(4\left(b^\lambda\right)^2 - 5b^\lambda - 1\right)\right.$$

$$\left. - \left(b^\lambda - 1\right)^2 \left(3b^\lambda + 1\right)\log b^\lambda + b^\lambda \left(b^\lambda + 1\right)\left(\log b^\lambda\right)^2\right),$$

i.e.,

$$f_b''(\lambda) = -\frac{g_b(\lambda)(\log b)^2}{\left(b^\lambda - 1\right)^4}\left(\left(b^\lambda - 1\right)^2 \left(4\left(b^\lambda\right)^2 - 5b^\lambda - 1\right)\right.$$

$$\left. - \left(b^\lambda - 1\right)^2 \left(3b^\lambda + 1\right)\log b^\lambda + b^\lambda \left(b^\lambda + 1\right)\left(\log b^\lambda\right)^2\right).$$

Let $b \geq 1$. Then, applying Lemma A.1,

$$\left(b^\lambda - 1\right)^2 \left(4b^{2\lambda} - 5b^\lambda - 1\right) - \left(b^\lambda - 1\right)^2 \left(3b^\lambda + 1\right)$$

$$\times \log b^\lambda + b^\lambda \left(b^\lambda + 1\right)\left(\log b^\lambda\right)^2$$

$$\geq \left(b^\lambda - 1\right)^2 \left(4b^{2\lambda} - 5b^\lambda - 1\right) - \left(b^\lambda - 1\right)^2 \left(3b^\lambda + 1\right)$$

$$\times \frac{b^\lambda - 1}{b^{\frac{\lambda}{2}}} + b^\lambda \left(b^\lambda + 1\right) \left(\frac{2\left(b^\lambda - 1\right)}{b^\lambda + 1}\right)^2$$

$$= \left(b^\lambda - 1\right)^2 \left(4b^{2\lambda} - 5b^\lambda - 1\right)$$

$$+ 4b^\lambda \frac{\left(b^\lambda - 1\right)^2}{b^\lambda + 1} - \left(b^\lambda - 1\right)^3 \frac{3b^\lambda + 1}{b^{\frac{\lambda}{2}}}$$

$$= \frac{\left(b^\lambda - 1\right)^2}{b^\lambda + 1} \left(4b^{3\lambda} - 5b^{2\lambda} - b^\lambda + 4b^{2\lambda} - 5b^\lambda - 1 + 4b^\lambda\right)$$

$$- \left(b^\lambda - 1\right)^3 \frac{3b^\lambda + 1}{b^{\frac{\lambda}{2}}}$$

$$= \frac{\left(b^\lambda - 1\right)^2}{b^\lambda + 1} \left(4b^{3\lambda} - b^{2\lambda} - 2b^\lambda - 1\right) - \left(b^\lambda - 1\right)^3 \frac{3b^\lambda + 1}{b^{\frac{\lambda}{2}}}$$

$$= \frac{\left(b^\lambda - 1\right)^2}{b^\lambda + 1} \left(b^{2\lambda}\left(b^\lambda - 1\right) + 2b^\lambda\left(b^\lambda - 1\right)\left(b^\lambda + 1\right)\right.$$

$$\left. + \left(b^\lambda - 1\right)\left(b^\lambda\right)^2 + b^\lambda + 1\right) - \left(b^\lambda - 1\right)^3 \frac{3b^\lambda + 1}{b^{\frac{\lambda}{2}}}$$

$$= \frac{\left(b^\lambda - 1\right)^3}{b^\lambda + 1} \left(b^{2\lambda} + 2b^{2\lambda} + 2b^\lambda + b^{2\lambda} + b^\lambda + 1\right)$$

$$- \left(b^\lambda - 1\right)^3 \frac{3b^\lambda + 1}{b^{\frac{\lambda}{2}}}$$

$$= \frac{\left(b^\lambda - 1\right)^3}{b^\lambda + 1} \left(4b^{2\lambda} + 3b^\lambda + 1\right) - \left(b^\lambda - 1\right)^3 \frac{3b^\lambda + 1}{b^{\frac{\lambda}{2}}}$$

$$= \frac{\left(b^\lambda - 1\right)^3}{\left(b^\lambda + 1\right)b^{\frac{\lambda}{2}}}\left(4b^{\frac{5\lambda}{2}} + 3b^{\frac{3\lambda}{2}} + b^{\frac{\lambda}{2}} - 3b^\lambda - 1 - 3b^{2\lambda} - b^\lambda\right)$$

$$= \frac{\left(b^\lambda - 1\right)^3}{\left(b^\lambda + 1\right)b^{\frac{\lambda}{2}}}\left(4b^{\frac{5\lambda}{2}} - 3b^{2\lambda} + 3b^{\frac{3\lambda}{2}} - 4b^\lambda + b^{\frac{\lambda}{2}} - 1\right)$$

$$= \frac{\left(b^\lambda - 1\right)^3}{\left(b^\lambda + 1\right)b^{\frac{\lambda}{2}}}\left(4b^\lambda\left(b^{\frac{3\lambda}{2}} - 1\right) - 3b^{\frac{3\lambda}{2}}\left(b^{\frac{\lambda}{2}} - 1\right) + b^{\frac{\lambda}{2}} - 1\right)$$

$$= \frac{\left(b^\lambda - 1\right)^3}{\left(b^\lambda + 1\right)b^{\frac{\lambda}{2}}}\left(4b^\lambda\left(b^{\frac{\lambda}{2}} - 1\right)\left(b^\lambda + b^{\frac{\lambda}{2}} + 1\right)\right.$$
$$\left. - 3b^{\frac{3\lambda}{2}}\left(b^{\frac{\lambda}{2}} - 1\right) + b^{\frac{\lambda}{2} - 1} - 1\right)$$

$$= \frac{\left(b^{\frac{\lambda}{2}} - 1\right)^4\left(b^{\frac{\lambda}{2}} - 1\right)^3}{\left(b^\lambda + 1\right)b^{\frac{\lambda}{2}}}\left(4b^{2\lambda} + 4b^{\frac{3\lambda}{2}} + 4b^\lambda - 3b^{\frac{3\lambda}{2}} + 1\right)$$

$$= \frac{\left(b^{\frac{\lambda}{2}} - 1\right)^4\left(b^{\frac{\lambda}{2}} - 1\right)^3}{\left(b^\lambda + 1\right)b^{\frac{\lambda}{2}}}\left(4b^{2\lambda} + b^{\frac{3\lambda}{2}} + 4b^\lambda + 1\right) \geq 0.$$

For $b \in (0, 1]$, we have, applying Lemma A.1,

$$\left(b^\lambda - 1\right)^2\left(4b^{2\lambda} - 5b^\lambda - 1\right) - \left(b^\lambda - 1\right)^2\left(3b^\lambda + 1\right)$$

$$\times \log b^\lambda + b^\lambda\left(b^\lambda + 1\right)\left(\log b^\lambda\right)^2$$

$$= \left(b^\lambda - 1\right)^2 \left(4b^{2\lambda} - 5b^\lambda - 1\right) + \left(b^\lambda - 1\right)^2 \left(3b^\lambda + 1\right)$$

$$\times \log\frac{1}{b^\lambda} + v\left(b^\lambda + 1\right)\left(\log\frac{1}{b^\lambda}\right)^2$$

$$\geq \left(b^\lambda - 1\right)^2 \left(4b^{2\lambda} - 5b^\lambda - 1\right) + \left(b^\lambda - 1\right)^2 \left(3b^\lambda + 1\right)\frac{2\left(\frac{1}{b^\lambda} - 1\right)}{\frac{1}{b^\lambda} - 1}$$

$$+ b^\lambda\left(b^\lambda + 1\right)\left(\frac{2\left(\frac{1}{b^\lambda} - 1\right)}{\frac{1}{b^\lambda} + 1}\right)^2$$

$$= \left(b^\lambda - 1\right)^2 \left(4b^{2\lambda} - 5b^\lambda - 1\right) + \left(b^\lambda - 1\right)^2 \left(3b^\lambda + 1\right)$$

$$\times \frac{2\left(1 - b^\lambda\right)}{b^\lambda + 1} + 4b^\lambda\frac{\left(b^\lambda - 1\right)^2}{b^\lambda + 1}$$

$$= \left(b^\lambda - 1\right)^2 \left(4b^{2\lambda} - 5b^\lambda - 1\right) + \frac{\left(b^\lambda - 1\right)^2}{b^\lambda + 1}$$

$$\times \left(6b^\lambda - 6b^{2\lambda} + 2 - 2b^\lambda + 4b^\lambda\right)$$

$$= \left(b^\lambda - 1\right)^2 \left(4b^{2\lambda} - 5b^\lambda - 1\right) + \frac{\left(b^\lambda - 1\right)^2}{b^\lambda + 1}\left(-6b^{2\lambda} + 8b^\lambda + 2\right)$$

$$= \frac{\left(b^\lambda - 1\right)^2}{b^\lambda + 1}\left(4b^{3\lambda} - 5b^{2\lambda} - b^\lambda + 4b^{2\lambda} - 5b^\lambda - 1 - 6b^{2\lambda} + 8b^\lambda + 2\right)$$

$$= \frac{\left(b^\lambda - 1\right)^2}{b^\lambda + 1}\left(4b^{3\lambda} - 7b^{2\lambda} + 2b^\lambda + 1\right)$$

$$= \frac{\left(b^\lambda - 1\right)^3}{b^\lambda + 1}\left(4b^\lambda - 3b^\lambda + 1\right)$$

$$= \frac{\left(b^\lambda - 1\right)^4}{b^\lambda + 1}\left(4b^\lambda + 1\right) \geq 0.$$

Therefore,

$$f_b''(\lambda) \leq 0, \quad b > 0, \quad \lambda \in \left[0, \frac{1}{2}\right].$$

Note that

$$f_b(0) = 0$$

and

$$f_b\left(\frac{1}{2}\right) = 2e\left(\frac{b-1}{2} + 1\right) - \left(\sqrt{b} + 1\right)\left(\sqrt{b}\right)^{\frac{\sqrt{b}}{\sqrt{b}-1}}$$

$$= e(b+1) - \left(\sqrt{b} + 1\right)\left(\sqrt{b}\right)^{\frac{\sqrt{b}}{\sqrt{b}-1}}$$

$$\geq 0,$$

where we have applied Lemma A.2. Therefore,

$$f_b(\lambda) \geq 0, \quad b > 0, \quad \lambda \in \left[0, \frac{1}{2}\right].$$

From here, we conclude that

$$2e\left((b-1)\lambda + 1\right) - \left(b^\lambda\right)^{\frac{b^\lambda}{b^\lambda - 1}}\left(b^\lambda - 1\right) \geq 0, \quad \lambda \in \left[0, \frac{1}{2}\right],$$

or

$$\frac{2e\left((b-1)\lambda + 1\right)}{\left(b^\lambda\right)^{\frac{b^\lambda}{b^\lambda - 1}}\left(b^\lambda + 1\right)} \geq 1, \quad b > 0, \ \lambda \in \left[0, \frac{1}{2}\right],$$

or

$$\frac{(b-1)\lambda + 1}{b^\lambda S\left(b^\lambda\right)} \geq 1, \quad b > 0, \ \lambda \in \left[0, \frac{1}{2}\right], \tag{A.6}$$

or

$$\lambda b + (1 - \lambda) \geq b^\lambda S\left(b^\lambda\right), \quad b > 0, \ \lambda \in \left[0, \frac{1}{2}\right].$$

Replacing b by $\frac{b}{a}$ in the above inequality, we find

$$\lambda \frac{b}{a} + (1 - \lambda) \geq \frac{b^\lambda}{a^\lambda} S\left(\left(\frac{b}{a}\right)^\lambda\right), \quad a, b > 0, \ \lambda \in \left[0, \frac{1}{2}\right],$$

or

$$(1 - \lambda)a + \lambda b \geq S\left(\left(\frac{b}{a}\right)^\lambda\right) a^{1-\lambda} b^\lambda, \quad a, b > 0, \ \lambda \in \left[0, \frac{1}{2}\right].$$

Now, from the inequality (A.6), replacing b with a, we get

$$\frac{(a-1)\mu + 1}{a^\mu S\left(a^\mu\right)} \geq 1, \quad \mu \in \left[0, \frac{1}{2}\right], \quad a > 0,$$

or

$$(a - 1)\mu + 1 \geq a^\mu S\left(a^\mu\right), \quad a > 0, \quad \mu \in \left[0, \frac{1}{2}\right].$$

Putting $\lambda = 1 - \mu$, we find

$$(1 - \lambda)a + \lambda \geq a^{1-\lambda} S\left(a^{1-\lambda}\right), \quad \lambda \in \left[\frac{1}{2}, 1\right], \quad a, b > 0.$$

Replacing in the last inequality a with $\frac{a}{b}$, we obtain

$$(1 - \lambda)a + \lambda b \geq a^{1-\lambda} b^\lambda S\left(\left(\frac{a}{b}\right)^{1-\lambda}\right)$$

$$= a^{1-\lambda} b^\lambda S\left(\left(\frac{b}{a}\right)^{1-\lambda}\right), \quad \lambda \in \left[\frac{1}{2}, 1\right], \quad a, b > 0.$$

This completes the proof. $\qquad\qquad\qquad\qquad\qquad\qquad\qquad\qquad$ $\square$

Corollary A.2. *For any positive numbers a, b and $\lambda \in [0,1]$, we have*

$$S\left(\left(\frac{a}{b}\right)^r\right)\left((1-\lambda)\frac{1}{a} + \lambda\frac{1}{b}\right)^{-1} \le a^{1-\lambda}b^\lambda,$$

where

$$r = \min\{\lambda, 1 - \lambda\}.$$

Proof. We replace a with $\frac{1}{a}$ and b with $\frac{1}{b}$ in Theorem A.7 and get

$$(1-\lambda)\frac{1}{a} + \lambda\frac{1}{b} \ge \left(\frac{1}{a}\right)^{1-\lambda}\left(\frac{1}{b}\right)^\lambda S\left(\left(\frac{a}{b}\right)^r\right)$$

or

$$S\left(\left(\frac{a}{b}\right)^r\right)\left((1-\lambda)\frac{1}{a} + \lambda\frac{1}{b}\right)^{-1} \le a^{1-\lambda}b^\lambda.$$

This completes the proof. □

A.4 Reverse Young Inequalities

Theorem A.8. *For any $a, b > 0$ and $\lambda \in [0,1]$, we have*

$$S(a)a^{1-\lambda} \ge (1-\lambda)a + \lambda \tag{A.7}$$

and

$$S\left(\frac{a}{b}\right)a^{1-\lambda}b^\lambda \ge (1-\lambda)a + \lambda b.$$

Proof. If $a = 1$, then the inequality (A.7) is evident. Let $a \ne 1$. Define the function

$$f_a(\lambda) = \frac{(1-\lambda)a + \lambda}{a^{1-\lambda}}, \quad \lambda \in [0,1].$$

We have

$$f_a(\lambda) = \left(\frac{1-a}{a}\lambda + 1\right) a^\lambda, \quad \lambda \in [0,1].$$

Now, we will find the maximum of the function f_a. For this purpose, we will compute its derivative. We have

$$f_a'(\lambda) = \frac{1-a}{a}a^\lambda + \left(\frac{1-a}{a}\lambda + 1\right) a^\lambda \log a$$

$$= \left(\frac{1-a}{a} + \left(\frac{1-a}{a}\lambda + 1\right)\log a\right) a^\lambda, \quad \lambda \in [0,1].$$

Hence,

$$f'(\lambda_a) = 0$$

$$\iff \frac{1-a}{a} + \lambda_1\frac{1-a}{a}\log a + \log a = 0$$

$$\iff \lambda_a\frac{1-a}{a}\log a = -\frac{1-a}{a} - \log a$$

$$\iff \lambda_a = -\frac{1}{\log a} + \frac{a}{a-1}.$$

By the Klein inequality, we have

$$\frac{a-1}{a} \leq \log a \leq a-1,$$

whereupon

$$\frac{1}{a-1} \leq \frac{1}{\log a} \leq \frac{a}{a-1}.$$

Therefore, $\lambda_a \geq 0$ and

$$\lambda_a = -\frac{1}{\log a} + 1 + \frac{1}{a-1} \leq 1.$$

Thus, $\lambda_a \in [0,1]$ and

$$\max_{\lambda \in [0,1]} f_a(\lambda) = f_a(\lambda_a)$$

$$= \left(\frac{1-a}{a}\left(\frac{a}{a-1} - \frac{1}{\log a}\right) + 1\right) a^{\frac{a}{a-1} - \frac{1}{\log a}}$$

$$= \left(-1 - \frac{1-a}{a\log a} + 1\right) a^{\frac{a}{a-1} - \frac{1}{\log a}}$$

$$= \frac{a-1}{a\log a} \cdot \frac{a^{\frac{a}{a-1}}}{a^{\frac{1}{\log a}}}$$

$$= \frac{\frac{a-a}{\log a}}{a^{-\frac{1}{a-1}} a^{\frac{1}{\log a}}}$$

$$= \frac{a^{\frac{1}{a-1}}(a-1)}{\log a\, a^{\frac{1}{\log a}}}$$

$$= b\frac{a^{\frac{1}{a-1}}(a-1)}{e\log a}$$

$$= \frac{a^{\frac{1}{a-1}}}{e\log a^{\frac{1}{a-1}}}$$

$$= S(a).$$

Consequently,

$$f_a(\lambda) \le S(a), \quad \lambda \in [0,1],$$

or

$$(1-\lambda)a + \lambda \le a^{1-\lambda} S(a).$$

Now, we replace a with $\frac{a}{b}$ in the last inequality and find

$$(1-\lambda)\frac{a}{b} + \lambda \le \frac{a^{1-\lambda}}{b^{1-\lambda}} S\left(\frac{a}{b}\right)$$

or

$$(1-\lambda)a + \lambda b \le a^{1-\lambda} b^{\lambda} S\left(\frac{a}{b}\right).$$

This completes the proof. $\square$

Theorem A.9. *For any $a, b > 0$ and $\lambda \in \left[0, \frac{1}{2}\right) \cup \left(\frac{1}{2}, 1\right]$, we have*

$$(1-\lambda)a + \lambda b - r\left(\sqrt{a} - \sqrt{b}\right)^2 \le S\left(\frac{a}{b}\right) a^{1-\lambda} b^{\lambda},$$

where

$$r = \min\{1 - \lambda, \lambda\}.$$

Proof.   Let $\lambda \in \left[0, \frac{1}{2}\right)$. Then $r = \lambda$ and applying Theorem A.8, we get

$$
\begin{aligned}
(1-\lambda)a + \lambda b - r\left(\sqrt{a} - \sqrt{b}\right)^2 &= (1-\lambda)a + \lambda b - \lambda\left(\sqrt{a} - \sqrt{b}\right)^2 \\
&= (1-\lambda)a + \lambda b - \lambda(a - 2\sqrt{ab} + b) \\
&= (1-\lambda)a + \lambda b - \lambda a + 2\lambda\sqrt{ab} - \lambda b \\
&= (1-2\lambda)a + 2\lambda\sqrt{ab} \\
&\leq S\left(\frac{a}{b}\right) a^{1-2\lambda}a^{\lambda}b^{\lambda} \\
&= S\left(\frac{a}{b}\right) a^{1-\lambda}b^{\lambda}.
\end{aligned}
$$

Let now, $\lambda \in \left(\frac{1}{2}, 1\right]$. Then $r = 1 - \lambda$ and applying Theorem A.8, we find

$$
\begin{aligned}
&(1-\lambda)a + \lambda b - r(\sqrt{a} - \sqrt{b})^2 \\
&= (1-\lambda)a + \lambda b - (1-\lambda)(\sqrt{a} - \sqrt{b})^2 \\
&= (1-\lambda)a + \lambda b - (1-\lambda)(a + b - 2\sqrt{ab}) \\
&= (1-\lambda)a + \lambda b - (1-\lambda)a - (1-\lambda)b + 2(1-\lambda)\sqrt{ab} \\
&= (2\lambda - 1)b + 2(1-\lambda)\sqrt{ab} \\
&\leq S\left(\frac{a}{b}\right) b^{2\lambda-1}a^{1-\lambda}b^{1-\lambda} \\
&= S\left(\frac{a}{b}\right) b^{\lambda}a^{1-\lambda}.
\end{aligned}
$$

This completes the proof. $\square$

Theorem A.10. *For any $a, b > 0$ and $\lambda \in \left[0, \frac{1}{2}\right) \cup \left(\frac{1}{2}, 1\right]$, we have*

$$
((1-\lambda)a + \lambda b)^2 - r^2(a-b)^2 \leq S\left(\frac{a}{b}\right)\left(a^{1-\lambda}b^{\lambda}\right)^2,
$$

where

$$
r = \min\{\lambda, 1 - \lambda\}.
$$

Proof. Let $\lambda \in \left[0, \frac{1}{2}\right)$. Then $r = \lambda$ and applying Theorem A.8, we get

$$
\begin{aligned}
((1-\lambda)a + \lambda b)^2 - r^2(a-b)^2 &= ((1-\lambda)a + \lambda b)^2 - \lambda^2(a-b)^2 \\
&= (1-\lambda)^2 a^2 + 2\lambda(1-\lambda)ab + \lambda^2 b^2 \\
&\quad - \lambda^2 a^2 + 2\lambda^2 ab - \lambda^2 b^2 \\
&= (1-2\lambda)a^2 + 2\lambda ab \\
&\leq S\left(\frac{a}{b}\right) a^{2(1-2\lambda)} a^{2\lambda} b^{2\lambda} \\
&= S\left(\frac{a}{b}\right) a^{2(1-\lambda)} b^{2\lambda} \\
&= S\left(\frac{a}{b}\right) \left(a^{1-\lambda} b^{\lambda}\right)^2.
\end{aligned}
$$

Let now, $\lambda \in \left(\frac{1}{2}, 1\right]$. Then $r = 1 - \lambda$ and applying Theorem A.8, we find

$$
\begin{aligned}
((1-\lambda)a + \lambda b)^2 - r^2(a-b)^2 &= ((1-\lambda)a + \lambda b)^2 - (1-\lambda)^2(a-b)^2 \\
&= (1-\lambda)^2 a^2 + 2\lambda(1-\lambda)ab + \lambda^2 b^2 \\
&\quad -(1-\lambda)^2 a^2 + 2(1-\lambda)^2 ab \\
&\quad -(1-\lambda)^2 b^2 \\
&= 2(1-\lambda)ab + (2\lambda - 1)b^2 \\
&\leq S\left(\frac{a}{b}\right) a^{2(1-\lambda)} b^{2(1-\lambda)} b^{2(2\lambda-1)} \\
&= S\left(\frac{a}{b}\right) a^{2(1-\lambda)} b^{2\lambda} \\
&= S\left(\frac{a}{b}\right) \left(a^{1-\lambda} b^{\lambda}\right)^2.
\end{aligned}
$$

This completes the proof. $\qquad\qquad\qquad\qquad\qquad\qquad\qquad\qquad\qquad$ □

A.5 The Hölder Inequality

Suppose that $\mathbb{T}$ is a time scale with forward jump operator and delta differentiation operator σ and Δ, respectively. Let also $a, b \in \mathbb{T}$, $a < b$.

Theorem A.11. Let $p_1, p_2 \in [1, \infty)$, $\frac{1}{p_1} + \frac{1}{p_2} = 1$, $f_1 \in L^{p_1}([a, b])$, $f_2 \in L^{P-2}([a, b])$. Then $f_1 f_2 \in L^1([a, b])$ and

$$\|f_1 f_2\|_1 \leq \|f_1\|_{p_1} \|f_2\|_{p_2}.$$

Proof. If $f_1 \equiv 0$ or $f_2 \equiv 0$, then the result is evident. Suppose that $f_1 \not\equiv 0$ and $f_2 \not\equiv 0$. We will consider the following cases:

1. Let $p_1 = 1$. Then $p_2 = \infty$. Hence,

$$\begin{aligned}
\|f_1 f_2\|_1 &= \int_a^b |f_1 f_2| \Delta \\
&\leq \|f_2\|_\infty \int_a^b |f_1| \Delta \\
&= \|f_1\|_1 \|f_2\|_\infty \\
&= \|f_1\|_{p_1} \|f_2\|_{p_2}.
\end{aligned}$$

2. Let $p_1 > 1$. Then $p_2 > 1$. We set

$$F_1 = \frac{f_1}{\|f_1\|_{p_1}},$$

$$F_2 = \frac{f_2}{\|f_2\|_{p_2}}.$$

By the Young inequality, we find

$$|F_1 F_2| \leq \frac{1}{p_1} |F_1|^{p_1} + \frac{1}{p_2} |F_2|^{p_2}$$

or

$$\frac{|f_1 f_2|}{\|f_1\|_{p_1} \|f_2\|_{p_2}} \leq \frac{|f_1|^{p_1}}{p_1 \|f_1\|_{p_1}^{p_1}} + \frac{|f_2|^{p_2}}{p_2 \|f_2\|_{p_2}^{p_2}},$$

whereupon

$$\begin{aligned}
\frac{1}{\|f_1\|_{p_1} \|f_2\|_{p_2}} \int_a^b |f_1 f_2| \Delta &\leq \frac{1}{p_1 \|f_1\|_{p_1}^{p_1}} \int_a^b |f_1|^{p_1} \Delta \\
&\quad + \frac{1}{p_2 \|f_2\|_{p_2}^{p_2}} \int_a^b |f_2|^{p_2} \Delta \\
&= \frac{1}{p_1} + \frac{1}{p_2} \\
&= 1.
\end{aligned}$$

From the last inequality, we get the desired result. This completes the proof.

□

Remark A.2. Note that Theorem A.11 can be generalized as follows. Let $p_j \in [1,\infty)$, $j \in \{1,\ldots,n\}$, $\sum_{j=1}^{n} \frac{1}{p_j} = 1$, $f_j \in L^{p_j}([a,b])$, $j \in \{1,\ldots,n\}$. Then $\prod_{j=1}^{n} f_j \in L^1([a,b])$ and

$$\left\| \prod_{j=1}^{n} f_j \right\|_1 \leq \prod_{j=1}^{n} \|f_j\|_{p_j}.$$

Theorem A.12. *Let* $p_1, p_2 \in [1,\infty)$, $r \in (0,\infty]$, $\frac{1}{p_1} + \frac{1}{p_2} = \frac{1}{r}$, $f_1 \in L^{\frac{p_1}{r}}([a,b])$, $f_2 \in L^{\frac{p_2}{r}}([a,b])$. *Then* $f_1 f_2 \in L^r([a,b])$ *and*

$$\|f_1 f_2\|_r \leq \|f_1\|_{p_1} \|f_2\|_{p_2}.$$

Proof. Note that

$$\frac{1}{\frac{p_1}{r}} + \frac{1}{\frac{p_2}{r}} = 1.$$

Then, applying the Hölder inequality, we get

$$\int_a^b |f_1 f_2|^r \Delta = \int_a^b |f_1|^r |f_2|^r \Delta$$

$$\leq \left(\int_a^b |f_1|^{p_1} \Delta \right)^{\frac{r}{p_1}} \left(\int_a^b |f_2|^{p_2} \Delta \right)^{\frac{r}{p_2}}$$

$$= \|f_1\|_{p_1}^r \|f_2\|_{p_2}^r$$

or

$$\|f_1 f_2\|_r^r \leq \|f_1\|_{p_1}^r \|f_2\|_{p_2}^r,$$

whereupon we get the desired result. This completes the proof. □

Remark A.3. Note that Theorem A.14 can be generalized as follows. Let $p_j \in [1,\infty)$, $r \in (0,\infty]$, $\sum_{j=1}^{n} \frac{1}{p_j} = \frac{1}{r}$, $f_j \in L^{\frac{p_j}{r}}([a,b])$, $j \in \{1,\ldots,n\}$. Then $\prod_{j=1}^{n} f_j \in L^r([a,b])$ and

$$\left\| \prod_{j=1}^{n} f_j \right\|_r \leq \prod_{j=1}^{n} \|f_j\|_{p_j}.$$

A.6   Reverse Hölder Inequalities

Let $\mathbb{T}$ be a time scale with forward jump operator and delta differentiation operator σ and Δ, respectively.

Theorem A.13. *Let $p_1, p_2 \in (1, \infty)$, $\frac{1}{p_1} + \frac{1}{p_2} = 1$, $f_1, f_2 : [a, b] \to (0, \infty)$, $f_1^{\frac{1}{p_1}} f_2^{\frac{1}{p_2}}$ be Δ-integrable on $[a, b]$. Then*

$$\|f_1\|_{p_1} \|f_2\|_{p_2} \leq \int_a^b S\left(\frac{f_1^{p_1} \|f_2\|_{p_2}^{p_2}}{f_2^{p_2} \|f_2\|_{p_2}^{p_2}}\right) f_1 f_2 \Delta.$$

Proof.   Let

$$a = \frac{f_1^{p_1}}{\|f_1\|_{p_1}^{p_1}}, \qquad b = \frac{f_2^{p_2}}{\|f_2\|_{p_2}^{p_2}}.$$

Now, we apply the reverse Young inequality for a and b and get

$$S\left(\frac{f_1^{p_1} \|f_2\|_{p_2}^{p_2}}{f_2^{p_2} \|f_2\|_{p_2}^{p_2}}\right) \frac{f_1}{\|f_1\|_{p_1}} \frac{f_2}{\|f_2\|_{p_2}} \geq \frac{f_1^{p_1}}{p_1 \|f_1\|_{p_1}^{p_1}} + \frac{f_2^{p_2}}{p_2 \|f_2\|_{p_2}^{p_2}},$$

whereupon

$$\frac{1}{\|f_1\|_{p_1} \|f_2\|_{p_2}} \int_a^b S\left(\frac{f_1^{p_1} \|f_2\|_{p_2}^{p_2}}{f_2^{p_2} \|f_2\|_{p_2}^{p_2}}\right) f_1 f_2 \Delta$$

$$\geq \frac{1}{p_1 \|f_1\|_{p_1}^{p_1}} \int_a^b f_1^{p_1} \Delta + \frac{1}{p_2 \|f_2\|_{p_2}^{p_2}} \int_a^b f_2^{p_2} \Delta$$

$$= \frac{1}{p_1} + \frac{1}{p_2}$$

$$= 1.$$

From here, we get the desired result. This completes the proof.   □

Theorem A.14. *Let $p_1, p_2 > 1$, $\frac{1}{p_1} + \frac{1}{p_2} = 1$, $f_1 \in L^{p_1}([a, b])$, $f_2 \in L^{p_2}([a, b])$, $f_1, f_2 : [a, b] \to (0, \infty)$ be such that*

$$0 < m \leq \frac{f_1^{p_1}}{f_2^{p_2}} \leq M < \infty.$$

Then

$$\left(\int_a^b f_1^{p_1} \Delta\right)^{\frac{1}{p_1}} \left(\int_a^b f_2^{p_2} \Delta\right)^{\frac{1}{p_2}} \leq \left(\frac{m}{M}\right)^{-\frac{1}{p_1 p_2}} \int_a^b f_1 f_2 \Delta.$$

Proof. Since

$$\frac{f_1^{p_1}}{f_2^{p_2}} \le M,$$

we get

$$f_2^{p_2} \ge \frac{1}{M} f_1^{p_1}$$

or

$$f_2 \ge \frac{1}{M^{\frac{1}{p_2}}} f_1^{\frac{p_1}{p_2}},$$

whereupon

$$f_1 f_2 \ge \frac{1}{M^{\frac{1}{p_2}}} f_1^{\frac{p_1}{p_2}+1}$$

$$= \frac{1}{M^{\frac{1}{p_2}}} f_1^{p_1}.$$

Hence,

$$\int_a^v f_1 f_2 \Delta \ge \frac{1}{M^{\frac{1}{p_2}}} \int_a^b f_1^{p_1} \Delta$$

or

$$\int_a^b f_1^{p_1} \Delta \le M^{\frac{1}{p_2}} \int_a^b f_1 f_2 \Delta$$

and

$$\left(\int_a^b f_1^{p_1} \Delta \right)^{\frac{1}{p_1}} \le M^{\frac{1}{p_1 p_2}} \left(\int_a^b f_1 f_2 \Delta \right)^{\frac{1}{p_1}}. \tag{A.8}$$

Next, by the inequality

$$\frac{f_1^{p_1}}{f_2^{p_2}} \ge m,$$

we get

$$f_1^{p_1} \ge m f_2^{p_2}$$

or

$$f_1 \ge m^{\frac{1}{p_1}} f_2^{\frac{p_2}{p_1}}$$

and

$$f_1 f_2 \geq m^{\frac{1}{p_1}} f_2^{\frac{p_2}{p_1}+1}$$
$$= m^{\frac{1}{p_1}} f_2^{p_2}.$$

Hence,

$$\int_a^b f_1 f_2 \Delta \geq m^{\frac{1}{p_1}} \int_a^b f_2^{p_2} \Delta$$

and

$$\left(\int_a^b f_1 f_2 \Delta\right)^{\frac{1}{p_2}} \geq m^{\frac{1}{p_1 p_2}} \left(\int_a^b f_2^{p_2} \Delta\right)^{\frac{1}{p_2}}.$$

By the last inequality and the inequality (A.8), we arrive at

$$M^{\frac{1}{p_1 p_2}} \int_a^b f_1 f_2 \Delta \geq m^{\frac{1}{p_1 p_2}} \left(\int_a^b f_1^{p_1} \Delta\right)^{\frac{1}{p_1}} \left(\int_a^b f_2^{p_2} \Delta\right)^{\frac{1}{p_2}}$$

or

$$\left(\frac{m}{M}\right)^{-\frac{1}{p_1 p_2}} \int_a^b f_1 f_2 \Delta \geq \left(\int_a^b f_1^{p_1} \Delta\right)^{\frac{1}{p_1}} \left(\int_a^b f_2^{p_2} \Delta\right)^{\frac{1}{p_2}}.$$

This completes the proof. □

Theorem A.15. *Suppose that $p_1, p_2, p_3 > 0$ and $\frac{1}{p_1} + \frac{1}{p_2} + \frac{1}{p_3} = 1$. If $f_1, f_2, f_3 : [a, b] \to (0, \infty)$ be such that*

(i) $0 < m \leq \frac{f_1^{p_1 s}}{f_2^{p_2 s}} \leq M < \infty$ *for some $sz > 0$ so that $\frac{1}{p_1} + \frac{1}{p_2} = \frac{1}{s}$*

and

(ii) $0 < m \leq \frac{(f_1 f_2)^s}{f_3^{p_3}} \leq M < \infty$ *for some $s > 0$,*

then

$$\left(\int_a^b f_1^{p_1} \Delta\right)^{\frac{1}{p_1}} \left(\int_a^b f_2^{p_2} \Delta\right)^{\frac{1}{p_2}} \left(\int_a^b f_3^{p_3} \Delta\right)^{\frac{1}{p_3}}$$
$$\leq \left(\frac{m}{M}\right)^{-\left(\frac{1}{s p_3} + \frac{s^2}{p_1 p_2}\right)} \int_a^b f_1 f_2 f_3 \Delta.$$

Proof. Let

$$\frac{1}{p_1} + \frac{1}{p_2} = \frac{1}{s}$$

or

$$\frac{1}{\frac{p_1}{s}} + \frac{1}{\frac{p_2}{s}} = 1 \tag{A.9}$$

and

$$\frac{1}{s} + \frac{1}{p_3} = 1.$$

Set

$$H = f_1 f_2.$$

Then, applying Theorem A.14, we get

$$\left(\int_a^b H^s \Delta\right)^{\frac{1}{s}} \left(\int_a^b f_3^{p_3} \Delta\right)^{\frac{1}{p_3}} \leq \left(\frac{m}{M}\right)^{-\frac{1}{sp_3}} \int_a^b H f_3 \Delta$$

$$= \left(\frac{m}{M}\right)^{-\frac{1}{sp_3}} \int_a^b f_1 f_2 f_3 \Delta,$$

which is equivalent to the inequality

$$\left(\int_a^b f_1^s f_2^s \Delta\right)^{\frac{1}{s}} \left(\int_a^b f_3^{p_3} \Delta\right)^{\frac{1}{p_3}} \leq \left(\frac{m}{M}\right)^{-\frac{1}{sp_3}} \int_a^b f_1 f_2 f_3 \Delta. \tag{A.10}$$

Now, using (A.9) and Theorem A.14, we find

$$\left(\int_a^b f_1^s f_2^s \Delta\right)^{\frac{1}{s}} \geq \left(\frac{m}{M}\right)^{\frac{1}{\frac{p_1}{s}\frac{p_2}{s}}} \left(\int_a^b f_1^{p_1} \Delta\right)^{\frac{1}{p_1}} \left(\int_a^b f_2^{p_2} \Delta\right)^{\frac{1}{p_2}}$$

$$= \left(\frac{m}{M}\right)^{\frac{s^2}{p_1 p_2}} \left(\int_a^b f_1^{p_1} \Delta\right)^{\frac{1}{p_1}} \left(\int_a^b f_2^{p_2} \Delta\right)^{\frac{1}{p_2}}.$$

Hence, by (A.10), we arrive at

$$\left(\frac{m}{M}\right)^{\frac{s^2}{p_1 p_2}} \left(\int_a^b f_1^{p_1} \Delta\right)^{\frac{1}{p_1}} \left(\int_a^b f_2^{p_2} \Delta\right)^{\frac{1}{p_2}} \left(\int_a^b f_3^{p_3} \Delta\right)^{\frac{1}{p_3}}$$

$$\leq \left(\int_a^b f_1^s f_2^s \Delta\right)^{\frac{1}{s}} \left(\int_a^b f_3^{p_3} \Delta\right)^{\frac{1}{p_3}}$$

$$\leq \left(\frac{m}{M}\right)^{-\frac{1}{s p_3}} \int_a^b f_1 f_2 f_3 \Delta,$$

whereupon

$$\left(\int_a^b f_1^{p_1} \Delta\right)^{\frac{1}{p_1}} \left(\int_a^b f_2^{p_2} \Delta\right)^{\frac{1}{p_2}} \left(\int_a^b f_3^{p_3} \Delta\right)^{\frac{1}{p_3}}$$

$$\leq \left(\frac{m}{M}\right)^{-\frac{1}{s p_3} - \frac{s^2}{p_1 p_2}} \int_a^b f_1 f_2 f_3 \Delta.$$

This completes the proof.   □

Appendix B

Jensen Inequalities

Theorem B.1 (Jensen's Inequality). *Let* $g \in \mathcal{C}([a, b])$ *and* $\Phi :$ $\mathbb{R} \to \mathbb{R}$ *be convex. Then*

$$\Phi\left(\frac{1}{b-a}\int_a^b g(t)\Diamond_\alpha t\right) \leq \frac{1}{b-a}\int_a^b \Phi(g(t))\Diamond_\alpha t.$$

Proof. Since Φ is convex, we have

$$\Phi\left(\frac{1}{b-a}\int_a^b g(t)\Diamond_\alpha t\right)$$

$$= \Phi\left(\frac{\alpha}{b-a}\int_a^b g(t)\Delta t + \frac{1-\alpha}{b-a}\int_a^b g(t)\nabla t\right)$$

$$\leq \alpha\Phi\left(\frac{1}{b-a}\int_a^b g(t)\Delta t\right) + (1-\alpha)\Phi\left(\frac{1}{b-a}\int_a^b g(t)\nabla t\right).$$

$$(B.1)$$

Let

$$x_0 = \frac{\int_a^b g(t)\Delta t}{b-a},$$

$$x = g(t).$$

Since Φ is convex, there exists a $\beta \in \mathbb{R}$ such that

$$\Phi(x) - \Phi(x_0) \geq \beta(x - x_0) \qquad (B.2)$$

317

or

$$\Phi(g(t)) - \Phi\left(\frac{1}{b-a}\int_a^b g(t)\Delta t\right) \geq \beta\left(g(t) - \frac{1}{b-a}\int_a^b g(t)\Delta t\right).$$

We integrate the last inequality with respect to t from a to b and find

$$\int_a^b \Phi(g(t))\Delta t - (b-a)\Phi\left(\frac{1}{b-a}\int_a^b g(t)\Delta t\right)$$

$$\geq \beta\left(\int_a^b g(t)\Delta t - \int_a^b g(t)\Delta t\right) = 0,$$

i.e.,

$$\Phi\left(\frac{1}{b-a}\int_a^b g(t)\Delta t\right) \leq \frac{1}{b-a}\int_a^b \Phi(g(t))\Delta t. \qquad \text{(B.3)}$$

As in the above,

$$\Phi\left(\frac{1}{b-a}\int_a^b g(t)\nabla t\right) \leq \frac{1}{b-a}\int_a^b \Phi(g(t))\nabla t.$$

By the last inequality and by (B.1), (B.3), we find

$$\Phi\left(\frac{1}{b-a}\int_a^b g(t)\Diamond_\alpha t\right) \leq \frac{\alpha}{b-a}\int_a^b \Phi(g(t))\Delta t + \frac{1-\alpha}{b-a}\int_a^b \Phi(g(t))\nabla t$$

$$= \frac{1}{b-a}\int_a^b \Phi(g(t))\Diamond_\alpha t.$$

This completes the proof. □

Theorem B.2 (The Generalized Jensen Inequality). *Let* $g, h \in \mathcal{C}([a,b])$ *with*

$$\int_a^b |h(t)|\Diamond_\alpha t > 0$$

and $\Phi : \mathbb{R} \to \mathbb{R}$ *be convex. Then*

$$\Phi\left(\frac{\int_a^b |h(t)|g(t)\Diamond_\alpha t}{\int_a^b |h(t)|\Diamond_\alpha t}\right) \leq \frac{\int_a^b |h(t)|\Phi(g(t))\Diamond_\alpha t}{\int_a^b |h(t)|\Diamond_\alpha t}.$$

Proof. We take

$$x_0 = \frac{\int_a^b |h(t)||g(t)| \Diamond_\alpha t}{\int_a^b |h(t)| \Diamond_\alpha t},$$

$$x = g(t)$$

in (B.2) and find

$$\int_a^b |h(t)| \Phi(g(t)) \Diamond_\alpha t - \left(\int_a^b |h(t)| \Diamond_\alpha t \right) \Phi \left(\frac{\int_a^b |h(t)||g(t)| \Diamond_\alpha t}{\int_a^b |h(t)| \Diamond_\alpha t} \right)$$

$$= \int_a^b |h(t)| \Phi(g(t)) \Diamond_\alpha t - \left(\int_a^b |h(t)| \Diamond_\alpha t \right) \Phi(x_0)$$

$$= \int_a^b |h(t)| \left(\Phi(g(t)) - \Phi(x_0) \right) \Diamond_\alpha t$$

$$\geq \beta \int_a^b |h(t)|(g(t) - x_0) \Diamond_\alpha t$$

$$= \beta \left(\int_a^b |h(t)||g(t)| \Diamond_\alpha t - x_0 \int_a^b |h(t)| \Diamond_\alpha t \right)$$

$$= \beta \left(\int_a^b |h(t)||g(t)| \Diamond_\alpha t - \int_a^b |h(t)||g(t)| \Diamond_\alpha t \right) = 0.$$

This completes the proof. $\qquad\square$

References

1. R. Agarwal, D. O'Regan and S. Saker. *Hardy Type Inequalities on Time Scales.* Springer, Switzerland, 2016.
2. G. Anastassiou. Principles of delta fractional calculus on time scales and inequalities. *Mathematical and Computer Modeling* 52(3–4), 556–566, 2010.
3. M. Awan, M. Noor and K. Noor. Hermite-Hadamard inequalities for exponential convex functions. *Applied Mathematics Information Sciences*, 12(2), 405–409, 2018.
4. M. Bohner and A. Peterson. *Dynamic Equations on Time Scales: An Introduction with Applications.* Birkhäuser, Boston, 2001.
5. M. Bohner and S. Georgiev. *Multivariable Dynamic Calculus on Time Scales.* Springer, Cham, 2017.
6. R. Castillo and E. Trousselot. Reverse generalized Hölder and Minkowski type inequalities and their applications. *Mathematics Bulletin*, 17(2), 137–142, 2010.
7. S. Furuichi. Refined Young inequalities with Specht's ratio. *Journal of the Egyptian Mathematical Society*, 20, 46–49, 2010.
8. S. Georgiev. *Fractional Dynamic Calculus and Fractional Dynamic Equations on Time Scales.* Springer, Berlin, 2016.
9. S. Hilger. Analysis on measure chains — A unified approach to continuous and discrete calculus. *Results in Mathematics*, 18, 18–56, 1990.
10. W. Liao, J. Wu and J. Zhao. New versions of reverse Young and Heinz mean inequalities with the Kantorovich constant. *Taiwanese Journal of Mathematics*, 19(2), 467–479, 2015.
11. D. Pachpatte. A generalized Gronwall inequality for Caputo fractional dynamic delta operator. *Progress in Fractional Differentiation and Applications*, 6(2), 129–136, 2020.

12. S. Rashid, M. Noor, K. Nisar, D. Baleanu and G. Rahman. A new dynamic scheme via fractional operators on time scales. *Frontiers in Physics*, 8(105), 1–10, 2020.

13. S. Saker, M. Kenawy, G. Alnemer and M. Zakarya. Some fractional dynamic inequalities of Hardy's type via conformable calculus. *Mathematics*, 8, 434, 2020.

14. M. Tominaga. Specht's ratio in the Young inequality. *Scientiae Mathematicae Japonicae Online*, 5, 525–530, 2001.

15. D. Ucar, F. Hatipoglu and A. Akincali. Fractional integral inequalities on time scales. *Open Journal of Mathematical Sciences*, 2(1), 361–370, 2018.

16. M. Zakarya, M. Altanji, C. Alnemer, H. El-Hamid, C. Cesarano and H. Rezk. Fractional Reverse Copson's inequalities via conformable calculus on time scales. *Symmetry*, 13(4), 542, 2021.

Index

Printed in the USA
CPSIA information can be obtained
at www.ICGtesting.com
JSHW011708050923
47823JS00003B/7